25th International Laser Physics Workshop (LPHYS'16)

Journal of Physics: Conference Series Volume 826

Yerevan, Armenia
11 – 15 July 2016

ISBN: 978-1-5108-3964-9
ISSN: 1742-6588

Printed from e-media with permission by:

Curran Associates, Inc.
57 Morehouse Lane
Red Hook, NY 12571

Some format issues inherent in the e-media version may also appear in this print version.

Printed by Curran Associates, Inc. (2017)

For permission requests, please contact the Institute of Physics
at the address below.

Institute of Physics
Dirac House, Temple Back
Bristol BS1 6BE UK

Phone: 44 1 17 929 7481
Fax: 44 1 17 920 0979

techtracking@iop.org

Additional copies of this publication are available from:

Curran Associates, Inc.
57 Morehouse Lane
Red Hook, NY 12571 USA
Phone: 845-758-0400
Fax: 845-758-2633
Email: curran@proceedings.com
Web: www.proceedings.com

Table of contents

Volume 826

25th Annual International Laser Physics Workshop (LPHYS'16)
11-15 July 2016, Yerevan, Armenia

Accepted papers received: 14 March 2017
Published online: 20 April 2017

Preface

011001
OPEN ACCESS
25th Annual International Laser Physics Workshop (LPHYS'16)

011002
OPEN ACCESS
Peer review statement

Papers

Modern Trends in Laser Physics

012001
OPEN ACCESS
Manifestations of induced molecular alignment in the interaction of ultrafast laser pulses with gases

A A Kolomenskii, N Kaya, G Kaya, M Sayrac, Y Boran, J Strohaber and H A Schuessler.....1

012002
OPEN ACCESS
Using Raman spectroscopy to determine the structure of copolymers and polymer blends

G Yu Nikolaeva, E A Sagitova, K A Prokhorov, P P Pashinin, P M Nedorezova, A N Klyamkina, M A Guseva and V A Gerasin.....11

012003
OPEN ACCESS
The use of Inductive Discharge for Laser Pumping of a Copper Vapor

M.A. Kazaryan, M.M. Malikov, V.T. Karpukhin and N.A. Lyabin.....21

012004
OPEN ACCESS
Theoretical treatment of the resonant hyper-Raman scattering in A^2B^6 semiconductors

L E Semenova, G Yu Nikolaeva and K A Prokhorov.....29

012005
OPEN ACCESS
Dielectric Anisotropy of Human Bone and CERABONE® in the Terahertz Spectral Range 0.2 to 2.5 THz

A.S. Nikoghosyan, H. Ting, J. Shen, R.M. Martirosyan, M.A. Kazaryan, M.Yu. Tunyan, A.V. Papikyan and A.A. Papikyan.....33

012006
OPEN ACCESS
New insights into the structure of polypropylene polymorphs and propylene copolymers probed by low-frequency Raman spectroscopy

E A Sagitova, P Donfack, G Yu Nikolaeva, K A Prokhorov, P P Pashinin, P M Nedorezova, A N Klyamkina and A Materny.....41

Strong Field & Attosecond Physics

012007
OPEN ACCESS
High-order harmonic generation by polyatomic molecules

S Odžak, E Hasović and D B Milošević.....50

012008
OPEN ACCESS
Above-threshold ionization in a bicircular field: quantum orbits unfolding in a plane

W Becker and D B Milošević.....56

012009
OPEN ACCESS
Electron Rescattering in a Bicircular Laser Field

E Hasović, W. Becker and D B Milošević.....64

012010
OPEN ACCESS
Ionization without interference

F. Cajiao Vélez, J. Z. Kamiński and K. Krajewska.....70

012011
OPEN ACCESS
Quantum-Plasma Model of Atomic Cluster Excitations

A. El-Khawaldeh and H.-J. Kull.....80

012012
OPEN ACCESS
Quantum Two-Stream Instability with Exchange Interaction

H.-J. Kull.....90

012013
OPEN ACCESS
Plasmon effects in photoemission

J. Z. Kamiński, F. Cajiao Vélez and K. Krajewska.....100

012014
OPEN ACCESS
Quantum-Mechanical Description of Ionization-Induced Generation of Tunable Mid-Infrared Pulses

A A Silaev, A A Romanov, V A Kostin and N V Vvedenskii.....110

Laser Biomedical Applications

012015
OPEN ACCESS
Comparative evaluation of the effects of high-intensity and low-intensity laser radiation on microcirculation among patients with knee arthritis

D B Kulchitskaya, T V Konchugova and N E Fedorova.....116

012016
OPEN ACCESS
New method in medical tomography based on vibrating wire: bench-test experiment on laser beam

M.A. Aginian, J Alonso, S.G. Arutunian, M. Chung, A.V. Margaryan, E.G. Lazareva, L.M. Lazarev and L.A. Shahinyan.....120

Nonlinear Optics & Spectroscopy

012017
OPEN ACCESS
THz Pulse Duration Influence on High Energy Level Excitation Due to Cascade Mechanism

Vyacheslav A. Trofimov, Dmitry Yu. Zagursky and Irina G. Zakharova.....131

012018
OPEN ACCESS
Analytical consideration and computer simulation of DFWM

Vyacheslav A. Trofimov and Igor E. Kuchik.....139

012019
OPEN ACCESS
Soliton self-formation in the medium with nanorods under conditions of the absorption spectrum finite width

Vyacheslav A. Trofimov, Tatiana M. Lysak and Igor E. Kuchik.....149

Physics of Cold Trapped Atoms

012020
OPEN ACCESS
Uncertainty product of an out-of-equilibrium Bose-Einstein condensate

S Klaiman, A I Streltsov and O E Alon.....159

012021
OPEN ACCESS
Entanglement production by evolution operator

V.I. Yukalov and E.P. Yukalova.....168

012022
OPEN ACCESS
Cavity resonances dominating the photon statistics in the non-equilibrium steady state

Felix Rüting and Christoph Weiss.....178

Quantum Information Science

012023
OPEN ACCESS
Peculiarities of excitation trapping in dense polyatomic ensemble in a Fabry-Perot cavity

A S Kuraptsev and I M Sokolov.....188

012024
OPEN ACCESS
Dynamic Entanglement Evolution of Multi-Qubits Systems

N S Ananikian, H A Lazaryan and M A Nalbandyan.....196

012025
OPEN ACCESS
Invariant Mass and Propagation Speed of Light Pulses in Vacuum

M V Fedorov and S V Vintskevich.....201

Fiber Optics

012026
OPEN ACCESS
Calculation of mutual information for nonlinear optical fiber communication channel at large SNR within path-integral formalism

A V Reznichenko, I S Terekhov and S K Turitsyn.....208

Extreme Light Technologies, Science, and Applications

012027
OPEN ACCESS
Conjecture of perturbative QED breakdown at $\alpha\chi^{2/3} \gtrsim 1$

Alexander Fedotov.....218

012028
OPEN ACCESS
Weighted Particle Monte Carlo Method for Effective Simulation of Electromagnetic Cascades

Nina Elkina.....226

012029
OPEN ACCESS
Observable Features of QED Cascades in Collisions of GeV Electrons with Intense Laser Pulses

Arseny Mironov, Alexander Fedotov and Nikolay Narozhny.....230

012030
OPEN ACCESS
Conversion of Transverse to Longitudinal Plasma Waves Induced by Radiation Reaction

E G Gelfer, A M Fedotov and N V Elkina.....236

EDITORIAL

Dear Readers,

The 25th annual International Laser Physics Workshop, LPHYS'16, took place in the city of Yerevan, Republic of Armenia. 319 participants from 32 countries attended the conference. It was hosted by the Yerevan Physics Institute, A Alikhanyan National Laboratory. This year the conference has been endorsed by the Optical Society (OSA), the year the society celebrates its 100th anniversary.

The LPHYS'16 Steering Committee and the Advisory & Program Committee would like to extend their sincere gratitude to Dr. Koryun B Oganesyan (Chair of the Local Organizing Committee) and to his team for the outstanding job performed on organizing, arranging, managing and putting in order the conference. Their combined effort lead to a successful result.

In this volume of the IOP Journal of Physics: Conference Series you will find selected proceedings of the Workshop in Yerevan.

Please make a note that the 26th annual International Laser Physics Workshop (LPHYS'17) will take place from July 17 to July 21, 2017 in the city of Kazan, Russia hosted by the Kazan E K Zavoisky Physical-Technical Institute (KPhTI).

With kind regards,

Steering and Advisory & Program committees

LPHYS'16

IOP Conf. Series: Journal of Physics: Conf. Series **826** (2017) 011002 doi:10.1088/1742-6596/826/1/011002

Peer review statement

All papers published in this volume of *Journal of Physics: Conference Series* have been peer reviewed through processes administered by the proceedings Editors. Reviews were conducted by expert referees to the professional and scientific standards expected of a proceedings journal published by IOP Publishing.

IOP Conf. Series: Journal of Physics: Conf. Series **826** (2017) 012001 doi:10.1088/1742-6596/826/1/012001

Manifestations of induced molecular alignment in the interaction of ultrafast laser pulses with gases

A A Kolomenskii[1], N Kaya[2,3], G Kaya[2], M Sayrac[1], Y Boran[1], J Strohaber[4] and H A Schuessler[1,2]

[1]Department of Physics and Astronomy, Texas A&M University, College Station, Texas 77843-4242, USA.
[2]Science Program, Texas A&M University at Qatar, Doha 23874, Qatar.
[3]Department of Physics, Giresun University, Giresun 28200, Turkey.
[4]Department of Physics, Florida A&M University, Tallahassee, Florida 32307, USA.

Abstract. The dynamics of rotational wave packets of laser-aligned linear molecules was studied with femtosecond laser pulses in a pump-probe configuration. The alignment produced with a pump beam creates a transiently anisotropic medium, which exhibits birefringence and also a dependence of ionization on the mutual orientation of the polarizations of the probe and pump beams. The induced molecular alignment was observed via variations of the yields of photoelectrons in strong field ionization. In addition, the induced transient birefringence due to alignment revivals altered the nonlinear interaction of the probe beam with the gas, resulting in delay dependent modulation of the probe beam. The revival signatures measured for N_2, CO and C_2H_2 gas molecules were in good agreement with the calculated molecular alignments taking into account the effects of electronic structure and molecular symmetry.

1. Introduction

In the nonadiabatic alignment, i.e. when the laser pulse duration is less than the characteristic molecular rotational period, the short laser pulse induces in molecules a rotational wave packet. The dipole moment due to the anisotropic polarizability of the molecule tends to align its axis of highest polarizability along the laser polarization direction [1]. The produced wave packet, i.e. a coherent superposition of the rotational states shows dephasing and re-phasing in time after the laser pulse has ended. Consequently, the gaseous medium exhibits anisotropic properties, which depend on the delay relative to the excitation pump pulse. Studies on molecular alignment and orientation in molecules (see [2, 3]) have attracted considerable interest due to the variety of applications and different manifestations of the alignment process, such as controlling chemical reactions [4], selectively-controlled alignment of isotopes [5], imaging of molecular structures [6], nanolithography with molecular beams [7], pulse compression [8], and quantum information processing [9]. Femtosecond laser technology allows aligning molecules and observing the molecular dynamics by using a pump-probe technique [10]. Alignment initiated by a pump pulse creates a dynamically anisotropic medium evolving in time, and hence, concurrent variations of the interaction with such a medium of a probe pulse can be observed by ionization [11], fragmentation [12, 13], high harmonic generation [14] or by spatial and spectral modulation of the probe pulse [15-17].

For many applications, it is important to align molecules outside the plane of polarization [18], which can be achieved by circularly polarized probe pulses. The circularly polarized aligning pulse produces a torque on the molecules by pulling the most polarizable axis into the plane of polarization. This creates

IOP Conf. Series: Journal of Physics: Conf. Series **826** (2017) 012001 doi:10.1088/1742-6596/826/1/012001

a rotational wavepacket that re-aligns under field-free conditions to the plane of polarization every revival period [2]. For a linear molecule, just before the revival, the molecular axis first aligns along the direction of laser propagation, which is called 'k-alignment' [19], then the molecule keeps rotating, and in a short time later the molecular axis is aligned with the field polarization plane.

Of particular interest is the effect of alignment on filament formation and white light generation. The nonlinear polarization exerted by a pump pulse yields an intensity-dependent refractive index and a transient birefringence as a result of the pulse interaction with the molecular gas. Then a probe pulse propagates in a medium with properties that depend on the delay in the wake of the pump pulse. For certain conditions, the intense probe pulse can experience self-focusing and spatially collapses due to the dynamic Kerr effect. This collapse however stops when the intensity becomes high enough to generate free electrons and nonlinear losses tending to defocus the beam resulting in filamentary propagation [20]. The spatial and temporal characteristics of the aligned molecules affect filamentation and white light generation [21], so these processes can be also used for probing the dynamics of molecular alignment. Here we present our recent work on different manifestations of molecular alignment in the interaction of intense femtosecond laser pulses with molecular gases. We monitored the effect of aligning by observing the strong field ionization yield of photoelectrons by using a pump-probe approach. The molecular ionization rate is often maximal when the maximum of the initial electron density distribution corresponds to the laser polarization direction of the applied field. However, not all diatomic molecules have their highest occupied molecular orbital (HOMO), which donates electrons most readily, aligned along the molecular axis. The measured molecular alignment is compared to the calculated time dependence of the molecular alignment parameter $\langle\langle\cos^2\theta\rangle\rangle$, and the effects of the electronic structure and symmetry of molecules on the signatures of alignment revivals were investigated. We study effects of molecular alignment with linear and circularly polarized laser pulses on strong field photoionization of electrons. We also observed transmission spectral modulation of a probe pulse due to the induced transient birefringence produced by alignment revivals. By changing the angle between polarizations of the pump pulse and the filament-producing probe pulse and performing measurements for a range of pump-probe delays around rotational revivals, we obtained a map of the rotational wave packet as a function of the polarization mismatch and the delay between the pump and probe pulses, which is also known as a "quantum carpet" [22]. The experimental results on the dynamics of rotational wave packets are compared with quantum-mechanically calculated angular and time dependencies of the molecular alignment parameter, showing good agreement.

2. Theory

The time-dependent Schrödinger equation describing a molecule in an intense laser field can be written based on the rigid rotor model as

$$i\hbar\frac{\partial}{\partial t}\Phi_{JM}(t)=\left(H_R^{(0)}+V_L(t)\right)\Phi_{JM}(t), \tag{1}$$

where $H_R^{(0)}=B\hat{J}^2$ is the field-free rotational Hamiltonian of the molecule with B and $\hat{J}$ being the rotational constant and the angular momentum operator, respectively. The interaction term is $V_L(t)=-\frac{1}{2}\ \Delta\alpha\cos^2\theta+\alpha_{\perp}\ \left\langle\varepsilon^2\ t\right\rangle$ for a linear symmetric top molecule subjected to the linearly polarized laser field $\varepsilon\ t$ of the field of the pump pulse; in case of a circular polarization the expression of the nonresonant dipole potential for circular polarization becomes $V\ t\ =-1/4\ \alpha_{\parallel}+\alpha_{\perp}-\Delta\alpha\cos^2\theta\ E^2\ t$ [23-26], where $\Delta\alpha=\alpha_{\parallel}-\alpha_{\perp}$ is the difference between the polarizability components parallel $\alpha_{\parallel}$ and perpendicular $\alpha_{\perp}$ to the molecular axis and θ is the angle between the molecular axis and the field propagation direction of the aligning pulse.

The rotational wave function $\Phi(t)$ can be expanded in a sum of free-field rotor eigenfunctions $|JM\rangle$ with eigenenergies $E_J = BJ(J+1)$, where J is an integer of the orbital momentum quantum number, and for each J the projection quantum number M takes the values $-J$, $-(J-1)$, ..., 0, ...$(J-1)$, J. Then the evolution of the initial state, $|\Phi_{J_0M_0}(t)\rangle$, created by the pump pulse is described by an expansion in the rotational wave packet states [2]:

$$|\Phi_{J_0M_0}(t)\rangle = \sum_{JM} d_{JM}^{J_0M_0}(t)\exp\left(-i\frac{E_J}{\hbar}t\right)|JM\rangle, \tag{2}$$

where the expansion coefficients $d_{JM}^{J_0M_0}(t)$ are determined by solving the set of differential equations [27] stemming from the time-dependent Schrödinger Eq. (1). The degree of alignment of an ensemble at temperature T can be found by averaging the different states over the Boltzmann distribution

$$\langle\langle\cos^2\theta\rangle\rangle(\mathbf{r},t) = \frac{\sum_{J_0}\sum_{M_0=-J_0}^{J_0} g_{J_0}\langle\cos^2\theta\rangle_{J_0M_0}(\mathbf{r},t)\exp\left[-BJ_0(J_0+1)/kT\right]}{\sum_{J_0}\sum_{M_0=-J_0}^{J_0} g_{J_0}\exp\left[-BJ_0(J_0+1)/kT\right]}, \tag{3}$$

where $\langle\cos^2\theta\rangle_{J_0M_0}(\mathbf{r},t) = \langle\Phi_{J_0M_0}(t)|\cos^2\theta|\Phi_{J_0M_0}(t)\rangle = \sum_{J',M'}\sum_{J,M} d_{J'M'}^{*J_0M_0} d_{JM}^{J_0M_0}\langle J',M'|\cos^2\theta|J,M\rangle\varphi_{J,J'}$ is the alignment degree for a single individual rotational state $|J_0M_0\rangle$, $\varphi'_{J,J\pm2} = \exp[-i(\pi\tau(4J+6)+\Delta)]$, where Δ is determined by the configuration of the HOMO orbital. We note that $\langle\cos^2\theta\rangle_{J_0M_0}(t)$ depends on coefficients $d_{JM}^{J_0M_0}(t)$, which in turn are determined by the interaction term that has certain spatial distribution over position vector $\mathbf{r}$. An additional factor g_{J_0} originates from the nuclear spin statistics. The alignment degree depends on molecular properties, i.e., the rotational constant B, statistical weight factors g_{J_0}, the ensemble temperature T and the polarizabilities $\Delta\alpha$ as well as on the laser fluence.

When molecules with anisotropic polarizability experience the aligning (pump) pulse in a non-adiabatic alignment regime, the refractive index changes both spatially and temporally. The nonadiabatic molecular alignment induces a periodic modulation of the refractive index $\Delta n(\mathbf{r},t)$ of the gas. For the probe pulse with the polarization direction of the aligning pulse the resulting birefringence can be characterized by the difference [28]

$$\Delta n(\mathbf{r},t) = \Delta n_{\parallel}(\mathbf{r},t) - \Delta n_{\perp}(\mathbf{r},t) = \frac{3\pi N}{n_0}\Delta\alpha\left(\langle\langle\cos^2\theta\rangle\rangle(\mathbf{r},t) - \frac{1}{3}\right), \tag{4}$$

where $\langle\langle\cos^2\theta\rangle\rangle(\mathbf{r},t)$ is the thermally averaged alignment expectation value given by Eq. (3), N is the molecular density, and n_0 is the linear refractive index. The spatial dependence, which follows the intensity distribution of the pump beam, creates an additional lensing effect for the probe beam due to the spatially inhomogeneous molecular alignment. We note that polarizabilities are modified when the field becomes sufficiently strong, in particular, absorption can include also the contribution from multiphoton processes and the changes of both the real and imaginary parts of the refractive index are proportional to the alignment factor, $\left(\langle\langle\cos^2\theta\rangle\rangle(\mathbf{r},t) - \frac{1}{3}\right)$.

3. Experimental setup

The pump-probe setup is depicted in Figure 1. A mode-locked Ti:sapphire laser oscillator provides ~20fs pulses at a repetition rate of 80 MHz. These pulses are seeded into a regenerative laser amplifier, which outputs ~50fs laser pulses at a repetition rate of 1 kHz and a central wavelength of 800 nm. Since shorter pulses have a higher peak intensity for a given pulse energy, temporal compression of the laser pulses in the focus was achieved by maximizing the integrated photoelectron yield, while adjusting the grating compressor in the laser amplifier. The maximum pulse energy from the amplified laser system was ~1 mJ.

In the experiments, different gases (N_2, CO or C_2H_2 from Matheson TRIGAS, with purities of 99.9995%, 99.9%, 99.6%, respectively) were admitted to the chamber through an adjustable leak valve. The pressure of the gases in the chamber was adjusted to 6×10^{-6} mbar, which was about three orders of magnitude higher than the background pressure. The linearly polarized incident light was split by a beam-splitter into two beams as pump and probe pulses with about equal intensities. A set consisting of a half-wave plate and a polarizer was placed into the pump arm assuring the proper polarization and intensity of the pump pulse.

The time delay between the pump and probe pulses was precisely adjusted using an optical delay line with a translational stage controlled by a stepping motor (GTS150, ESP300, Newport), which provided a high temporal resolution of ~0.67 fs. Another beam-splitter was used to recombine the two beams. The laser beam was focused in the chamber filled with gas by a 20cm achromatic lens. The pump pulse creates a rotational wave packet, while the probe pulse, with an adjustable time delay with respect to the pump pulse, ionizes molecules in the focus, producing electrons. The ionized electrons, ejected along the polarization direction of the probe laser field were detected by the MCP detectors. The electrons traveled within a μ-metal TOF tube in a magnetic field-free region. The signals from the MCP detectors were amplified by a preamplifier (ZKL-2 Mini-Circuits). A FAST ComTech MCS6 multiple-event time digitizer with 100 ps time resolution was used for data acquisition. Laser pulses were detected behind one of the mirrors on the probe arm by a photodiode, and the signal was used to trigger the data-acquisition software. The total number of counts, depending on the pump-probe delay, was registered as the output signal.

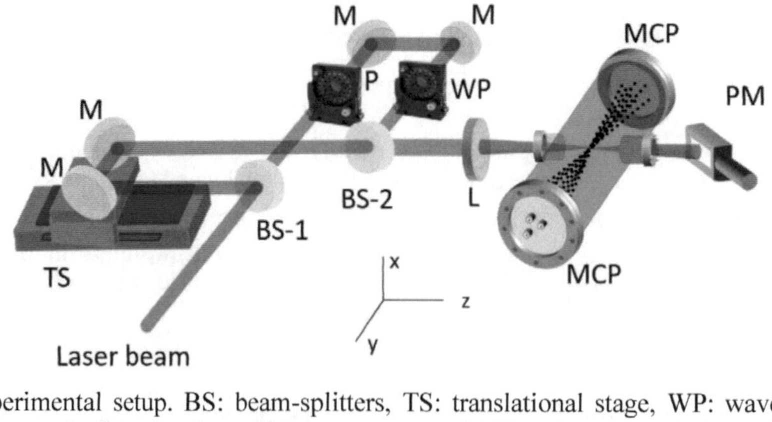

Figure 1. Experimental setup. BS: beam-splitters, TS: translational stage, WP: wave plate, M: flat mirrors, L: achromatic focusing lens, PM: power meter, MCP: microchannel plates. The dots depict electrons flying toward the MCP detectors.

When studying induced gas birefringence we used a gas cell at relatively high pressure (~4bar) instead of the vacuum chamber utilized in experiments with photoelectrons. In addition, the power measurement of the probe pulse variations were performed by using a photodiode power meter (Ophir NOVA II with PD300-UV head), the spectral measurements were performed with an Ocean Optics USB-2000 spectrometer and the pulse duration measurements were done with a correlator Grenouille (8-20, Swamp Optics).

4. Linear molecules aligned with linearly polarized laser pulses

We have observed alignment effects [29] with different gases: N_2, CO, and C_2H_2. The choice of the molecules in our study was determined by their different configurations of HOMO orbitals. We present

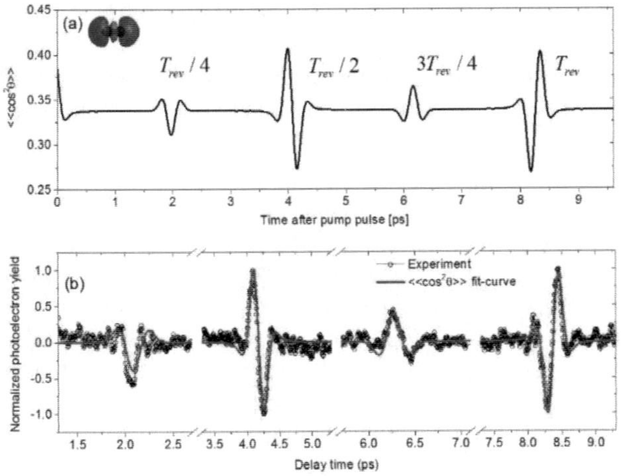

Figure 2. (a) The calculated time dependence of the molecular alignment parameter induced by the pump pulse in N_2; the configuration of the HOMO is also shown; (b) the time evolution of the measured SFI signal for N_2 molecules and the fit curve. Collinearly polarized pump and probe pulses were used with peak intensities of $2x10^{13}$ W/cm^2 and $7.8x10^{13}$ W/cm^2, respectively.

observed temporal signatures of the alignment and compare them to the calculated time evolution of the alignment parameter $\langle\langle \cos^2\theta \rangle\rangle$, which depends on the temperature, molecular properties and laser parameters. To demonstrate the effect of the HOMO configuration we present the results for N_2 and C_2H_2.

Figure 2(a) shows the calculated alignment parameter $\langle\langle \cos^2\theta \rangle\rangle$ at temperature T=300 K with our laser parameters and the properties of N_2 molecules with rotational constant B=1.9896cm^{-1} [30, 31]. Figure 2(b) depicts the measured signal of the SFI yield for N_2 molecules aligned by a linearly polarized pump pulse. The photoelectron yield is smaller at $T_{rev}/4$ and $3T_{rev}/4$ and larger at $T_{rev}/2$ and T_{rev}. For the interpretation of the results, we consider the total wave function of a linear molecule, which according to the Born-Oppenheimer approximation can be factorized, $\Psi_{tot} = \psi_{el} \times \psi_{vib} \times \psi_{rot} \times \psi_{ns}$, as a product of the electronic wave function ψ_{el}, the vibrational wave function ψ_{vib}, the rotational wave function ψ_{rot}, and the nuclear spin wave function ψ_{ns}. The observed alignment signatures can be explained in the following way. $^{14}N_2$ is a boson, and therefore the total wave function Ψ_{tot} is symmetric [32]. At room temperatures, the electronic ψ_{el} ($^3\sum_g^+$), and the vibrational ψ_{vib} wave functions of ^{14}N correspond to the ground state, and both are symmetric [32]. Therefore, to understand the ψ_{rot} behavior, we should consider the symmetric and antisymmetric forms of the nuclear spin wave function ψ_{ns} of $^{14}N_2$. For Ψ_{tot} to be symmetric, both ψ_{rot} and ψ_{ns} must be symmetric or antisymmetric. For ^{14}N the nuclear spin is I=1 and the nuclear spin of a N_2 molecule takes only values I_{tot}=0, 1, or 2. Consequently, N_2 exists in even (I_{tot}=0, 2) and odd (I_{tot} =1) forms. A state degeneracy with I_{tot} is 2 I_{tot} +1, so the even-N_2 and odd-N_2 statistical weights are 6 and 3, respectively. Due to the relative abundance of the even-N_2 versus odd-N_2, 2:1 for even/odd states, their opposed contributions to the $T_{rev}/4$ revival do not cancel completely, however the signal has a reduced amplitude compared to the T_{rev} and $T_{rev}/2$ revivals, as is observed in the experiment (Fig. 2(b)). Consequently, the measured yield in Fig. 2(b) varies in phase with the time dependence of the molecular alignment parameter of Fig 2(a). In Fig. 2(b) we also presented the corresponding fit curve based on Eq. (3) for the alignment parameter $\langle\langle \cos^2\theta \rangle\rangle$. Thus we can conclude that the phase shift $\Delta = 0$ for the N_2.

The molecular ionization rate is therefore maximal when molecules are aligned along the laser polarization direction and the configuration of the HOMO orbital is parallel to the molecular axis, because such configuration makes ejection of an electron easier. This agrees with the σ_g HOMO symmetry of the neutral nitrogen molecule [33], where the molecular ionization rate is maximal, when the molecules are aligned along the probe polarization direction. As seen in the inset of Fig. 2(a), N_2 has its maximum electron density along the internuclear (molecular) axis in its σ_g HOMO configuration. By fitting the curve within one full revival period and taking the same laser parameters as experimental ones, we also found the characteristic rotational constant and the corresponding revival time for nitrogen gas. We note that no offset for the time-axis was introduced. The theoretical values of the rotational

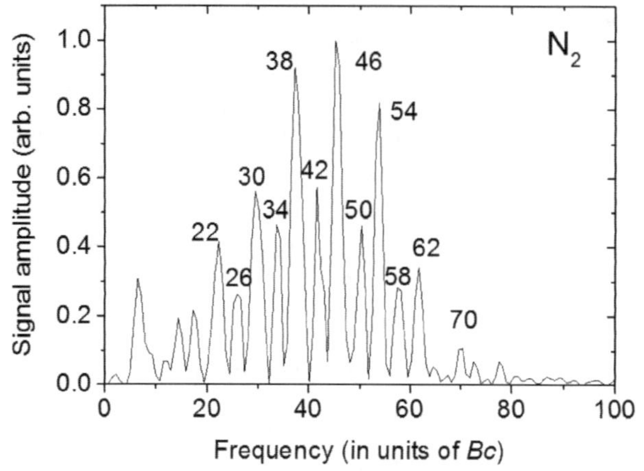

Figure 3. Frequency spectrum amplitudes of the measured time-dependent SFI yield for N_2 shown in Fig. 2(b). The numbers on spectral peaks, which correspond to the allowed transitions, show the frequencies in terms of $4J$+6. The series (6, 14, 22, 30, 38, 46, 54, ...)Bc and (10, 18, 26, 34, 42, 50, 58, ...)Bc come from even and odd values of J's, respectively.

constant 1.9896cm^{-1} [30, 31] and the revival time 8.383ps agree well with the experimental values of the rotational constant B=(1.999±0.009)cm^{-1} and the corresponding revival time 8.342ps.

In Fig. 3, we show the corresponding frequency spectrum amplitudes of the measured time-dependent SFI yield for N_2 obtained by a Fourier transform (FT) analysis. The spectrum reveals two sequences of allowed Raman transitions obtained from the matrix elements of $\langle\langle\cos^2\theta\rangle\rangle$, which are calculated as $(E_{J+2}-E_J)=(4J+6)hBc$ with the selection rule $\Delta J=\pm 2$. Namely, these sequences are: (6, 14, 22, 30, 38, ...)hBc for even J's and (10, 18, 26, 34, 42,...) hBc for odd J's. Recall that the nuclear spin of the nitrogen nucleus is 1, and both even and odd rotational states are permitted for the nitrogen molecule [5]. The relative ratio of the even and odd J states is 2:1, which follows from the nuclear spin statistics and gives an additional factor of ~2 for even J's in the spectrum [34].

Figure 4 (a) shows the alignment parameter $\langle\langle\cos^2\theta\rangle\rangle$ calculated for a temperature T=300 K with our laser parameters for acetylene (C_2H_2) molecules with the rotational constant B=1.1766 cm^{-1} [35] and measured time trace of the ionization yield with variations due to alignment of C_2H_2 molecules. Acetylene has both even and odd J-states, which are populated with the ratio 1:3 [36]. As a result of this fact, even and odd wave packets partially cancel each other, and some alignment and antialignment are observed at quarter revivals as seen in Fig. 4(a,b). The experimental results on molecular revivals of C_2H_2 molecules at quarter revival, half revival, three-quarter revival, and full revival are in good agreement with the calculated time dependence of molecular alignment $\langle\langle\cos^2\theta\rangle\rangle$ parameter, but with the inverted polarity. This can be understood from the configuration of HOMO for acetylene, which is dominated by a π_u orbital [37]. Molecular orbitals of π_u type have higher electron density above/below the internuclear axis, with a node along the internuclear axis, as shown in the inset of Fig. 4(a). Consequently, the temporal structure of C_2H_2 revivals in Fig. 4(b) has inverted polarity as compared to the calculated molecular alignment parameter $\langle\langle\cos^2\theta\rangle\rangle$ in Fig. 4(a). With $\Delta=\pi$ the temporal signature of the molecular alignment parameter $\langle\langle\cos^2\theta\rangle\rangle$ is inverted, as seen in Fig. 4 (b), and the calculated function gives the shape similar to the experimental one. The theoretical value of the

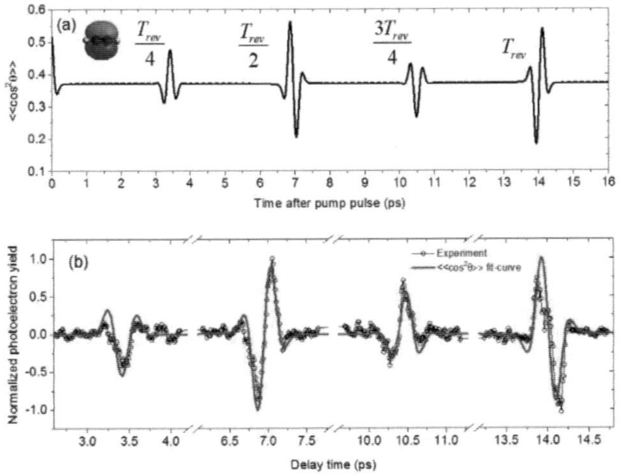

Figure 4. (a) The calculated for C_2H_2 time dependence of the molecular alignment parameter; the configuration of the HOMO is also shown. (b) The time evolution of the measured SFI signal for C_2H_2 and the fit curve. We used linearly polarized pump and probe pulses with peak intensities of $3x10^{13}$ W/cm^2 and $7.6x10^{13}$ W/cm^2, respectively.

molecular constant for C_2H_2, B=1.1766cm^{-1} [35], and the respective revival time 14.175ps match well the characteristic rotational constant B=(1.180±0.003) cm^{-1} and the revival period 14.133ps following from the fitting procedure.

The frequency spectrum of the time-dependent ionization yield for C_2H_2 is shown in Fig. 5. The number on each spectral peak shows the frequency in terms of $4J+6$ which corresponds to allowed Raman transitions, given by the series (6, 14, 22, 30, 38, 46, 54, …)hBc for even J and by the series (10, 18, 26, 34, 42, 50, 58, …)hBc for odd J. The relative ratio of the amplitudes of the even and odd spectral components is close to 1:3, following from the nuclear spin statistics.

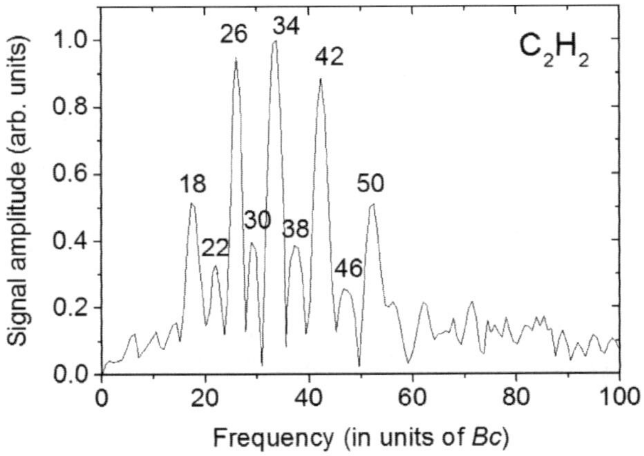

Figure 5. Frequency spectrum amplitudes of the measured time-dependent photoelectron yield for C_2H_2, shown in Fig. 4(b). The numbers on the spectral peaks show frequencies in terms of 4J+6, which correspond to allowed Raman transitions from the series (6, 14, 22, 30, 38, 46, 54, …)hBc for even J and the series (10, 18, 26, 34, 42, 50, 58, …)hBc for odd J. The relative ratio of the magnitudes of the spectral harmonics corresponding to even and odd J correlates with the nuclear spin statistics ratio 1:3.

5. Impulsive molecular alignment by circularly polarized laser pulses

A combination of a polarizer and a quarter-wave plate placed into the pump arm (Fig. 1) assured the proper circular polarization and power of the pump pulse. A pump pulse with circular polarization produces a net alignment along the laser pulse propagation axis at certain phases of the evolution. This gives the possibility to control alignment of linear molecules outside the plane of polarization, which can provide new capabilities for molecular imaging. As an example we show results for CO, which contains nonidentical nuclei, and therefore no additional factor arises from the nuclear spin statistics [38-40]. Thus, the revivals at odd multiples of $T_{rev}/4$ are completely cancelled, whereas only the

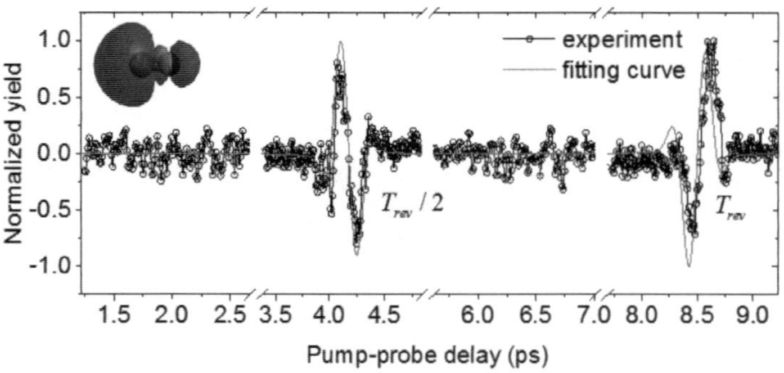

Figure 6. Strong-field ionization yield of impulsively aligned CO gas molecules along with the fitting curve.

revivals at multiples of $T_{rev}/2$ remain, as presented in Fig. 6. From the fitting function, we found $\Delta = 0$ for CO, which is similar to the N_2 measurements. CO has σ HOMOs, so there is no nodal plane along the internuclear axis, and therefore the calculated time dependency of molecular alignment is in phase with the measured strong-field photoelectron yield. For CO, we calculated the characteristic rotational constant as $1.9553 cm^{-1}$, and the corresponding revival period is 8.529ps.

6. Alignment with arbitrary relative polarizations of pump and probe pulses

In order to investigate how the structure of the rotational full revival changes with polarization angle α, we ran the experiment for a series of pump-probe delay scans with different linear polarization angles of the pump pulse with respect to the horizontally polarized probe pulse. For this experiment we used a cell filled with N_2 gas at typical pressure of 4bar. The created alignment was monitored by the nonlinear spectral modulation of the probe pulse. By blocking the probe pulse, we checked that no spectral modification was produced with the pump pulse only. The resulting structure of such a mapping of the rotational wave packet the so-called "quantum carpet" [22] is shown in Fig. 7: experimental results in

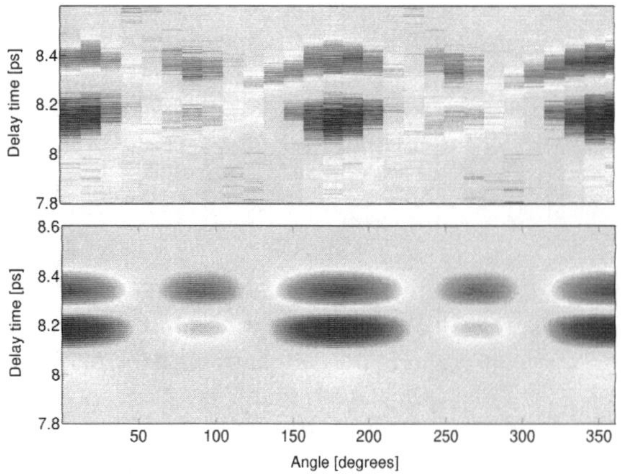

Figure 7. Experimental (top) and theoretical (bottom) quantum carpets of a rotational full revival around 8.3ps.

the top and the theoretical calculation in the bottom. The calculation is performed based on Eq. (3) using the relation: $\langle\langle\cos^2\theta'\rangle\rangle\ t = \frac{1}{2}\ 3\cos^2\alpha - 1\ \langle\langle\cos^2\theta\rangle\rangle\ t + \frac{1}{2}\sin^2\alpha$, where θ' is the angle between molecular axis and polarization of the probe beam, θ is the angle between the molecular axis and the field propagation direction of the aligning pulse and α is the relative angle between the polarizations of pump and probe pulses. The angle and delay dependent modulation (Fig. 7) shows that the transmitted power of the probe beam closely correlates with the calculated alignment parameter. The enhancement

/reduction of the alignment is shown by red/blue colours, respectively. We note the existence of angles $\alpha_c \approx 55°$, 125°, 235°, 305°, where the signal is independent of the time delay, which is a specific signature of the σ_g symmetry of the active molecular orbital, characteristic of the nitrogen molecule [41].

7. Conclusions

The revival signatures of the linear molecules produced by linearly and circularly polarized femtosecond pump pulses have been studied under nonadiabatic alignment conditions and measured by the detection of the strong field ionization photoelectron yield in a pump-probe setup with a variable delay. The revivals at odd multiples of $T_{rev}/4$ are strongly affected by the nuclear spin-statistics factors. The effect of the configuration of the HOMO electron density distribution on the shape of the temporal dependence of the photoelectron yields measured with aligned molecules was discussed. N_2 and CO have their maximum electron densities along the internuclear (molecular) axis due to σ_g and σ HOMOs, respectively. The molecular ionization rate is maximal when molecules are aligned along the laser polarization direction, because such configuration makes ejection of an electron easier. Consequently, in case of N_2 and CO our measured photoelectron yields were modulated in phase with the time dependency of the molecular alignment parameter. For the C_2H_2, due to the π_u HOMO symmetry of C_2H_2, which has higher electron density above/below the internuclear axis with a node along the direction of this axis, our experimental photoelectron yields have a polarity-inverted dependence compared to the molecular alignment parameter (a (-π) shift). By taking into account this additional phase shift a good agreement between calculated and measured alignment signatures was obtained. In addition, the spectra of revivals reveal peaks corresponding to sequences of transitions between even and odd rotational states with amplitude ratios in agreement with the population ratio following from the nuclear spin statistics. Thus, we have demonstrated that the implemented alignment monitoring by measuring photoelectron yields presents a viable and efficient way of studying rotational molecular dynamics. The impulsive alignment of CO molecules with circularly polarized pump pulses was demonstrated. The ability of controlling the alignment of linear molecules outside the plane of polarization can open new capabilities for molecular imaging.

Using an N_2-filled gas cell at relatively high pressure (~4bar) we carried out an experiment to study the molecular alignment wakes of femtosecond laser pulses by modulation of the probe pulse. The variations of the transmitted probe pulse were explained by the induced intensity-dependent variations of the nonlinear absorption. In addition, we mapped ultrafast rotational wave packets as a function of the pump-probe delay and the angle between their polarizations (i.e. a "quantum carpet" was experimentally measured).

Acknowledgments

This work was supported by the Robert A. Welch Foundation Grant No. A1546, and the Qatar Foundation under the grant NPRP 6-465-1-091.

References

[1] Xu N, Li J, Li J, Zhang Z and Fan Q 2011 *Lasers - Applications in Science and Industry*, ed K. Jakubczak, p 229
[2] Stapelfeldt H and T Seideman T 2003 *Rev. Mod. Phys.* **75** 543
[3] Seideman T and Hamilton E 2005 *Adv. At., Mol., Opt. Phys.* ed P R Berman and C C Lin (Cambridge: Academic Press) **52** 289
[4] Henrik S 2004 *Phys. Scr.* **2004** 132.
[5] Fleischer S, Averbukh I S and Prior Y 2006 *Phys. Rev.* A **74** 041403
[6] Itatani J, Levesque J, Zeidler D, Niikura H, Pepin H, Kieffer J C , Corkum P B and Villeneuve D M 2004 *Nature* **432** 867
[7] Gordon R J, Zhu L, Schroeder W A and Seideman T 2003 *J. Appl. Phys.* **94** 669
[8] Bartels R A, Weinacht T C, Wagner N, Baertschy M, Greene C H, Murnane M M and Kapteyn H C 2001 *Phys. Rev. Lett.* **88** 013903 (2001).

[9] Lee K F, Villeneuve D M, Corkum P B and Shapiro E A, 2004 *Phys. Rev. Lett.* **93**, 233601
[10] Zewail A H 2000 *J. Phys. Chem.* A **104**, 5660
[11] Pavičić D, Lee K F, Rayner D M, Corkum P B and Villeneuve D M 2007 *Phys. Rev. Lett.* **98**, 243001

[12] Péronne E, Poulsen M D, Bisgaard C Z, Stapelfeldt H and Seideman T 2003 *Phys. Rev. Lett.* **91**, 043003
[13] Nevo I, Holmegaard L, Nielsen J H, Hansen J L, Stapelfeldt H, Filsinger F, Meijer G and Kupper J 2009 *Phys. Chem. Chem. Phys.* **11** 9912
[14] Vozzi C, Calegari F, Benedetti E, Caumes J P, Sansone G, Stagira S, Nisoli M, Torres R., Heesel E, Kajumba N, Marangos J P, Altucci C and Velotta R 2005 *Phys. Rev. Lett.* **95** 153902
[15] Ripoche J-F, Grillon G, Prade B, Franco M, Nibbering E, Lange R and Mysyrowicz A 1997 *Opt. Commun.* **135** 310
[16] Calegari F, Vozzi C, Gasilov S, Benedetti E, Sansone G, Nisoli M, De Silvestri S.and Stagira S 2008 *Phys. Rev. Lett.* **100** 123006 (2008).
[17] Kaya N, Kaya G, Sayrac M, Boran Y, Anumula S, J. Strohaber, A. A. Kolomenskii and H. A. Schuessler, *Opt. Express* **24** 2562 (2016).
[18] Smeenk C T L, Arissian L, Sokolov A V, Spanner M, Lee K F, Staudte A, Villeneuve D M and Corkum P B 2014 *Phys. Rev. Lett.* **112** 253001
[19] Smeenk C T L and Corkum P B 2013 *J. Phys. B At. Mol. Opt. Phys.* **46** 201001
[20] Couairon A and Mysyrowicz A 2007 *Phys. Rep.* **441** 47
[21] Chen Y H , Varma S and Milchberg H M 2008 J. Opt. Soc. Am. B **25** B122
[22] Bryan W A, English E M L, McKenna J, Wood J, Calvert C R, Turcu I C E, Torres R, Collier J L , Williams I D and Newell W R 2007 *Phys. Rev.* A **76**, 023414
[23] Seideman T 1997 *J. Chem. Phys.* **107** 10420
[24] Graybeal J D *Molecular Spectroscopy* 1988 (McGraw-Hill: New York)
[25] Purcell S M and Barker P F 2010 *Phys. Rev. A* **82** 033433
[26] Purcell S M Barker P F 2009 *Phys. Rev. Lett.* **103** 153001
[27] Ortigoso J, Rodríguez M, Gupta M and Friedrich B 1999 *J. Chem. Phys.* **110** 3870.
[28] Yuan S, Li M, Feng Y, Li H, Zheng L, Chin S L and Zeng H, *J. Phys. B At. Mol. Opt. Phys.* **48**, 094018 (2015)
[29] Kaya G, Kaya N, Strohaber J, Hart N A, Kolomenskii A A and Schuessler H A *Appl. Phys.* B **122** 288 (2016)
[30] Haynes W M L D R 2011 *CRC handbook of chemistry and physics: a ready-reference book of chemical and physical data* (Boca Raton: CRC Press)
[31] Bendtsen J 1974 *J. Raman Spectrosc.* **2** 133-45.
[32] Atkins P, de Paula J and Friedman R 2009 *Quanta, Matter, and Change: A Molecular Approach to Physical Chemistry* (Oxford: Oxford University Press)
[33] Tong X M , Zhao Z X and Lin C D 2002 *Phys. Rev.* A **66** 033402
[34] Faisal F H M, Abdurrouf A , Miyazaki K and Miyaji G 2007 *Phys. Rev. Lett.* **98** 143001
[35] Herman M, Campargue A, El Idrissi M I and Vander Auwera J 2003 J. *Phys. Chem. Ref. Data* **32**, 921
[36] Li H, Li W, Feng Y, Liu J, Pan H and Zeng H 2012 *Phys. Rev.* A **85** 052515
[37] Torres R, Kajumba N, Underwood J G, Robinson J S, Baker S, Tisch J W G, de Nalda R, Bryan W A, Velotta R, Altucci C, Turcu I C E and Marangos J P 2007 *Phys. Rev. Lett.* **98** 203007
[38] R B Philip and Per J 2006 *Molecular Symmetry and Spectroscopy, 2nd Ed* (Ottawa: NRC Research Press)
[39] McQuarrie, D A and J D Simon 1997 *Physical Chemistry: A Molecular Approach* (Herndon: University Science Books)
[40] Kaya N, Kaya G, Strohaber J, Kolomenskii A A and Schuessler H A 2016 *Eur. Phys. J.* D **70** 224
[41] Abdurrouf A and Faisal F H M 2009 *Phys. Rev.* A **79** 023405

Using Raman spectroscopy to determine the structure of copolymers and polymer blends

G Yu Nikolaeva[1], E A Sagitova[1], K A Prokhorov[1], P P Pashinin[1], P M Nedorezova[2], A N Klyamkina[2], M A Guseva[3] and V A Gerasin[3]

[1] Prokhorov General Physics Institute, Russian Academy of Sciences, Vavilov St. 38, 119991, Moscow, Russia
[2] Semenov Institute of Chemical Physics, Russian Academy of Sciences, Kosygin St. 4, 119991, Moscow, Russia
[3] Topchiev Institute of Petrochemical Synthesis, Russian Academy of Sciences, Leninsky Pr. 29, 119991, Moscow, Russia

Email: nikolaeva@kapella.gpi.ru

Abstract. We present Raman structural study of two grades of random propylene/1-octene copolymers with low and high molecular weights and the 1-octene content up to 4.5 mole %. The copolymer spectra are compared with the spectra of the α, γ, and smectic modifications of isotactic polypropylene. Raman investigation has showed that the degree of crystallinity and conformational order of the copolymer macromolecules slightly decrease with the growth of the 1-octene content. The degree of crystallinity is slightly higher for the samples of the high-molecular-weight grade compared to the low-molecular-weight one. Furthermore, we present Raman spectra of polyethylene/polypropylene (90/10) blends, mixed in the polyethylene melt at three different temperatures, corresponding to three different states of polypropylene macromolecules. It was concluded that the degree of crystallinity and conformational order of the polyethylene macromolecules in the blends are the highest for the temperature, at which polypropylene macromolecules have lost their packing and conformational order.

1. Introduction

Along with a filler loading, copolymerization and blending are effective and inexpensive ways of polymer structure modification. In principle, these three ways allow purposeful and wide-range variation in the material properties and, as a consequence, creation of state-of-the-art functional materials for new high-tech applications. To make development of a new material maximally efficient, researchers need an informative, non-destructive and fast method of quantitative structural characterization. Raman spectroscopy is exactly the method, which meets all these requirements. Its main advantage is the ability to provide quantitative information on the macromolecules in different configurational, conformational, and phase states, including description of various crystalline modifications and amorphous regions with different state of order.

In this contribution we put an emphasis on Raman structural study of random propylene copolymers and polyethylene/polypropylene (PE/PP) blends, which are practically important materials due to their excellent physical-chemical properties and relatively low cost. Interpretation of the copolymer spectra was based on comparison with Raman spectra of various polymorphs of isotactic PP with the degree of crystallinity, varied in a wide range (from 0 to 70%).

2. Experimental

We study the nascent form of the random propylene/1-octene copolymers with the 1-octene content of 0 mole % (α-iPP1), 0.7 mole % (C1-0.7), and 4.4 mole % (C1-4.4) (Series 1, table 1). The copolymers were synthesized in the medium of liquid propylene over the homogeneous metallocene catalyst of C_2-symmetry $Me_2SiInd_2ZrCl_2$ by using the method, described in [1]. We compare the Raman spectra of these copolymers with the spectra of the nascent form of the random propylene/1-octene copolymers, synthesized in the medium of liquid propylene over the homogeneous metallocene catalyst of C_2-symmetry *rac*-$Me_2Si(4\text{-}Ph\text{-}2\text{-}MeInd)_2ZrCl_2$ (Series 2, table 1, samples α-iPP2 and C2-4.5). Method of the synthesis as well as differential scanning calorimetry (DSC), X-ray analysis, and mechanical testing data for the copolymers from the Series 2 were published in [1]. Raman spectra of the copolymers of the Series 2 with the 1-octene content up to 24 mole % are analyzed in detail in [2,3].

Two series of the copolymers differ mainly in molecular weight (M_w), measured by gel permeation chromatography (table 1, [1]). Table 1 also summarizes data on the degree of crystallinity, determined by X-ray analysis, DSC and Raman spectroscopy (for details see Results and Discussion section). The mole content of the comonomers in the copolymers was measured by ^{13}C NMR spectroscopy [1].

Table 1. The degree of crystallinity of the propylene/1-octene copolymers and neat PP samples, measured by X-ray analysis, DSC and Raman spectroscopy.

Sample / Molecular weight	Crystallinity, %		
	X-ray analysis	DSC	Raman spectroscopy I_{809}/I_{ref} (I_{809}/I_{1463})
Propylene/1-octene copolymers, Series 1, catalyst $Me_2SiInd_2ZrCl_2$			
α-iPP1 (isotactic), M_w=49 800	68 (α phase)	65	53 (0.95)
copolymer (C1-0.7) 0.7 % of 1-octene, M_w=45 200	65 (α phase)	63	52 (0.93)
copolymer (C1-4.4) 4.4 % of 1-octene, M_w=30 800	60 (α phase)	57	50 (0.87)
Propylene/1-octene copolymers, Series 2, catalyst *rac*-$Me_2Si(4\text{-}Ph\text{-}2\text{-}MeInd)_2ZrCl_2$			
α-iPP2 (isotactic), M_w=742 400	68 (α phase)	66	56 (1.09)
copolymer (C2-4.5) 4.5 % of 1-octene, M_w=220 500	54 (42% α phase + 12% γ phase)	39	51 (spectrum was recorded using 473 nm wavelength)
Neat PP samples			
s-iPP3 (isotactic)	smectic	smectic	43 (1.50)
PP4 (isotactic with a large amount of stereodefects)	0	0	0 (0)
γ -iPP5 (isotactic with a regulated amount of stereodefects)	10 (γ phase)	no data	10 (0.42)
γ-iPP6 (isotactic with a regulated amount of stereodefects)	20 (γ phase)	no data	29 (0.54)
γ-iPP7 (isotactic with a regulated amount of stereodefects)	35 (γ phase)	no data	34 (0.70)
γ-iPP8 (isotactic)	70 (γ phase)	no data	53 (1.06)
sPP9 (syndiotactic)	30	no data	-

All the samples under study were in the nascent form. The only exception was the smectic modification (s-iPP3), which was prepared in a form of thick and transparent film by melting of the PP granules at 200°C and subsequent pressing and quenching in icy water.

The fully-amorphous PP sample (PP4) and the PP samples of the γ modification with the degree of crystallinity of 10% (γ-iPP5), 20% (γ-iPP6), and 35% (γ-iPP7) were synthesized in the medium of liquid propylene over the homogeneous metallocene catalyst of C_1 - symmetry $C_2H_4(Flu)(5,6cyclopenta\text{-}2\text{-}MeInd)MtCl_2$ in accordance with the method [4,5]. The method allows to produce an isotactic PP with a regulated amount of stereodefects and, thus, to get the neat γ modification with variable degree of crystallinity. X-ray analysis, DSC, and mechanical data for these samples are also presented in [4,5].

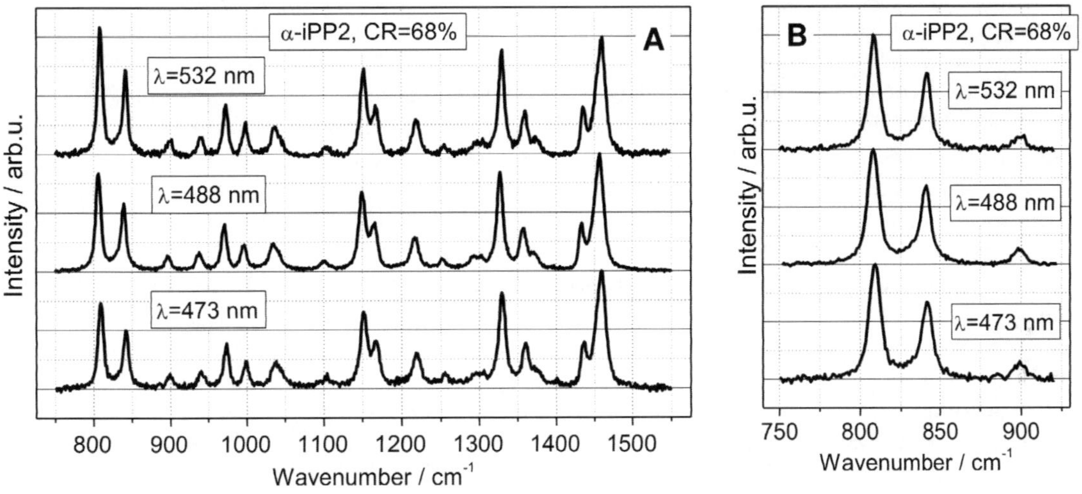

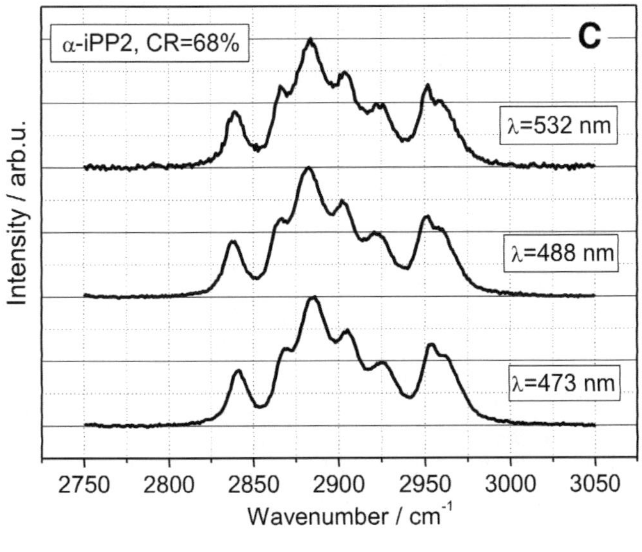

Figure 1. Raman spectra of the nascent powder of α-iPP2: A – the region 750 – 1550 cm^{-1}, B – the region 750 – 920 cm^{-1}, C - the region 2750 – 3050 cm^{-1}. The spectra are normalized to the peak intensity of the line at 1463 cm^{-1} (A), 809 cm^{-1} (B) and 2883 cm^{-1} (C). λ – the excitation wavelength.

The method of synthesis of the neat γ modification of isotactic PP with the degree of crystallinity of 70% (γ-iPP8) and the corresponding X-ray analysis data are described in [6]. The same information for the syndiotactic PP (sPP9) is presented in [7].

Isotropic films of the blends of 90 mass % of PE and 10 mass % of isotactic PP were produced by mixing in the PE melt at three temperatures of 155, 165 and 190°C with subsequent pressing and quenching in room-temperature water. The films of the neat PE and neat PP were prepared at 150 and 190°C, respectively. The melting temperatures of the neat PE and neat PP were 128 and 165°C respectively. The relative content of the blend components was chosen in accordance with the works [8-10], where new technology for producing the all-polymer composites was proposed.

The Raman setup consisted of a Sapphire SF 532 laser (Coherent Inc., USA), a double monochromator (U1000, Jobin Yvon, France) and Peltier-element cooled photomultiplier detector, operating in photon counting regime. The excitation wavelength was 532 nm. Spectra were recorded at 90°-scattering with spectral resolution of 5 cm^{-1}.

In the course of preliminary measurements with different excitation wavelengths (473 and 488 nm from Ar^+-Kr^+ Stabilite 2018 laser (Spectra Physics, USA) and 532 nm from Sapphire SF 532 laser (Coherent Inc., USA)) we found out that the relative intensities of the isotactic PP Raman lines can be affected by the choice of the excitation wavelength. In particular, it can be seen from figure 1A that the intensities of the lines at 809 and 841 cm^{-1} increase noticeably relative to the intensity of the line at 1463 cm^{-1} with growth of the excitation wavelength. Integral intensities of the lines at 809 and 841 cm^{-1} are used for evaluation of the degree of crystallinity and conformational order of macromolecules of isotactic PP [11], while the line at 1463 cm^{-1} (the deformation vibration of CH_3 groups) often serves as a reference line due to the fact, that its intensity is less dependent on the degree of crystallinity and conformational order of macromolecules (for example, [12,13]).

The relative intensities in the narrow spectral region 750 – 920 cm^{-1} (figure 1B) as well as in the region 2750 – 3050 cm^{-1} (figure 1C) are independent of the excitation wavelength. Thus, care should be taken when comparing the intensities of the isotactic PP Raman lines in the region 750 – 1550 cm^{-1}, if the spectra are recorded using different excitation wavelengths. However, the degree of crystallinity and conformational order of isotactic PP macromolecules, evaluated as the ratios of integral intensities of the lines in the region 750 – 920 cm^{-1} [11], do not depend on the excitation wavelength.

It is important that the line at 809 cm^{-1}, which integral intensity was proposed to be a measure of the degree of crystallinity of isotactic PP [11], is observed in a Raman spectrum of smectic isotactic PP [14]. Thus, the degree of crystallinity, evaluated from Raman spectra using the method [11], accounts for the smectic phase or similar states of isotactic PP macromolecules. Besides, it should be noted that the degree of crystallinity, evaluated from data of different techniques, should not coincide exactly because of different nature of physical effects, which form the basis of the methods. But the trends in the changes of the degree of crystallinity, evaluated by different techniques, should be the same. The error of measurement of the degree of crystallinity is similar for all the methods, used in this work, and is about 5 – 15%, depending on the absolute value of this parameter.

3. Results and Discussion

3.1. Propylene/1-octene copolymers and polymorphs of isotactic PP

Assignment of the Raman-active lines of isotactic and syndiotactic PP can be found in [11,14-19] and in [20-25] respectively. The relationships between appearance of the regularity bands in the isotactic PP vibrational spectrum and the length n of sequences of monomer units in the helical conformation have been established [13,26]. The minimum values of n, at which the bands at 973, 998, 841, and 1220 cm^{-1} appear in the isotactic PP vibrational spectrum, are 2-4, 5-10, 12-14, and about 14 monomer units respectively (see detailed literature review in [27]). Appearance of the line at 809 cm^{-1} was attributed to the minimal length n of about 7 [28-30] and of about 15 monomer units [31,32].

The line at 809 cm^{-1} was assigned to the vibrations of isotactic PP helical chains in the crystalline phases with the monoclinic, hexagonal, and orthorhombic crystalline lattices and in the smectic phase [11,14]. Ratio of integral intensity I_{809} of this line and the total integral intensity I_{ref} in the spectral region from 750 to 890 cm^{-1} is used as a measure of the degree of crystallinity of isotactic PP ([11], table 1). The line at 841 cm^{-1} was assigned to the vibrations of isotactic PP helical chains with significant amount of conformational defects, localized in the amorphous phase [11].

The Raman spectra of the copolymers from the Series 1 as well as the spectra of the polymorphs of isotactic PP and the spectrum of the syndiotactic PP are presented in figures 2ABC. The spectrum of the syndiotactic PP looks very different from the spectra of the isotactic PPs, especially in the region 750 – 1550 cm^{-1}, where practically all the regularity bands of isotactic PP are localized. The spectrum of the smectic modification differs from the spectrum of the α modification mainly in the enhancement of the relative intensities of the lines at 809 and 841 cm^{-1} and in the form of the intensity distribution in the region 2750 – 3050 cm^{-1}.

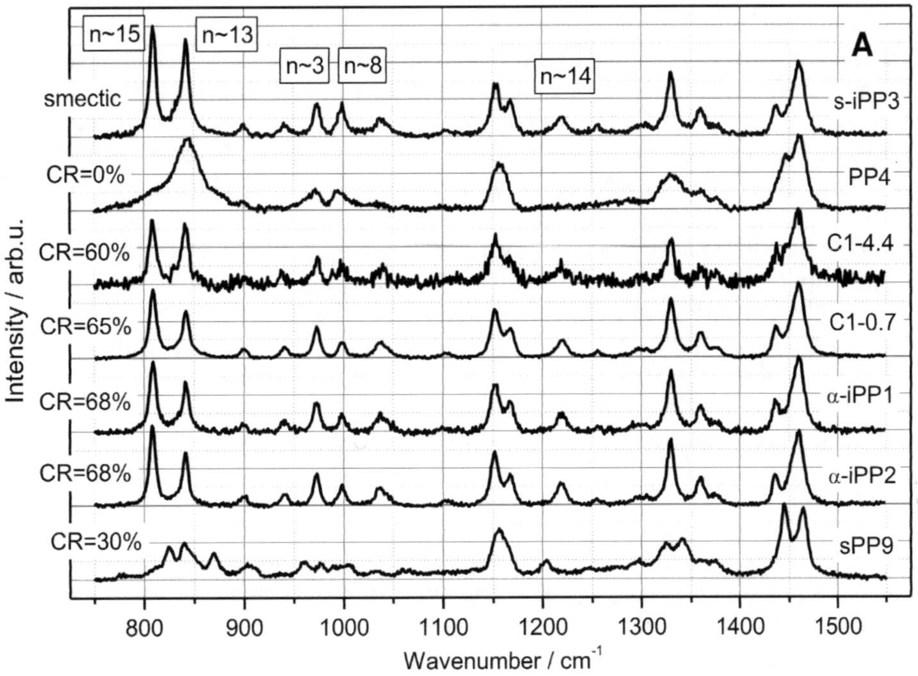

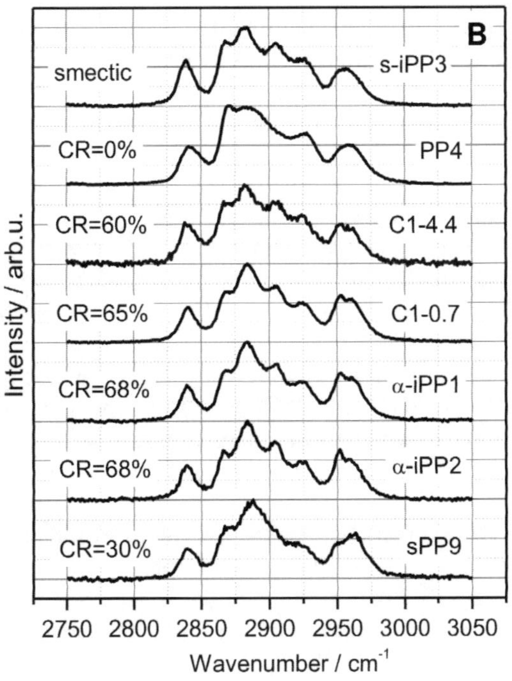

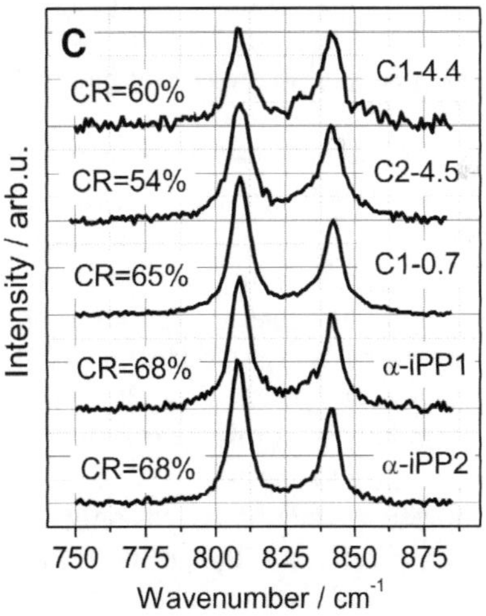

Figure 2. Raman spectra of the propylene/1-octene copolymers and neat PPs in the regions 750 – 1550 cm^{-1} (A), 2750 – 3050 cm^{-1} (B), and 750 – 885 cm^{-1} (C). The spectra are normalized to the peak intensity of the line at 1463 cm^{-1} (A), 2883 cm^{-1} (B) and 841 cm^{-1} (C). The numbers on the left side are the degree of crystallinity, measured by X-ray analysis.

When comparing the copolymer spectra, the spectrum of the α modification (with the degree of crystallinity of 68%) and the spectrum of the fully-amorphous PP, it is clear that the intensities of the regularity bands at 809, 841, 973, 998, and 1220 cm^{-1} monotonously decrease with the increase in the 1-octene content (and/or with the reduction of the degree of crystallinity) with the most pronounced effect, observed for the lines at 809 and 1220 cm^{-1}, which are related to the highest n values.

The intensity of the line at 809 cm^{-1}, normalized to the intensity of the line at 841 cm^{-1}, is slightly higher in the spectra of the neat PP and the copolymer from the Series 2 in comparison with the spectra of the samples from the Series 1 (figure 2C). It means that the degree of crystallinity is slightly higher for the samples of the high-molecular-weight grade as compared to the low-molecular-weight one.

Figures 3AB show the Raman spectra of various polymorphs of isotactic PP. The spectra of the α modification with the degree of crystallinity of 68% and the γ modification with the degree of crystallinity of 70% are very similar. The most significant difference is observed in the range at about 2960 cm^{-1} for the doublet of the lines, assigned to the asymmetric stretching vibrations of CH_3 groups [33]. As in the case of the α modification, for the γ modification the intensities of the regularity bands at 809, 841, 973, 998, and 1220 cm^{-1} decrease with the reduction of the degree of crystallinity with the most pronounced effect, observed for the lines at 809 and 1220 cm^{-1}.

3.2. PE/PP blends

Raman spectrum of PE is well investigated, and almost all the PE Raman-active lines are reliably assigned [34-39]. The Raman-active regularity bands for a PE macromolecule in the *trans*-conformation are observed at 1062, 1130, 1170, 1295, 2850, and 2881 cm^{-1} and are assigned to the sequences of *trans*-conformers with the minimal length n of 6–8, 18, 18, 18, 6-8, and 18 CH_2 groups respectively [40]. The line at 1417 cm^{-1} belongs to the vibrations of the macromolecules in the *trans*-conformation in the orthorhombic crystalline phase [35]. Structure of PE/PP blends was analyzed by Raman spectroscopy in a number of works [10,41-52]. However, Raman investigation of structure of PE/PP blends as a function of temperature of the polymer mixing was not carried out before.

Figures 4AB show Raman spectra of the neat PE and the blends of 90 mass % of PE and 10 mass % of isotactic PP, produced by mixing in the PE melt at three temperatures of 155, 165, and 190°C. The spectra were normalized to the peak intensity of the line at 2850 cm^{-1} [40].

As it was indicated above, the melting temperatures of the neat PE and the neat PP were 128°C and 165°C respectively. Thus, the blends were mixed at three temperatures, at which the PE was in the state of a melt, where the orthorhombic crystalline phase and the *trans*-conformation of the PE macromolecules were destroyed. At 155°C PP macromolecules should retain both the helical conformation and packing in the crystalline lattice. At 165°C we expect destruction of the PP crystallites, but retention of the helical conformation of the PP macromolecules for the most part. At 190°C all the PP crystallites should be destroyed, and the helical conformation of the PP macromolecules should be strongly disturbed. Thus, we expect the different structure and, as a consequence, different Raman spectra for the blends, mixed at these three temperatures.

The spectra of the blends turned out to be similar to the spectrum of the neat PE. This is due to the low PP content in the blends and, thus, to weakness of the PP Raman lines. For the blend, prepared at 155°C, we observed decrease in the intensities of all the PE regularity bands as well as of the line at 1417 cm^{-1}. It means that the degree of crystallinity and conformational order of the PE macromolecules for this blend are lower compared with other two blends and the neat PE. Thus, the solid PP particles more strongly disturb the conformational ordering and crystallization of the PE macromolecules during cooling compared with the separated PP macromolecules uniformly-distributed over the blend.

The peak intensities of all the PE regularity bands as well as of the line at 1417 cm^{-1} are slightly higher for the blend, prepared at 190°C, compared with the blend, mixed at 165°C. This means that the most perfect structure of PE is obtained at the mixing temperature of 190°C.

Figure 5 presents the temperature-dependent Raman spectra of the isotactic PP α-iPP2 (with the melting temperature of 164°C) in the region 750 – 1550 cm^{-1}. The intensities of all the isotactic PP regularity bands are lower at 175°C compared with the spectrum, measured at 25°C. It is an indication of disruption of the macromolecular packing and conformational order.

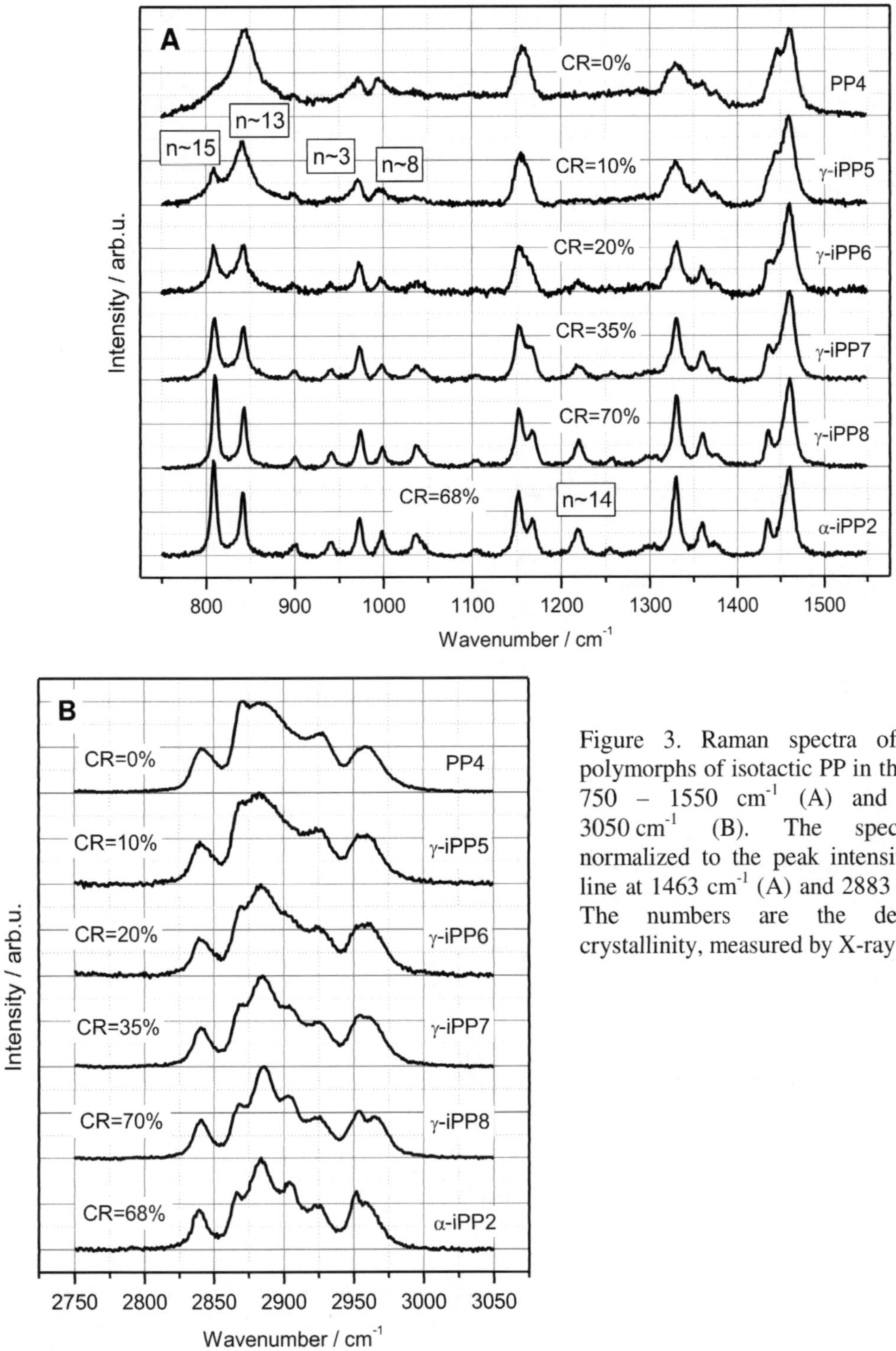

Figure 3. Raman spectra of various polymorphs of isotactic PP in the regions 750 – 1550 cm^{-1} (A) and 2750 – 3050 cm^{-1} (B). The spectra are normalized to the peak intensity of the line at 1463 cm^{-1} (A) and 2883 cm^{-1} (B). The numbers are the degree of crystallinity, measured by X-ray analysis.

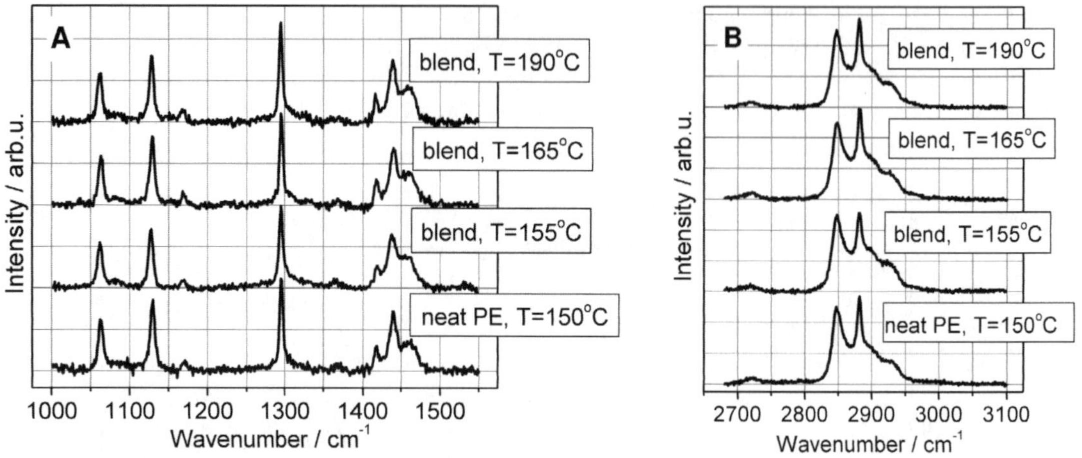

Figure 4. Raman spectra of the PE/PP blends in the regions 750 – 1550 cm^{-1} (A) and 2750 –3050 cm^{-1} (B). The spectra are normalized to the peak intensity of the line at 2850 cm^{-1}. The numbers are the temperatures, at which the polymers were mixed.

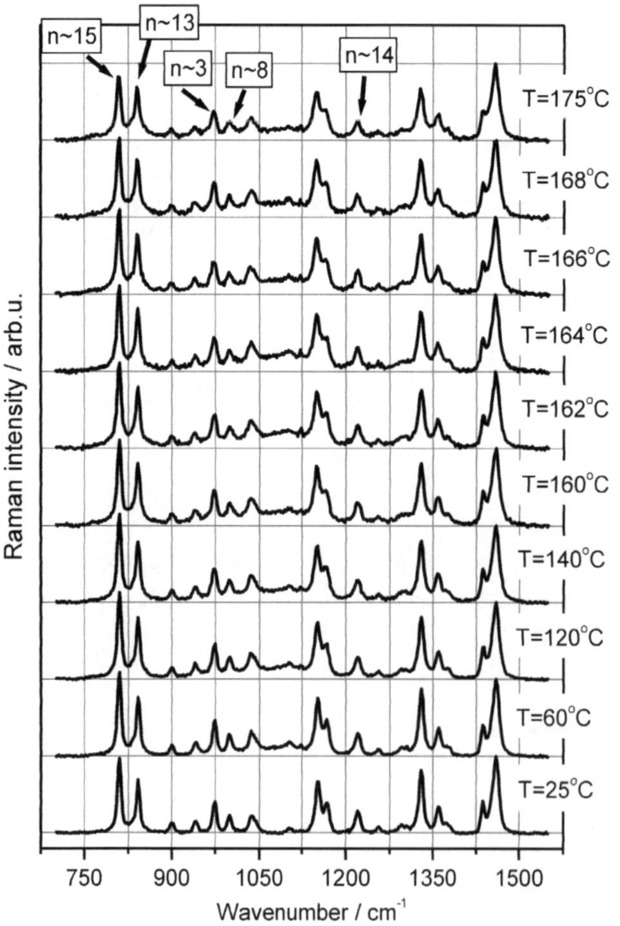

Figure 5. Raman spectra of the isotactic PP α-iPP2 with the melting temperature of 164°C in the region 750 – 1550 cm^{-1} at various temperatures. The spectra are normalized to the peak intensity of the line at 1463 cm^{-1}.

Significant amount of the conformationally-ordered segments is still present at temperatures, exceeding the melting point (figure 5). In the temperature region from 60 to 166°C the intensity of the line at 809 cm^{-1} (relative to the intensity of the line at 1463 cm^{-1}) is higher compared with the spectrum, recorded at 25°C. We explain this fact by the increase in the content of the smectic modification, which is formed due to destruction of the PP crystallites. This explanation is confirmed by higher intensity of the line at 809 cm^{-1} in the spectrum of the smectic modification compared to the spectrum of the α and γ modifications, measured at room temperature (figures 2A, 3A).

4. Conclusion

Based on Raman spectroscopy data we found out that the degree of crystallinity and conformational order of macromolecules of the propylene/1-octene copolymers decrease with the growth of the 1-octene content. The degree of crystallinity is slightly higher for the neat polypropylene and the copolymer of the high-molecular-weight grade compared to the low-molecular-weight one. Raman study of the polyethylene/polypropylene blends revealed that the degree of crystallinity and conformational order of the polyethylene macromolecules are the highest for the mixing temperature, at which the polypropylene macromolecules completely lost their packing and conformational order.

Acknowledgments
This work is supported by the Russian Foundation for Basic Research (14-02-00832) and Program III.2 of Russian Academy of Sciences. The authors are very grateful to Dr. B.F. Shklyaruk (Topchiev Institute of Petrochemical Synthesis, Russian Academy of Sciences) for providing with the sample of the γ modification with the degree of crystallinity of 70% as well as for the X-ray analysis and DSC data for all the samples under study. The authors are very grateful to Dr. N.N. Mel'nik (Lebedev Physical Institute, Russian Academy of Sciences) for carrying out the temperature measurements.

References

[1] Nedorezova P M, Chapurina A V, Koval'chuk A A, Klyamkina A N, Aladyshev A M, Optov V A and Shklyaruk B F 2010 *Polym. Sci.* B **52** 15-25
[2] Chernyshov K, *et al.* 2010 *Macromol. Symp.* **296** 505-16
[3] Gen D E, *et al.* 2010 *Laser Phys.* **20** 1354-67
[4] Nedorezova P M, *et al.* 2005 *Russ. Chem. Bull.* **54** 400–13
[5] Nedorezova P M, Veksler E N, Aladyshev A M, Tsvetkova V I, Optov V A, Brusova G P and Lemenovskii D A 2006 *Kinet. Catal.* **47** 245–50
[6] Shklyaruk B F, Dreval' V E, Bocharova V A, Kulichikhin V G, Kuleznev V N and Antipov E M 2004 *Polym. Sci.* A **46** 605-12
[7] Nedorezova P M, Tsvetkova V I, Aladyshev A M, Savinov D V, Klyamkina A N, Optov V A and Lemenovskii D A 2001 *Polym. Sci.* A **43** 356–65
[8] Krajenta J, Pawlak A and Galeski A 2015 *Polimery* **60** 664–6
[9] Li Zh, Shi Yu-J, Sun Ch-X, Zhang Q and Fu Q 2015 *Compos. Sci. Technol.* **115** 34–42
[10] Cardenas M A, Perera R, Villarreal N, Rosales C and Pastor J M 2007 *J. Appl. Polym. Sci.* **106** 2298-312
[11] Nielsen A S, Batchelder D N and Pyrz R 2002 *Polymer* **43** 2671-6
[12] Zhao J, Peng Zh, Zhang J and Li G 2011 *Polym. Bull.* **67** 1649–59
[13] Chen J, Cao Ya, Kang J and Li H 2010 *J. Macromol. Sci.* B **50** 248– 65
[14] Chalmers J M, Edwards H G M, Lees J S, Long D A, Mackenzie M W and Willis H A 1991 *J. Raman Spectrosc.* **22** 613–8
[15] Fraser G V, Hendra P J, Watson D S, Gall M J, Willis H A and Cudby M E A 1973 *Spectrochim. Acta* A **29** 1525–33
[16] Snyder R G and Schachtschneider J H 1964 *Spectrochim. Acta* **20** 853–69
[17] Arruebarrena de Baez M, Hendra P J and Judkins M 1995 *Spectrochim. Acta* A **51** 2117–24
[18] Tadokoro H, Kobayashi M, Ukita M, Yasufuku K, Murahashi S and Torii T 1965 *J. Chem. Phys.* **42** 1432–49
[19] Vasko P D and Koenig J L 1970 *Macromolecules* **3** 597–601
[20] Gatos K G, Kandilioti G, Galiotis C and Gregoriou V G 2004 *Polymer* **45** 4453–64

[21] Hahn T, Suen W, Kang S, Hsu S L, Stidham H D and Siedle A R 2001 *Polymer* **42** 5813–22
[22] Hsu S L, Hahn T, Suen W, Kang S, Stidham H D and Siedle A R 2001 *Macromolecules* **34** 3376–83
[23] Lee K-H, Ohsawa O, Watanabe K, Kim I-S, Givens S R, Chase B and Rabolt J F 2009 *Macromolecules* **42** 5215–8
[24] Masetti G, Cabassi F and Zerbi G 1980 *Polymer* **21** 143–52
[25] Sevegney M S, Kannan R M, Siedle A R, Naik R and Naik V M 2006 *Vib. Spectrosc.* **40** 246–56
[26] Kissin Yu V and Rishina L A 1976 *Eur. Polym. J.* **12** 757–9
[27] Prokhorov K A, Nikolaeva G Yu, Sagitova E A, Pashinin P P, Nedorezova P M and Klyamkina A N 2016 *Vib. Spectrosc.* **85** 22–8
[28] Ruiz-Orta C and Alamo R G 2012 *Polymer* **53** 810–22
[29] Arranz-Andres J, de la Orden M U, Martinez Urreaga J, Gomez-Elvira J M, Perez E and Cerrada M L 2015 *Eur. Polym. J.* **63** 227–36
[30] Ruiz-Orta C, Fernandez-Blazquez J P, Pereira E J and Alamo R G 2011 *Polymer* **52** 2856–68
[31] Khafagy R M 2006 *J. Polym. Sci.* B **44** 2173–82
[32] Brookes A, Dyke J M, Hendra P J, Meehan S 1997 *Spectrochim. Acta* A **53** 2313–21
[33] Gen D E, Prokhorov K A, Nikolaeva G Yu, Sagitova E A, Pashinin P P, Shklyaruk B F and Antipov E M 2011 *Laser Phys.* **21** 125–9
[34] Naylor C C, Meier R J, Kip B J, Williams K P J, Mason S M, Conroy N and Gerrard D L 1995 *Macromolecules* **28** 2969-78
[35] Strobl G R and Hagedorn W 1978 *J. Polym. Sci.: Polym. Phys. Ed.* **16** 1181–93
[36] Kurelec L, Rastogi S, Meier R J and Lemstra P J 2000 *Macromolecules* **33** 5593–601
[37] Abbate S, Zerbi G and Wunder S L 1982 *J. Phys. Chem.* **86** 3140–9
[38] Gall M J, Hendra P J, Peacock O J, Cudby M E A and Willis H A 1972 *Spectrochim. Acta* A **28** 1485–96
[39] Boerio F J and Koenig J L 1970 *J. Chem. Phys.* **52** 3425–31
[40] Sagitova E A, *et al.* 2016 *Vib. Spectrosc.* **84** 139-45
[41] Shemouratov Yu V, Prokhorov K A, Nikolaeva G Yu, Pashinin P P, Kovalchuk A A, Klyamkina A N, Nedorezova P M, Demidenok K V, Lebedev Yu A and Antipov E M 2008 *Laser Phys.* **18** 554-67
[42] Shemouratov Yu V, Prokhorov K A, Sagitova E A, Nikolaeva G Yu, Pashinin P P, Lebedev Yu A and Antipov E M 2009 *Laser Phys.* **19** 2179-83
[43] Zavgorodnev Yu V, Sagitova E A, Nikolaeva G Yu, Prokhorov K A, Pashinin P P, Novokshonova L A, Ushakova T M, Starchak E E and Antipov E M 2012 *Proc. Int. Conf. on Advanced Laser Technologies (Sept. 2 - 6, 2012, Thun)* vol. 1, ed B Neuenschwander, *et al.* (Bern: Bern Open Publishing) 79 pp 1-5
[44] Barnes S E, Brown E C, Sibley M G, Edwards H G M, Scowen I J and Coates P D 2005 *Appl. Spectrosc.* **59** 611–9
[45] Coates P D, Barnes S E, Sibley M G, Brown E C, Edwards H G M and Scowen I J 2003 *Polymer* **44** 5937–49
[46] Furukawa T, Sato H, Kita Ya, Matsukawa K, Yamaguchi H, Ochiai S, Siesler H W and Ozaki Yu 2006 *Polym. J.* **38** 1127–36
[47] Garton A, Batchelder D N and Cheng C 1993 *Appl. Spectrosc.* **47** 922–7
[48] Loos J, Bonnet M and Petermann J 2000 *Polymer* **41** 351-6
[49] Lopez Quintana S, Schmidt P, Dybal J, Kratochvıl J, Pastor J M and Merino J C 2002 *Polymer* **43** 5187–95
[50] Schmidt P, Dybal J, Scudla J, Raab M, Kratochvil J, Eichhorn K-J, Lopez Quintana S and Pastor J M 2002 *Macromol. Symp.* **184** 107–22
[51] Tselios Ch, Bikiaris D, Maslis V and Panayiotou C 1998 *Polymer* **39** 6807–17
[52] Tselios Ch, Bikiaris D, Savidis P, Panayiotou C and Larena A 1999 *J. Mater. Sci.* **34** 385-94

The use of Inductive Discharge for Laser Pumping of a Copper Vapor

M.A. Kazaryan [1], M.M. Malikov[2], V.T.Karpukhin[2], N.A. Lyabin[3]

Results of numerical simulation re presented for copper vapor laser (CVL) excitation by a pulse-periodical inductive discharge. A CVL realization with a coaxial discharge chamber is studied. It was shown that a coaxial chamber is better matched with specific features of inductive pumping method than an ordinary cylindrical chamber. High lasing output characteristics are obtained in numerical experiments, which confirms possibility of efficient pumping CVL by a new inductive method.

Key words: copper vapor laser, inductor, transformer, inductive discharge, coaxial chamber, numerical simulation, laser kinetics.

INTRODUCTION

Study of various methods for exciting an active medium of lasers on self-contained transitions of metal atoms [1] is associated with a tendency to enhance output characteristics and obtain certain technical and operational advantages. A pulse-periodical inductive discharge of the transformer type was suggested long ago for exciting gas lasers [2-4]. This method was successively employed in [5-7] and other works for pumping lasers on atom and molecular working media. For the first time the inductive method was tested for exciting copper vapor laser in experimental work [8], and, although lasing was not observed, this work stimulated further investigations. First results of numerical experiments [9, 10] on exciting CVL by an inductive discharge confirmed the possibility of creating such lasers. Note that the nonelectrode inductive excitation is interesting from fundamental point of view as well. For example, in certain conditions, pulsed magnetic fields arising in plasma may change spectroscopic and electric characteristics of discharge plasma. In this work we present results of numerical simulation of operation of an inductive copper vapor laser (ICVL) with a coaxial chamber. The results are compared to those of work [10] where a cylindrical chamber was used. Physical conditions and constructive parameters are discussed, in which efficient lasing may be reached.

PHYSICAL MODEL AND PROBLEM STATEMENT

A simplest ICVL construction with pulse-periodic pumping by inductive discharge is shown in Fig. 1a. Storage capacitor C_s discharges through switch to inductor *1,* which is a continuous turn around the heat-insulating volume *2*. The active volume (copper vapor and neon) is arranged in a coaxial gap $\Delta r_{пл} = (r_2 - r_1)$ between a central ceramic insertion *3* of radius r_1 and external wall of the discharge chamber *4* of radius r_2. A pulsed magnetic field of the inductor creates a vortex electric field $E_\varphi(r,t)$ and azimuthal conduction current in plasma of working mixture. In this case, the conductor is considered as a primary coil of transformer (without core) and the plasma turn is a secondary coil. A specific feature of ICVL distinct from induction lasers [5-7] mentioned above is a thick heat insulator, which increases a distance from the plasma turn to the inductor and substantially reduces the coupling coefficient K_c of a transformer negatively affecting operation of the latter. Other features of ICVL are: high pulse repetition frequency of pumping (dozens kHz) and comparatively low pre-pulse resistance of plasma ~1 Ohm·cm. To produce inverted population of working levels it is necessary to rapidly deposit electric energy into plasma for a time lapse of ~10 – 100 ns. The temperature of walls *3* and *4* of the discharge tube should be 1900 K, which is needed for maintaining the copper vapor pressure. A pressure of buffer neon gas is 01-1 atm.

In Fig. 1b one can see a simplified schematic diagram of the pulsed inductive pumping of the transformer type (with purely inductive coupling of coils). Here L_1 and J_1 are the inductance and current of the primary circuit, L_2 и J_2 are the inductance and current of the plasma turn. The values of L_1, L_2 and mutual inductance M were calculated by the formulae from [11]. The electrical resistance of the switch $R_{sw}(t)$ was described by the model function capable of specifying the switching duration and Joule loss in the primary circuit of the transformer. The ohmic resistance R_I of inductor was taken into account. The initial voltage $U_{in}(0)$ across the storage capacitor and the initial inductance L_{in} were prescribed.

From known differential transformer equations [12] and the Kirchoff laws one can derive the system of equations for the scheme in Fig. 1b.

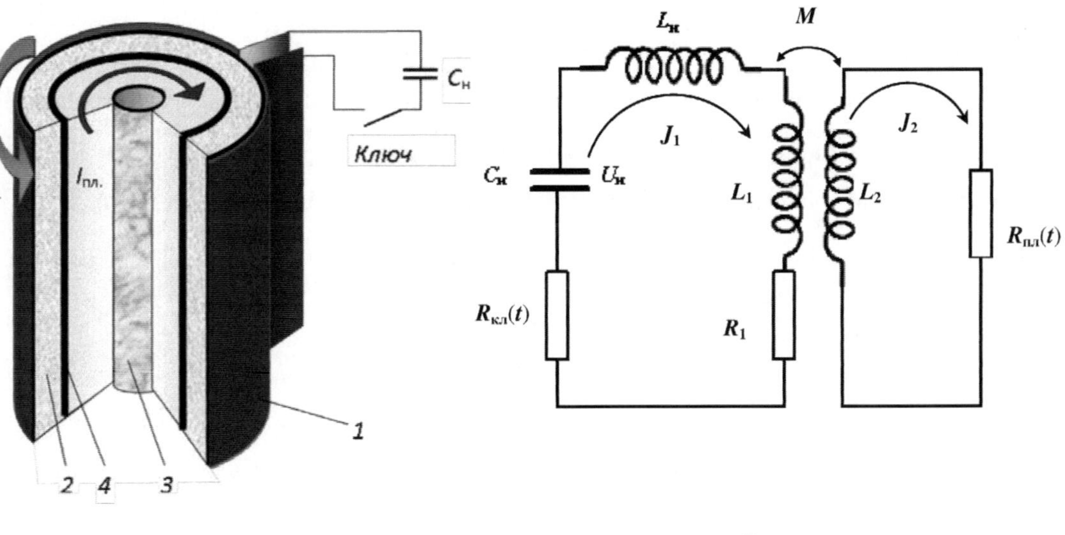

Fig.1. a – construction of ICVL with a coaxial discharge chamber, b – electric scheme of pulse-periodic inductive pumping.

$$
\begin{aligned}
U_{н} &= \left(L_{н} + L_1\right)\frac{dJ_1}{dt} - M\frac{dJ_2}{dt} - J_1\left(R_{кл}(t) + R_1\right) \\
0 &= R_{пл}(t)J_2 + L_2\frac{dJ_2}{dt} - M\frac{dJ_1}{dt} \\
\frac{dU_{н}}{dt} &= -\frac{J_1}{C_{н}}
\end{aligned}
\qquad (1)
$$

Resistance of plasma turn $R_{pl}(t)$ depends on the specific electric conductivity $\sigma(r,t)$ and the radial current density distribution in plasma $j(r,t)$. All these values including $R_{pl}(t)$ sharply vary during the exciting pulse. For finding $R_{pl}(t)$ system of equations (1) should be solved in conjunction with equations describing physical processes and $\sigma(r,t)$ in non-equilibrium plasma of working medium. For this purpose we used the model and software of conventional CVL developed earlier [13-15]. A basis of the CVL model was the system of equations describing kinetics of the population of excited atomic levels, electron concentration, balance of electron energy. Also, equations of developing induced emission in an optical plane resonator were used. The model comprised a specially developed method for calculating thermal parameters of working medium and high-temperature elements of laser construction. A large number of energy levels of copper and neon atoms were taken into account as well as a series of elementary processes. Note that in a steady state (with respect to a pulse repetition frequency) regime, the self-consistent values of all plasma parameters were numerically calculated, as well as specific conductivity $\sigma(t)$, current and voltage (in the primary and secondary circuits). The power of Joule heat release was determined, as well as electron and gas temperatures, radiation power, and laser efficiency. Note that equations in model [13, 14] are written in zero-dimensional approximation, that is, all plasma parameters including σ(t) are averaged over plasma volume, and transport processes are taken into account approximately. Hence, into the balance of electron energy we put an average power of Joule heat release w equal to

$$w = \frac{1}{V_\text{p}} \int \left(\sigma E_\varphi^2\right) dV_\text{p} \approx \frac{J_2^2 \cdot R_\text{пл}(t)}{V_\text{p}} \tag{2}$$

(here V_p is the working volume). On the other hand, in order to establish relationship between the effective resistance of plasma turn $R_\text{pl}(t)$ and specific average conductance σ(t) by using (2) one should know the radial dependence of vortex electric field and conductance in a cylindrical volume. There are many works (e.g., [16-18]) devoted to calculation of radial distribution of electric parameters in plasma of inductive discharge. For a conductive solid cylinder the induced vortex field $E_\varphi(r,t)$ in plasma has a maximum near external wall and falls to zero at the cylinder axis. In the case of a hollow cylinder (coaxial plasma volume, see Fig. 1a), the value of $E_\varphi(r,t)$ does not fall to zero on internal wall ($r = r_1$) and in the case of weak skin-effect ($\delta \gg (r_2 - r_1)$) weakly varies [17] (here, δ is the depth of field penetration into a conductive medium). Hence, in a coaxial discharge chamber, more homogeneous plasma can be produced than in a cylindrical chamber, which is a positive factor for lasing. In addition, in this case, description of plasma processes by zero-dimensional model is more reasonable. It will be shown below that employment of coaxial chamber gives a chance to substantially increase the coupling coefficient K_c of transformer.

In [17, 18], formulae for calculating $R_\text{pl}(t)$ are given as applied to solid or hollow cylinders with uniform conductivity σ and arbitrary penetration depth δ. The expressions comprise complicated combinations of Bessel functions of the first- and second kinds. Those are rather cumbersome and we omit them here. At $\delta \gg (r_2 - r_1)$ the expressions for $R_\text{pl}(t)$ take an obvious form [17]. For the solid and hollow cylinders we have, respectively

$$R_\text{пл}(t) \approx 4\pi/\sigma\ell_\text{тр}\,, \qquad R_\text{пл}(t) \approx \frac{4\pi}{\sigma\ell_\text{тр}} \frac{\left(r_2^2 + r_1^2\right)}{\left(r_2^2 - r_1^2\right)} \tag{3}$$

Here, ℓ_tube is the length of plasma volume (tube).

In real conditions, plasma conductance is not uniform. In this case, there are no analytical solutions for distributions of electrical parameters. Values of fields $E_\varphi(r,t)$, current densities $j_\varphi(r,t)$, σ(r,t) and total resistance $R_\text{pl}(t)$ are found numerically by solving combined non-stationary one-dimensional electromagnetic problem and non-stationary one-dimensional equations describing non-equilibrium double-temperature plasma. A detailed review of such calculation models of inductive plasma heating one can find in [16, 18]. In [18], authors recommend to employ formulae similar to (3) as a first approximation. In the present work we will limit ourselves, as was mentioned, to zer-dimensional model of plasma processes and equations (2), (3) combining plasma conductivity with system (1) for electric transformer design.

In numerical simulations we have considered two variants of ICVL construction with coaxial chambers, differing in dimensions and coupling coefficients K_c (see Table). For the basis in the first variant of ICVL we have taken parameters corresponding to ordinary commercial active element LT-30Cu [19]: the working volume is $V_w \approx 280$ cm^3, diameter of external envelope is ~10 cm, nominal value of storage capacitor is C_{st}=0.5 nF, and initial voltage across it $U_{in}(0)$=28 kV. The second variant differ by greater geometrical dimensions, volume of the coaxial discharge chamber ($V_w \approx 1410$ cm^3) and values of storage capacitor C_{st}. In both the variants, the pressure of neon equal to 250 millimeters of mercury, temperature of internal wall *4* (Fig. 1a) T_4=1823 ^{o}K, copper vapor concentration $1{,}5 \cdot 10^{15}$ см$^{-3}$ were similar and approximately corresponded to the parameters of working medium in LT-30Cu. The pumping pulse repetition frequency was 10 kHz. Inductance L_{sw} of switch circuit was taken $0.5 \cdot L_1$, resistance R_1 of the primary transformer circuit with the skin-effect for copper conductor taken into account was 10^{-3} Ohm. The coupling coefficient was calculated by the known formula $K_c = M/\sqrt{L_1 L_2}$. The duration of pumping current pulses τ_p was determined in calculations by the instant when 98-99% of the initial (stored in C_s) energy was converted to Joule heat in plasma and on resistances of the switch and primary circuit. In this case the balance equation for electric energy was used [12]:

$$\frac{C_н U_н^2(0)}{2} = \left[\frac{(L_н + L_1) J_1^2}{2} + \frac{L_2 J_2^2}{2} - M J_1 J_2 + \frac{C_н U_н^2}{2} \right]_t + \int_0^t \left[J_2^2 \cdot R_{пл} + J_1^2 \cdot (R_{кл} + R_1) \right] dt \qquad (4)$$

The value of $R_{pl}(t)$ was chosen such that 30-40% of stored energy was lost in the primary circuit and in switch, and 70-60% in plasma. Such a proportion is typical for exciting an ordinary CVL with a thyratron switch. Hence, in numerical experiments with inductive pumping we have provided pulsed specific energy deposition into plasma approximately equal to that of ordinary CVL. This approach was chosen to objectively compare CVLs of both the types. In the present work we consider a solid single-turn inductor shown in Fig. 1a.

RESULTS OF NUMERICAL SIMULATION

IOP Conf. Series: Journal of Physics: Conf. Series **826** (2017) 012003 doi:10.1088/1742-6596/826/1/012003

In the first variant of ICVL (see Table), the inductor of radius r_{ind} directly covers the external envelope of the active element of LT-30Cu and the thickness of het insulation $\Delta r_{hi}= (r_{ind}-r_2)$ of active element is reduced from 4 cm to 2,5 cm; in this case the value of K_c is 0.47. Note that in using LT-30Cu with ordinary cylindrical working volume without changing the heat insulator thickness; the value of K_c was only 0.18 [10]. In the second variant, the thickness of heat insulation remains 2.5 cm, however, the inductor radius r_{ind}, radius of coaxial plasma volume r_2, and the lengths of tube (and inductor) ℓ_{tube} are increased. As one can see from Table, this also increases K_c to 0.53.

Table. 1. Variants of geometrical and electrical-technical parameters of ICVL.

Var. No.	r_1 cm	r_2 cm	Δr_{pl} cm	r_{ind} cm	Δr_t cm	ℓ_{tube} cm	V_w cm3	L_1 nH	L_2 nH	M nH	K_c
1	2	2,5	0,5	5	2,5	40	280	25	3,6	4,4	0,47
2	2	3,0	1,0	5,5	2,5	90	1410	13	1,9	2.6	0,53

DISCUSSION

Calculations and results presented [10] show that in the inductive excitation of CVL, the pumping current pulse is a train of high-frequency oscillations with the period of 10-30 ns. whereas in ordinary CVL the pulse of current has a shape close to aperiodic discharge with a duration of 150-200 ns. High frequency of current oscillations is related with that the inductance of plasma turn, inductance of the inductor, and the coefficient of coupled induction are by two-three orders of magnitude less than the inductance of the discharge circuit in conventional CVL with longitudinal discharge between electrodes (at similar V_w). Note that the effective resistance of plasma R_{pl} is less by approximately two orders of magnitude than the resistance of longitudinal discharge in ordinary CVL (at similar values of σ). The latter is related with a large transversal cross-section of azimuth current and short length along current (for a single azimuth run). By these parameters the considered coaxial ICVLs are closer to so-called CVL with transversal discharge (see, e.g., [20]).

CONCLUSION

Results of numerical study [9, 10] and the present work confirm possibility of creating efficient CVL excited by pulse-periodic inductive (non-electrode) discharge. Preliminary calculations show that at a higher coupling coefficient one can reach the power and efficiency of emission typical for conventional CVLs.

ICVL with coaxial discharge chambers may have even higher coupling coefficients. In the numerical experiments discussed above, high radiation power was obtained at the level of 100 W, substantial efficiency of about 5% and technical efficiency of 1.5-3% were attained. These calculated efficiencies are higher than efficiencies of ordinary CVLs operated on pure mixtures of copper and neon. Analysis performed in the present work shows that in the whole coaxial discharge chambers are better suited for creating ICVLs on this bases than cylindrical chambers.

For estimating prospects for creating efficient ICVLs and employing them in practice it is necessary to thoroughly study operation of such lasers and optimize them with respect to the main physical and technical prescribed parameters with the aim of determining limiting output characteristics. Employment of inductive pumping, probably, will help solving the problem of long-term service time and reliability of sealed-off laser elements. Creation of efficient ICVLs substantially expands the range of CVL applications in industry, precious micro-treatment of materials [19], in selective technologies, physical investigations, diagnosing multi-phase gas flows [21, 11], medicine etc.

Authors are grateful to V.M. Mkhitryan for the help and interest to the work.

REFERENCES

1. *Batenin V.M.,Boichenko A.M., Buchanov V.V., Kazaryan M.A., Klimovskii I.I., Molodykh E.I.* Lazery na samoogranichennykh perekhodakh atomov metallov -2 [Lasers on Self-Contained Transitions of Metal Atoms – 2], v.1. Moscow: FIZMATLIT, 2009, 544 pp.
2. V.M. Mkhitaryan, Laser-Inform No. **15–16,** p. 18 (2004).
3. A.M. Razhev, V.M. Mekhitarian, D.S. Churkin, JETP Lett. **82**, 290 (2005).
4. A.M. Razhev, D.S. Churkin, JETP Lett. **86**, 479 (2007).
5. A. M. Razhev, V. M. Mekhitarian, A. A. Zhupikov, D. S. Churkin, Proc. SPIE **6263**, 8 (2006)
6. A. M. Razhev, D. S. Churkin, A. A. Zhupikov, Quantum Electron. **39**, 901 (2009)
7. A M. Razhev, D S Churkinand E S Kargapol'tsev, Laser Phys. **24**, 074004 (5pp) (2014)
8. R.G. Manucharyan, V. M. Mekhitarian, A.A. Isaev, MA. Kazaryan et al, Simposium "Copper Vapor Lasers (LPM-2006)", Rostov-Don, p. 14 (2006).
9. M.M. Malikov, V.T. Karpukhin, M.A. Kazaryan. Simposium "Copper Vapor Lasers (LPM-2014)", Rostov-Don, p. 64 (2014).

10. М. М. Маликов, М. А. Казарян, В. Т. Карпухин. О возможности эффективной накачки лазеров на парах меди импульсно-периодическим индукционным разрядом // Краткие сообщения по физике, …. (2015)
11. P.L. Kalantarov, L.A. Tseitlin. Raschet inductivnistei [Calculation of Inductances] Leningrad: Energiya, 1970.
12. V.P. Popov. Osnovy teorii tsepei [Foundations of Chain Theory] Moscow: M.: Vyssh. Shkola, 2007
13. L.B. Direktor, M.M. Malikov. Physiacl Model and Method of Calculating Parameters of Copper Vapor Laser. PreprintNo.5-249 IVTAN. Moscow, 1988, 52 p.
14. M.M. Malikov. "Eksperimental'noe i raschetno-teoreticheskoe issledovanie fizicheskikh protsessov v lasernykh sistemakh na parakh medi c tsel'yu uluchsheniya udel'nykh vykhodnykh kharakteristik" [Experimental and calculation-theoretical study of physical processes in copper vapor laser systems aimed at improving specific output characteristics] PhD thesis Moscow, 2011.
15. L. B. Direktor, V. T. Karpukhin, and M. M. Malikov, High Temperature, **52**, 428, (2014).
16. Yu.P. Raizer, High-Frequency High-Pressure Induction Discharge and electrodeless plasmotron. // Uspekhi Fiz. Nauk **99**, 687, (1969)
17. Ustanovka induktsionnogo nagreva [Induction Heating Installation]. Ed. A.E. Slukhotskii, Leningrad: Energoizdat. 1981
18. S.V. Dresvin. Osnovy teorii i rascheta vysokochastotnykh plazmotronov [Basics of Theory and Calculation of High-Frequency Plasmotrons] Leningrad: Energoatomizdat, 1991.
19. A.G. Grigoryants, M.A. Kazaryan, N.A. Lyabin. Lazery na parakh metallov: konstuktsiya, kharakteristiki i primeneniya [Copper Vapor Lasers: Construction, Characteristics, Applications. Moscow: Fizmatlit, 2005. 312 p.
20. *Borovich B.L.* Transverse-discharge copper-vapor laser. In Metal Vapor Lasers and Their Applications. CIS Selected Papers. G. Petrash Editor.// Proc. SPIE. 1993. V. 22110. P. 46-63.
21. A. Yu. Varaksin, M.V. Protasov, Yu.S. Teplitskii, High Temperature, **52**, 554, (2014).
22. A. Yu. Varaksin, High Temperature, **52**, 753, (2014).

[1]Lebedev Physical institute, Leninskii pr., 53, 119991 Moscow, Russia
[2]Joint Institute of High Temperature, Russian Academy of Sciences, ul. Izhorskaya, 13, build. 2, 125412 Moscow, Russia e-mail: mmalikov@oivtran.ru, phone: +7 (916) 735-61-66
[3]SPC "ISTOK", Vokzalnaya ul., 2, Fryazino, Moscow reg., Russia

Theoretical treatment of the resonant hyper-Raman scattering in A^2B^6 semiconductors

L E Semenova, G Yu Nikolaeva and K A Prokhorov

Prokhorov General Physics Institute, Russian Academy of Sciences, Vavilov St. 38, 119991, Moscow, Russia

E-mail: sl@kapella.gpi.ru

Abstract. In this work the theoretical treatment of the hyper-Raman scattering of light by LO-phonons is given for A^2B^6 semiconductors of the wurtzite structure, taking into account the influence of the complex top valence band which consists of the three closely-lying subbands.

1. Introduction

The hyper-Raman scattering (HRS) of light by phonons was observed in a number of semiconductors under two-photon excitation near the absorption edge [1-12]. In publications [13-17], devoted to theoretical investigations of the resonant HRS, the scattering process was considered within the framework of the model taking into account the hydrogen-like Wannier excitons as intermediate virtual states. This model enables one to explain many features observed in the HRS. In particular, an abrupt increase in the hyper-Raman intensity near the two-photon resonance with the lowest exciton state was reported to be found under experimental investigations of the resonant HRS by longitudinal optical (LO) phonons in a CdS crystal [11]. This phenomenon was explained by a scattering mechanism including two-photon dipole transitions to excitonic states of the *s*-type [16].

In this work the resonant HRS in direct-band hexagonal semiconductor crystals A^2B^6 is considered, taking into account the influence of the complicated top valence band structure.

2. Basic formulas

From a microscopic point of view the Stokes resonant HRS process can be described as follows: 1) absorption of two incident photons; 2) a creation of a phonon; 3) emission of a scattered photon [13,15]. In the initial and final states an electronic system of a semiconductor is assumed to be in the ground state. And intermediate virtual states are the excitons formed by a transition of an electron from the degenerate (or almost degenerate) valence band v to the excitonic level E_λ near the degenerate conduction band c. The wave functions of the relative electron-hole motion $\tilde{\chi}^{(\lambda)}_{j_e j_h}(\boldsymbol{r})$ are assumed to satisfy a system of the effective mass approximation equations [18-20]. Here the indices $j_e=1,2,\ldots s_c$ and $j_h=1,2,\ldots,s_v$ denote the states in the conduction band c and the valence band v, and s_c and s_v are the appropriate degeneracy multiplicities. Owing to small magnitudes of photon wave vectors the HRS process is supposed to occur in the center of a zone of Brillouin. Hence in this paper the excitons with the wave vector K=0 are considered.

In hexagonal semiconductors A^2B^6 the top valence band is known to be split into three closely-lying doubly degenerate subbands Γ_9, Γ_7 and Γ_7 by the crystal structure and the spin-orbit interaction

[21,22]. The transitions from them to the excitonic levels near the lowest doubly degenerate conduction band Γ_7 lead to the three hydrogen-like series A, B and C in the absorption spectra. If the scalar mass $\bar{\mu}_{j_e j_h}$ is introduced, then for $\tilde{\chi}^{(\lambda)}_{j_e j_h}(\boldsymbol{r})$ in these crystals the equation system can be written as

$$\left(\frac{1}{2}\frac{1}{\bar{\mu}_{j_e j_h}}\left(\hat{p}_x^2+\hat{p}_y^2+\hat{p}_z^2\right)+U_{eh}(\boldsymbol{r})\right)\tilde{\chi}^{(\lambda)}_{j_e j_h}(\boldsymbol{r})+\sum_{j'_h=1}^{s_v}\hat{V}_{j'_h j_h}\tilde{\chi}^{(\lambda)}_{j_e j'_h}(\boldsymbol{r})=\left(E_\lambda-\varepsilon_{j_e j_h}\right)\tilde{\chi}^{(\lambda)}_{j_e j_h}(\boldsymbol{r}) \qquad (1)$$

where $\hat{p}_\alpha=-i\hbar\,\mathrm{d}/\mathrm{d}r_\alpha$, $U_{eh}(\boldsymbol{r})$ is the Coulomb energy of the electron-hole interaction, $\varepsilon_{j_e j_h}$ is the energy gap. The operator $\hat{V}_{j'_h j_h}$ has a form

$$\hat{V}_{j'_h j_h}=\delta_{j'_h j_h}\frac{1}{2}\left\{\left(\frac{1}{\mu^{\perp}_{j_e j_h}}-\frac{1}{\bar{\mu}_{j_e j_h}}\right)\left(\hat{p}_x^2+\hat{p}_y^2\right)+\left(\frac{1}{\mu^{\parallel}_{j_e j_h}}-\frac{1}{\bar{\mu}_{j_e j_h}}\right)\hat{p}_z^2\right\}+\left(\delta_{j'_h j_h}-1\right)D^{\beta\alpha}_{j'_h j_h}\hat{p}_\alpha\hat{p}_\beta\,. \qquad (2)$$

Here the effective mass parameters $D^{\alpha\beta}_{jj'}$ are defined by [18,20]

$$D^{\alpha\beta}_{jj'}=\frac{1}{2m}\delta_{jj'}\delta_{\alpha\beta}+\frac{1}{m^2}\sum_n\frac{\pi^\alpha_{jn}\pi^\beta_{nj'}}{\varepsilon_{jn}} \qquad (3)$$

where $\boldsymbol{\pi}_{jn}$ is the interband matrix element of the momentum operator.

The band structure of the considered crystals is mainly known [22]. For them the dipole transitions from the top valence band v and the lowest conduction band c to the higher-lying conduction band are possible. Besides, the selection rules allow the dipole transitions between the top and lower-lying valence bands. Thus, one can determine the nonzero matrix elements of the dipole transitions between these bands. This enables one to find the nonzero parameters $D^{\alpha\beta}_{jj'}$.

If the anisotropy influence on $U_{eh}(\boldsymbol{r})$ and $\mu_{j_e j_h}$ and the terms associating equations for different subbands in Eq. (1) are neglected ($\hat{V}_{j'_h j_h}\approx 0$), the equation system is reduced to the individual hydrogen-like equations [18,19] for which the wave functions $\chi^{(\xi\ell m)}_{j_e j_h}(\boldsymbol{r})$ and energies are well known [23]. Disregarded terms can be taken into account with the use of the perturbation theory.

Thus, considering $\hat{V}_{j'_h j_h}$ in the equation system (1) as perturbation, the first-order corrections for the lowest discrete energy levels $A_{n=1}$ and $B_{n=1}$ were found

$$E^{(1)}_{A_{n=1}}=\frac{2}{3}R_A\left\{2\left(\frac{\bar{\mu}_A}{\mu_A^{\perp}}-1\right)+\left(\frac{\bar{\mu}_A}{\mu_A^{\parallel}}-1\right)\right\} \qquad (4)$$

$$E^{(1)}_{B_{n=1}}=\frac{2}{3}R_B\left\{2\left(\frac{\bar{\mu}_B}{\mu_B^{\perp}}-1\right)+\left(\frac{\bar{\mu}_B}{\mu_B^{\parallel}}-1\right)\right\} \qquad (5)$$

where $\mu_{A(B)}$ and $R_{A(B)}$ are the effective mass and the exciton Rydberg for the A (B) excitonic series. Besides, the wave functions in the first approximation were obtained for these states. But appropriate expressions are not presented in this work because they too cumbersome.

3. The hyper-Raman scattering

In order to estimate the role of the first order corrections obtained the HRS by LO-phonons under two-photon excitation near the $B_{n=1}$ level was considered for the scattering geometry *z(xxz)y*.

As it is known, the two-photon dipole transition to *s*-exciton state can be described with the use of the three-band as well as two-band models, and their contributions are opposite in sign [15,24]. In accordance with the two-band model absorption of an photon is accompanied by the weakly-forbidden dipole transition from the ground state to an exciton of the *p*-type, another photon causes the intraband dipole transition between excitons [24,25]. In this work allowance was made for the dipole transitions between excitonic states of the different series too. The mechanism of the intraband Fröhlich exciton-lattice interaction is considered in the zero phonon wave vector approximation. Under calculations the dominant scattering mechanism including the two-photon transition to the $B_{n=1}$ exciton state was only taking into account. The Green's function approach was used under summation over intermediate excitonic states [15,26].

Taking into account the published data [22,27], the rough estimations of the frequency dependence of the resonant HRS cross section were performed for a CdS crystal. For comparison figure 1 demonstrates the frequency dependence of hyper-Raman cross section calculated for the wave functions and the energy in a first approximation and with regard for the dipole transitions between excitons relating to different subbands (1) and for unperturbed wave functions (2). As can be seen from the figure, the inclusion of these transitions and the first order correction can give rise to some increase in the cross section.

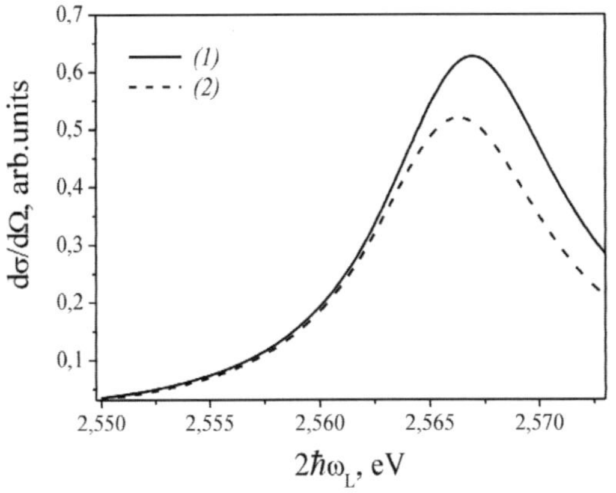

Figure 1. The HRS cross section calculated for the scattering mechanism including the two-photon transition to the $B_{n=1}$ exciton state in CdS as a function of the doubled energy of incident photons.

4. Conclusion

Taking into account the influence of anisotropy and the complicated structure of the top valence band with the use of the perturbation theory, for the hexagonal semiconductors A^2B^6 the wave functions and the energies in the first approximation were obtained for the $A_{n=1}$ and $B_{n=1}$ excitonic states. The HRS by LO-phonons near the two-photon resonance with the $B_{n=1}$ exciton level was considered for a CdS crystal. Rough estimations performed have shown that allowance made for the first order correction and the dipole transitions between the excitons of the different series do not lead to considerable change of the frequency dependence of the HRS cross section.

Acknowledgments
This study was supported by the Program III.2 of the Russian Academy of Sciences "Fundamental problems of laser technologies".

References

[1] Polivanov Yu N and Sayakhov R Sh 1979 *JETP Lett.* **30** 580-3
[2] Watanabe K and Inoue K 1989 *J. Phys. Soc. Jap.* **58** 726-32
[3] Inoue K and Watanabe K 1989 *Phys. Rev.* B **39** 1977-9

[4] Inoue K, Yoshida K, Minami F and Kato Y 1992 *Phys. Rev.* B **45** 8807-10
[5] Inoue K, Minami F, Kato Y, Yoshida K and Era K 1992 *J. Cryst. Growth* **117** 738-41
[6] Matsushita S, Minami F, Yoshida K, Inoue K and Akinaga H 1996 *J. Luminesc.* **66&67** 414-7
[7] Watanabe K and Inoue K 1990 *Phys. Rev.* B **41** 7957-60
[8] Watanabe K, Inoue K and Minami F 1992 *Phys. Rev.* B **46**, 2024-33
[9] Kono S, Naka N, Hasuo M, Saito S, Suemoto T and Nagasawa N 1996 *Solid State Commun.* **97** 455-8
[10] Zubkova L E, Ondriash K K, Polivanov Yu N and Prokhorov K A 1993 *JETP Lett.* **57** 348-51
[11] Maslov V A, Ondriash K K, Polivanov Yu N, Prokhorov K A, Semenova L E and Chuzavkov Yu L 1996 *Las. Phys.* **6** 132-43
[12] Filippidis L, Siegle H, Hoffman A and Thomsen C 1999 *Phys. Stat. Sol.* (b) **212** R1-2
[13] Garcia-Cristobal A, Cantarero A, Trallero-Giner C and Cardona M 1998 *Phys. Rev.* B **58** 10443-51
[14] Garcia-Cristobal A, Cantarero A, Trallero-Giner C and Cardona M 1999 *Physica* B **263-264** 809-12
[15] Semenova L E and Prokhorov K A 2003 *JETP* **96** 922-31
[16] Semenova L E and Prokhorov K A 2004 *Las. Phys. Lett.* **1** 253-8
[17] Semenova L E and Prokhorov K A 2005 *Las. Phys. Lett.* **2** 262-6
[18] Dresselhaus G 1956 *J. Phys. Chem. Solids* **1** 14-22
[19] Makarov V P 1971 *Phys. Stat. Sol.* (b) **44** 475-88
[20] Denisov M M and Makarov V P 1973 *Phys. Stat. Sol.* (b) **56** 9-59
[21] Reynolds D C 1969 Excitons in II-VI compounds *Optical Properties of Solids* ed S Nudelman and S S Mitra (New York: Plenum Press) chapter 10 pp 239-86
[22] Sobolev V V 1980 *Bands and Excitons of the* $A^{II}B^{VI}$ *Group Compounds* (Chisinau: Shtiintsa)
[23] Landau L D and Lifshitz E M 1989 *Course of Theoretical Physics* vol 3 (Moscow: Nauka)
[24] Rustagi K, Pradere F and Mysyrowicz A 1973 *Phys. Rev.* B **8** 2721-32
[25] Mahan G D 1968 *Phys. Rev.* **170** 825-38
[26] Hostler L 1964 *J. Math. Phys.* **5** 591-611
[27] Nguyen B A, Nguyen V H, Nguyen T T and Nguyen A V 1982 *Phys. Rev.* B **25** 4075-80

Dielectric Anisotropy of Human Bone and CERABONE® in the Terahertz Spectral Range 0.2 to 2.5 THz

A.S. Nikoghosyan[1], H. Ting [2], J. Shen[2], R.M. Martirosyan[1], M.A. Kazaryan[4] M.Yu.Tunyan[3], A.V. Papikyan[3], A.A. Papikyan[3]

[1]Yerevan State University, 1 Alex Manoogian, 0025 Yerevan, Armenia
[2]Capital Normal University, Key Laboratory for Terahertz Spectroscopy and Imaging, Beijing, 100048 China
[3]Yerevan State Medical University, Yerevan, Armenia
[4]Optics, Lebedev Physical Institute, Moscow, Russia

E-mail: nika@ysu.am

Abstract.
Terahertz time-domain spectroscopy (THz-TDS) was applied to study properties of a human jawbone in transmission geometry. The fiber femtosecond laser (Fx-100, IMRA) with a pulse width of 113 fs, a central wavelength of 800 nm and an average power of 120 mW was used as a laser source for pumping and detecting terahertz pulses. The refractive indices $n(\omega)$ and absorption coefficients $\alpha(\omega)$ of the human jawbone and bone substitute Cerabone® were determined in vitro by the THz-TDS in a wide frequency range from 0.2 to 2.5 THz. It is shown that the refractive index of the human jawbone changes between the values of 2.24 and 2.36, and Cerabone® between 2.4 and 2.65. The absorption coefficient of the human jawbone depending on frequency increases from 1.7 cm^{-1} to 178.5 cm^{-1}, showing several resonance absorption lines after 1.6 THz. The absorption coefficient of Cerabone® increases from zero to 80 cm^{-1}, and the resonance absorption occurs at 1.7 THz. The experimental results indicate that $n(f)$ and $\alpha(f)$ of a human jawbone change with the alteration of the direction of the linear polarization vector of the electric field of THz pulse relative to the axis of the plate of the human jawbone. The obtained results allowed us to determine the proximity of the physical properties of the Cerabone® with the natural bone matrix.

PACS: ***http://www.aip.org/pacs***
Keywords: terahertz time domain spectroscopy, dielectric anisotropy, human bone, terahertz range, Cerabone

1. Introduction

A bone material is composed of an organic matrix of collagen fibers and mineral hydroxyapatite ($Ca_5(PO4)_3OH$) nanoparticles. An average tooth dentin contains 70% hydroxyapatite crystals, 20% collagen (e.g., proteins), and 10% water. The organic constituents provide flexibility, whereas the mineral provides strength and quality. Due to the specific arrangement of nanosized mineral platelets and collagen fibrils with respect to the main axis in case of a long bone (figure 1(a)), signals relative to vibrational units of both mineral and collagen can result highly anisotropic. Up to now, very few studies have been conducted concerning the study of the human bones [1,2] and its anisotropy by THz radiation [3].

Many biological tissues (eye cornea, retina, dura mater, teeth, tendon, cartilage, muscle, myocardium, artery wall, nerve, etc.) are structurally anisotropic. Tissue birefringence results from the linear anisotropy of fibrous structure. The refractive index of a medium is higher along the length of fibres than along their cross section [3]. A tissue structure is a system composed of parallel cylinders that create a uniaxial birefringent medium with the optic c-axis parallel to the fibrils (ciliders) axes, as shown in figure 1(b) [4, 5].

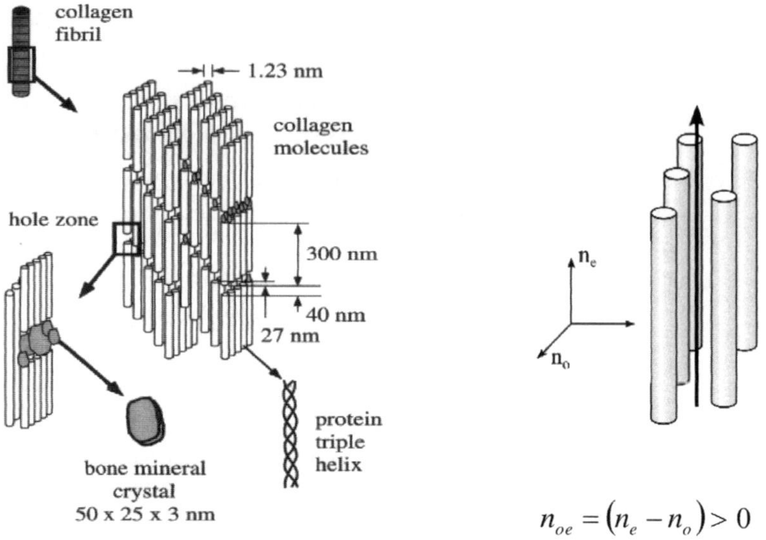

Figure 1. System of long dielectric cylinders array (mineralized collegen fibrils of turkey) with major components

A structure of parallel dielectric fibrils immersed in isotropic homogeneous ground substance behaves as a positive uniaxial birefringent medium $n_e - n_o > 0$, where n_e is the refractive index of extraordinary ray and n_o the refractive index of ordinary ray. The hidroxyapatite can be found in space between fibrils - in layers that transverse across the fibrils and both within and outside the collagen fibrils [6, 7]. The collagen fibrils periodically separated by a tiny gap from 1 nm to 67 nm. The results of electron diffraction and X-ray diffraction, carried out on samples, support the concept of organized intra-fibril mineral crystals within the organic collagen matrix.

THz wave (frequency range from 0.1 to 10 THz) found the most widespread practical use in THz time-domain spectroscopy [8,9] and in the THz imaging [10,11]. It is known that the time of vibrational motion of biological molecules have the order of picoseconds, and therefore, the frequency of vibrations is in the terahertz frequency range. Intermolecular interactions are usually weaker than the intramolecular and only THz spectroscopy in the time domain is sensitive to resolve their spectrum in the THz range. THz wave is non-invasive (since the photon energy of terahertz waves at several orders of magnitude smaller than the photon energy of the x-ray wavelength) and the non-contactin nature, can penetrate into the non-conductive materials and to provide additional spectroscopic data for the accurate diagnosis and analysis of the material.

In this paper we report the results obtained from a study of dielectric anisotropic properties of a human jawbone in transmission geometry using THz time-domain spectroscopy (THz -TDS).

2. Experimental technique

A schematic arrangement of the THz-TDS system, demonstrating the principal configuration used to obtain THz spectral data, is shown in figure 2. The fiber femtosecond laser (Fx-100, IMRA) with a pulse duration of 112 fs, a central wavelength of 800 nm and an average power of 120 mW was used as a laser source for pumping and detecting terahertz pulses.

The laser output, in the form of high repetition frequency - 75 MHz, is divided into two optical paths - the pump and probe beams. The pump beam is focused into a gap between biased electrodes deposited upon the surface of gallium arsenide. The pumping pulses have photon energy (1.43 eV) above the direct band-gap of the GaAs this induced conductivity changes. Electron–hole pairs are created by each laser pulse in a semiconductor, which, when accelerated by the bias field, act as a transient current source. The transient current radiates a sub-picosecond, single-cycle coherent THz electric pulse, as shown in figure.3a. The THz pulse from photoconductive antenna (PCA) is emitted in a dipole like pattern. The resulting radiation is polarized along the direction of the bias field. The polarization ratio is usually better than 10:1.

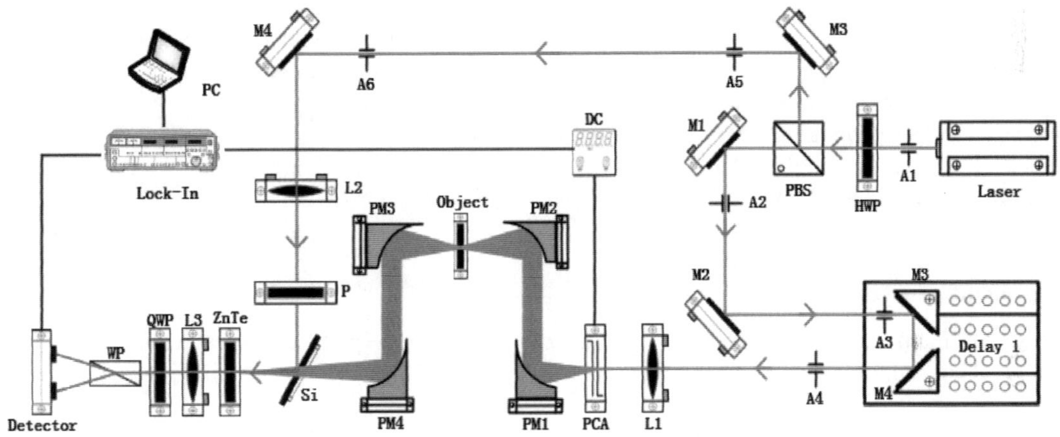

Figure 2. Experimental setup of the THz spectrometer.

Figure 2 is experimental setup of the THz spectrometer, where L is the femtosecond laser Fx-100, IMRA; A1–A6 the diafragms; M1–M6 the mirrors deflecting the optical beam; DL optical delay line; PM1–PM4 the parabolic mirrors; HWP the half-wave (λ/2) plate; PBS the polarizing beam splitter; L1–L3 the focusing lenses; PCA the GaAs photoconductive antenna; DC the data acquisition software; P the polarizer; Si the silicon wafer which unites the THz and optical beams; ZnTe the crystall; QWP the quarter-wave (λ/4) plate; WP the Wollaston prism; S the test material; D the detecting diodes; LA the lock-in amplifier; PC the personal computer.

The THz pulse after PCA is collected and directed to the test material using parabolic mirrors PM1 and PM2. The THz radiation transmitted through the test material is then directed by two parabolic mirrors PM3 and PM4 towards a detector. The subpicosecond THz pulse and optical femtosecond pulse were combined by the thin plate of Si inside the ZnTe crystal.

A small fraction of the pump optical beam - probe beam, was used for coherent detection of transmitted radiation from the test material (jawbone) using a dynamic free-space electro-optic cell, which consists of an electro-optic crystal ZnTe with a (110) orientation and the thickness 1 mm, of the quarter-wave plate providing optical bias, and of the Wollaston prism (WP) as the analyzer. The probe beam has controlled the detector, whose response was proportional to the amplitude and the sign of the electric field of the THz pulse. Thus, the whole system represents a version of a coherent pump-probe spectroscopic setup [12].

The electro-optic sampling method is widely applied for the coherent detection of THz pulse due to its short response time, high sensitivity and wide bandwidth. For the electro-optic sampling, the phase-matching condition requires the group velocity of probe pulse equal to the phase velocity of THz pulse. The phase mismatch is defined as [13]

$$\Delta k = \frac{f_{THz}}{c}\left(n_{THz}\left(f_{THz}\right) - n_g\left(f_{opt}\right)\right) \qquad (1)$$

The amplitude of detection response is proportional to the thickness of crystal ZnTe at the phase-matching condition, but a thicker crystal results in the reduction of bandwidth of THz pulse. The compromise should be made between the detection response and bandwidth. We used a ZnTe crystal of 1 mm thickness, ipso facto to provide the coherent detection of THz pulse in spectral range from 0.2 THz to 2.7 THz.

The temporal form of the THz pulse E(t) was determined by the change in the time delay between pump and probe beams using the optical delay line (DL). The balanced electro-optic detection method provides an excellent signal to noise ratio >1500 (to a noise-limited frequency of 3 THz) by the use of phase sensitive amplification with a lock-in amplifier. The amplified signal from a lock-in amplifier is transmitted to the computer through the analog to digital converter. Two spectra, THz field E(f) and phase Φ(f), are processed via a fast Fourier transform in the program Origin from originally obtained time-domain dependence E(t).

3. Results and discussion

THz time-domain transmission spectroscopic properties of the jawbone is presented in figure 3 and figure 4. The temporal forms of the THz pulses transmitted through the air (reference pulse) and human jawbone and the corresponding THz field spectra obtained after the fast Fourier transform are shown in figures 3a,b,c. The phase-frequency dependences $\Phi_1(\omega)$ and $\Phi_2(\omega)$ are shown in figure 3d, and the refractive indices and absorption coefficients at figure 3e, f. The laser polarization is parallel to the bone axis. To demonstrate the reproducibility of the experiment, on figure 3 (b, c, e, f) the curves of three consecutive measurements are given. The thickness of the jawbone was 0.44 mm and the cross-section 2×1 mm^2. During the measurements, the air temperature was 21°C, the humidity $<$ 1.5%, the dynamic range was more than 1000, and the signal/noise ratio at the peak position was about 400.

Temporal waveforms of THz pulses transmitted through the jawbone, THz amplitude spectra of the fields after a FFT , the index of refraction **n(ω)** and absorption coefficient **α(f ω)** for different directions of the laser polarization are depicted in figures 4 a-d. Figure 4(c) shows that the index of refraction n(f) of the jawbone and therefore the velocity of the THz pulse is different for different directions of the THz polarization with respect to the axis of the plate of the human jawbone. The dispersion of the refractive index of jawbone is weak for all directions of polarization, however, $n_{45}(f)$ is greater than the value $n_{ver}(f)$ and $n_{par}(f)$ at any frequency in the spectral range of 0.2-1.5 THz.

IOP Conf. Series: Journal of Physics: Conf. Series **826** (2017) 012005 doi:10.1088/1742-6596/826/1/012005

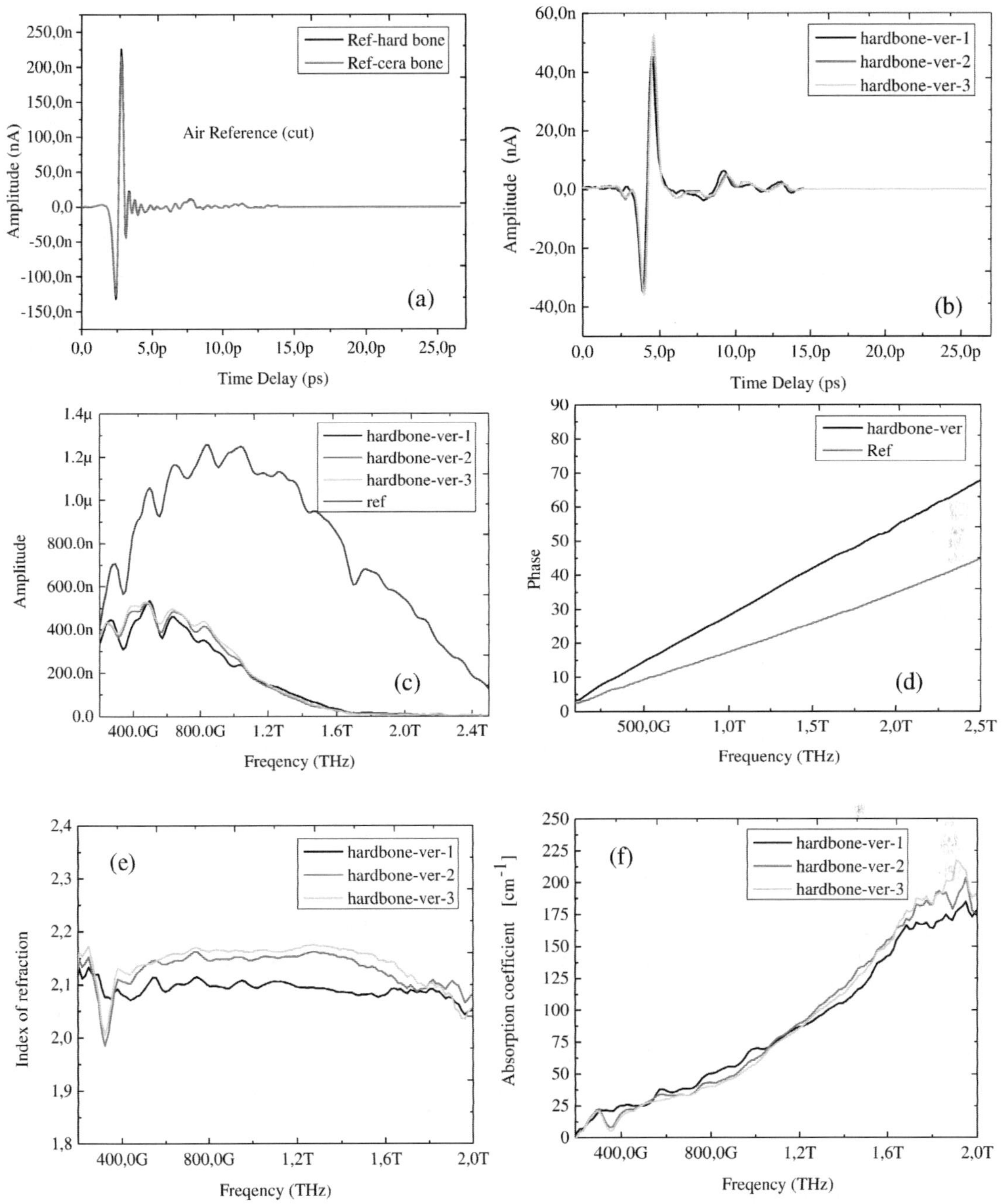

Figure 3. Temporal waveforms of THz pulses transmitted through (a) the air and (b) jawbone, THz amplitude spectra of the fields after a FFT (c); phase-frequency response (d); refractive index (e) and absorption coefficient human jawbone (f). The THz pulse polarization is parallel to the bone axis.

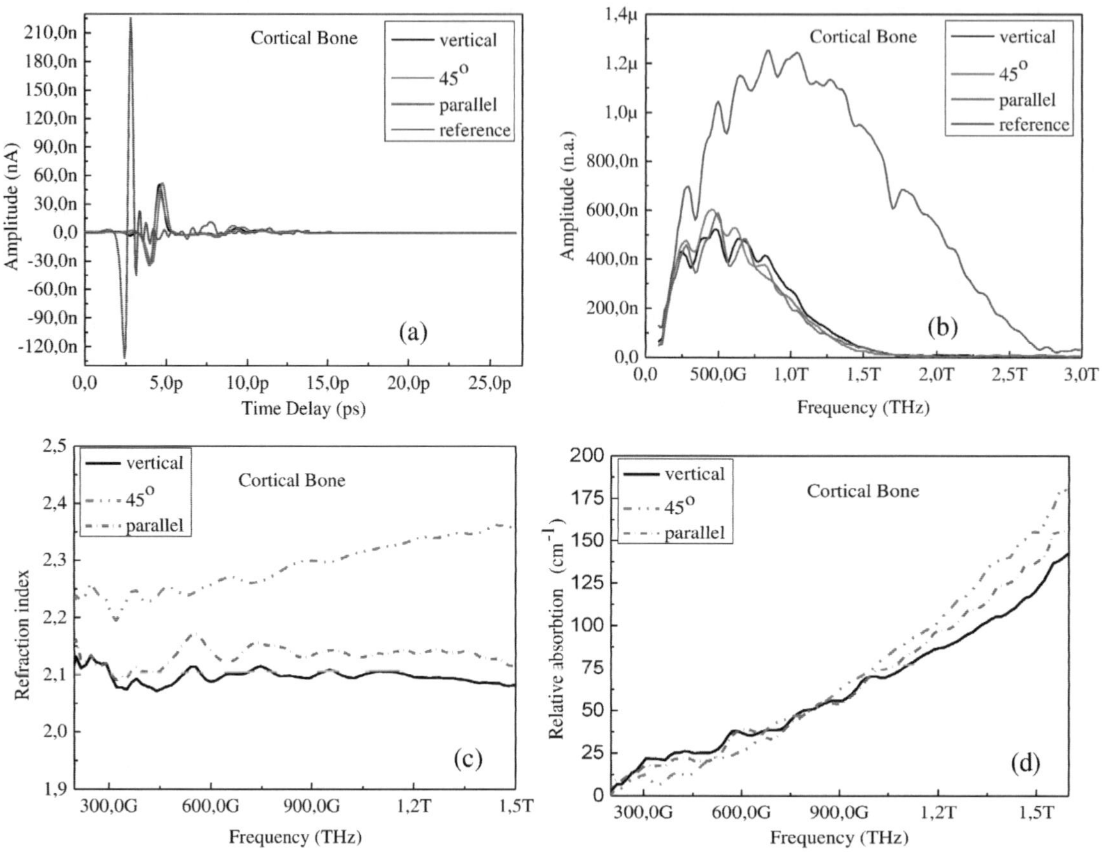

Figure 4. The index of refraction n(f) and absorption coefficient α(f) for different directions of the THz polarization (vertical, 45°, and parallel to the bone axis), in the spectral range of 0.2-2.5 THz.

The difference between the absorption coefficient α(f) for different directions of the THz pulse polarization depend on the relationship between the sizes and the properties of the cylinders or plates, can take positive or negative values. For example, only at frequencies 250 GHz and 830 GHz the absorption coefficient α(f) for different directions of the THz pulse polarization is same, $\Delta\alpha = 0$, but at frequencies above 830 GHz $\Delta\alpha$ is positive, and below 830 GHz negative.

Table 1. Optical parameters n(f) and α(f) of the jawbone depending on the orientation of the THz pulse polarization

Polarization	**n_{max}(f)**	**n_{min}(f)**	**Δn(f)**	**α_{min}(f), cm^{-1}**	**α_{max}(f)**	**Δα(f)**
Horizontal	2.1725	2.0896	0.0829	7.132	157.514	150.38
Vertical	2.1332	2.0711	0.0621	3.956	142.200	138.24
45°	2.3631	2.1941	0.169	1.717	178.487	176.77

It is known that the variation amount of trace elements in bone causes osteomalacia, osteoporosis and other diseases. Crystals form mineralized bone varies from plate to the needle. In shape and size of the crystals can be affected by age, state of health. Throughout life, all bones are in a dynamic process of growth and resorption called reconstruction. Inhibition or lack of bone reconstruction accompanied by a deterioration of the mechanical properties and results in macro- and microfractures. All these factors can influence the anisotropy of the bone. The information about the tissue, obtained by THz pulses orthogonal polarizations will improve also the diagnosis of tissue - whether it is cancerous or not. Cancer cells produce large heterogeneity in healthy tissues in shape and chemical composition.

The use of orthogonally polarized pulses in THz imaging of tissues provides unprecedented potential for nondestructive mapping of anisotropy in tissues with cancer. It is well known that enlarged nuclei of the cell are primary indicators of cancer

Conclusions

Dielectric anisotropy of a human jawbone has been studied using THz time-domain transmission spectroscopy in the wide frequency range 0.2–2.5 THz in vitro. The refractive index n(f) and absorption coefficient α(f) for different directions of the THz pulse polarization were measured. As the results of studies, it was found that the refractive index of the human jawbone changes between the values of 2.07 and 2.36 depending on the THz pulse polarization, and Cerabone® – between 2.4 and 2.65 [12]. The absorption coefficient of the human jawbone increases with the frequency from 1.7 cm^{-1} to a value of 178.5 cm^{-1}. The absorption coefficient of Cerabone® increases to 80 cm^{-1}, and the resonance absorption occurs at 1.7 THz [12]. The difference n(ω) and α(ω) in different directions at any frequency can be associated with the structural anisotropy of a bone that is, both with different dimensions of the bone particles and a specific tissue structure.

It was found that before and after the 830 GHz $\Delta\alpha = \alpha_{ver} - \alpha_{par}$ and $\Delta\alpha = \alpha_{ver} - \alpha_{45^0}$ changes sign.

This is the first representation of the frequency-dependent refractive index n(f) and absorption coefficient α(f) of human jawbone.

Acknowledgments

This research was supported by the National Instrumentation Program (Grant No.2012YQ140005), the National Nature Science Foundation of China (Grant Nos. 61505125), the Nature Science Foundation of Beijing (Grant No. 4144069) and Science and Technology Project of Beijing Municipal Education Commission (Grant No. KM201410028004).

References

[1] Almany Magal R. Reznikov N. Shahar R. Weiner S 2014 Three-dimensional structure of minipig fibrolamellar bone: Adaptation to axial loading J. Struct. Biol. **186** 253–264

[2] Berry E, Fitzgerald A J,. Zinov'ev N N, Walker G C, Homer-Vanniasinkam S, Sudworth C D, Miles R E, Chamberlain J M, Smith MA 2003 Optical properties of tissue measured using Terahertz pulsed imaging Proceedings of SPIE: Medical Imaging: Physics of Medical Imaging **5030** 459-470

[3] Tuchin V V 2015 Tissue Optics and Photonics: Biological Tissue Structures J. of Biomedical Photonics & Eng, **1** no.1 3-21

[4] Reznikov N, Shahar R, Weiner S, 2014 Acta Biomater **10** 3815
[5] Larry Arsenault A 1991 Image Analysis of Mineralized and Non-Mineralized Type I Collagen Fibrils. J. of Electron Microscopy Technique **18**, 262-268
[6] Svetoslav Nikolov and Dierk Raabe Hierarchical June 2008 Modeling of the Elastic Properties of Bone at Submicron Scales: The Role of Extrafibrillar Mineralization. Biophysical Journal **94** 4220–4232
[7] ErtsL D, Gathercole E J, Atkins D T 1994 Scanning probe microscopy of intrafibrillar crystallites in calcified collagen J. of Materials Science: Materials in Medicine **5** 200–206
[8] Fitzgerald A J., Berry E, Zinov'ev N N, Homer-Vanniasinkam S, Miles RE, Chamberlain J M, and Smith M A, 2003 *J. Biol. Phys.* **29** 123.
[9] Berry E, Fitzgerald A J, Zinov'ev N N, Walker G C, Homer-Vanniasinkam S, Sudworth C D, Miles R E, Chamberlain J M, and Smith M A, 2003 Proceedings of SPIE. **5030** 459
[10] Zinov'ev N N, Nikoghosyan A S, and Chamberlain J M 2006 *Proceedings of SPIE*, **6257** 62570P1
[11] Pickwell E, Wallace V P, Cole B E, Ali S, Longbottom C, Lynch R and Pepper M 2007 *Caries Res.*, **41** 4955.
[12] Nikoghosyan A S, Ting H, Shen J, Martirosyan R M, Tunyan M Yu, Papikyan A V, Papikyan A A 2016 Optical properties of human jawbone and human bone sabstitute CERABONE® in the terahertz range Journal of Contemporary Physics (Armenian Academy of Sciences) **51** No. 3 256-264
[13] A. Yariv 1983 Inroduction to Optical Electronics Moscow, Высшая школа

Figure Caption

Figure 1. System of long dielectric cylinders array (mineralized collegen fibrils of turkey) with major components.

Figure 2. Experimental setup of THz spectrometer

Figure 3. Temporal waveforms of THz pulses transmitted through (a) the air and (b) jawbone, THz amplitude spectra of the fields after a FFT (c); phase-frequency response (d); refractive index (e) and absorption coefficient human jawbone (f). The THz pulse polarization is parallel to the bone axis.

Figure 4. The index of refraction n(f) and absorption coefficient α(f) for different directions of the THz polarization (vertical, 45°, and parallel to the bone axis), in the spectral range of 0.2-2.5 THz.

New insights into the structure of polypropylene polymorphs and propylene copolymers probed by low-frequency Raman spectroscopy

E A Sagitova[1,2], P Donfack[2], G Yu Nikolaeva[1], K A Prokhorov[1], P P Pashinin[1], P M Nedorezova[3], A N Klyamkina[3] and A Materny[2]
[1] Prokhorov General Physics Institute, Vavilov St. 38, 119991, Moscow, Russia
[2] Department of Physics and Earth Sciences, Jacobs University Bremen, Campus Ring 1, 28759, Bremen, Germany
[3] Semenov Institute of Chemical Physics, Kosygin St. 4, 119991, Moscow, Russia

E-mail: sagitova@kapella.gpi.ru

Abstract. We present a low-frequency (or low-wavenumber) Raman spectroscopy study of isotactic polypropylene, namely the different modifications - α, γ, and smectic, with a degree of crystallinity ranging from 0 to 70 %, as well as of random propylene/1-butene and propylene/1-octene copolymers with the incorporated monomer contents of 5.3 and 19.5 mol % for 1-butene and 4.5 and 24.0 mol % for 1-octene. In the Raman spectra of the propylene/1-octene copolymers and the neat isotactic polypropylene, we have found that the band at 105 cm^{-1}, which is connected to the torsion vibrations of CH_2-$CHCH_3$ groups in the polymer backbone, can be assigned to the vibrations of helical macromolecules in the α and γ crystallites and the smectic phase. If the smectic phase is absent in the corresponding polymer system, then the intensity of the 105 cm^{-1} band is related to the molecular vibrations in the crystallites. We have also observed a broadening of the band at 398 cm^{-1} accompanied by the appearance of a high-wavenumber shoulder in the Raman spectra of the copolymers with the increase in the incorporated monomer content. This experimental finding can be explained by the reduction of the degree of crystallinity, conformational disordering or by the increase in the dimensions of the crystal unit cell.

1. Introduction

Polypropylene (PP) exhibits a complex molecular and supramolecular structure, which has a major impact on its physical properties. There are three stereoisomeric forms of PP: the isotactic, syndiotactic and atactic configurations, and the macromolecules in each of these configurations can adopt different conformations. Besides PP, random copolymers of propylene with α-olefins (ethylene, 1-butene, 1-octene, etc.), which can be considered to be materials with PP-like structure, are widely used in industry [1-4].

The molecular chains of most commercially-important isotactic PP in the helical conformation can crystallize in the α, β, and γ modifications, which have the monoclinic, hexagonal, and orthorhombic lattices, respectively [1, 2, 5]. The different crystalline modifications can coexist in an isotactic PP sample, for which the total degree of crystallinity does not exceed 75%. Usually, non-crystalline regions of isotactic PP contain macromolecules in both the helical and non-helical conformations [6]. However, there is also the smectic or pseudocrystalline modification, in which the macromolecules of isotactic PP do not form crystals but exhibit the helical conformation as well [5, 7]. Often, the smectic modification is considered to be an intermediate form between the amorphous and the crystalline states of isotactic PP.

Raman spectroscopy can potentially provide quantitative information about the configurational, conformational, and phase states of macromolecules, the content and length distribution of conformationally-ordered macromolecular segments, and about the sizes of the macromolecular crystallites. The spectral range 600–1600 cm^{-1} in the Raman and IR spectra of isotactic PP is the most important for a quantitative analysis of the phase and conformational compositions. This spectral region was investigated in numerous works, and the obtained results are summarized in our recent papers [8, 9]. However, a number of spectral lines in this region are still not precisely assigned. For example, for the line at 809 cm^{-1}, whose intensity is often used to calculate the degree of crystallinity, contradictory assignments in the Raman and IR spectra can be found [8, 9]. While the range 2700–3100 cm^{-1} is less suited for quantitative structural analysis of propylene-containing materials because of strong overlap of numerous bands with large bandwidths, the range 0–600 cm^{-1} contains a number of well-separated and structure-sensitive Raman-active lines. Nevertheless, the 0–600 cm^{-1} region has been less investigated for the phase and conformational composition analysis, probably due to the relatively more demanding Raman measurements in this very low-frequency region.

The low-frequency Raman spectra of isotactic PP at room and cryogenic temperatures were studied by G.V. Fraser *et al.* [2] for the α-modification, by P.D. Vasko and J.L. Koenig [7] for the α and smectic modifications, and by J.M. Chalmers *et al.* [5] for the α, β, γ, and smectic modifications. It should be noted that the sample referred to as γ PP in the study by Chalmers *et al.* [5] was in reality a mixture of the α and γ modifications with equal contents of α and γ phases and with a total degree of crystallinity of 52%. According to the works in references [2, 5, 7], the bands at approximately 106, 171, 250, 320, 398, 460, and 530 cm^{-1} are found in the Raman spectra of all the crystalline modifications of isotactic PP (α, β, and γ). Also, the band at approximately 50 cm^{-1} and the longitudinal acoustic mode, whose peak position is inversely proportional to the length of isotactic segments in the helical conformation [10, 11], are observed in the Raman spectra of all the crystalline forms.

In the publication by Chalmers *et al.* [5] however, the reported Raman spectrum of highly-crystalline α PP contains the band doublet at approximately 460 cm^{-1} with the peak positions at 453 and 460 cm^{-1}. Unfortunately, for the β, γ, and smectic forms, these authors presented only tabulated data of peak positions, without any information about the intensities and widths of the Raman bands and without reporting the Raman spectra themselves. The peak positions reported for room-temperature Raman measurements in this work showed a noticeable difference for the smectic modification as compared to the other modifications. On the contrary, P.D. Vasko and J.L. Koeing [7] reported that these peak positions in the Raman spectra of the α and smectic modifications are identical. The only difference was an additional weak band at 417 cm^{-1} observed in the Raman spectra of the smectic modification at -160°C. Chalmers *et al.* [5] also observed a band at 413 cm^{-1} in the Raman spectra of the γ modification of PP at cryogenic temperature, but they did not observe this band in the case of the smectic PP.

The studies of the low-frequency Raman spectra of molten isotactic PP were reported by P.D. Vasko and J.L. Koeing [7] and by R.M. Khafagy [12] as well as by V.M. Hallmark *et al.* [13] and T. Hahn *et al.* [14], who carried out numerical calculations for simulating the isotropic Raman spectra of isotactic PP. On the one hand, the band at 398 cm^{-1} is observed in solid, smectic and melt states of isotactic PP [7]. This band is assigned either to the umbrella vibration about the tertiary carbon atom [13] in helix segments with length of at least $n = 6$ or 9 monomer units (two or three complete turns in the helix) [14] or to the superposition of the wagging vibration of CH_2-groups ($\omega(CH_2)$) and the bending vibration of CH-groups ($\delta(CH)$) [12]. On the other hand, the calculated spectrum contains a broad shoulder band on the high-wavenumber side of the 398 cm^{-1} band [13, 14]. This shoulder was assigned to the vibrations of conformational defects in the macromolecules [13].

IOP Conf. Series: Journal of Physics: Conf. Series **826** (2017) 012006 doi:10.1088/1742-6596/826/1/012006

However, only a little shift (~ 2 cm^{-1}) of this band without any appearance of a high-wavenumber broad shoulder was found in the Raman spectra recorded when solid isotactic PP was heated up to and above the melting temperature in the experimental work by R.M. Khafagy [12]. Moreover, this author suggested that the 398 cm^{-1} band depends only on the molecular configuration. Also, the high-wavenumber shift of the band at 398 cm^{-1} was observed in the Raman spectra of isotactic PP-halloysite nanocomposites upon increase in the halloysite content, and this was explained by a drop in the degree of crystallinity of the polymer upon loading it with the filler [15].

Modern developments in PP synthesis allow to produce PP-based functional materials with different structural characteristics and with properties varying in a wide range. In particular, samples with 100 % content of the γ modification in the crystalline phase can be synthesized [8].

In the present work, we have studied the low-frequency Raman spectra of the α, γ, and smectic PP polymorphs, fully amorphous PP, and random copolymers, which include propylene/1-butene and propylene/1-octene copolymers with various comonomer content.

2. Experimental

Table 1. Degree of crystallinity of the investigated copolymers and PP samples

Polymer system	Content of 1-octene or 1-butene (mol %)	Degree of crystallinity (CR) based on X-ray analysis (%)
	Polymer films	
propylene/1-octene copolymers (CPOs)	24.0	0
	4.5	54 (42% of α phase + 12% of γ phase)
propylene/1-butene copolymers (CPBs)	19.5	55 (α phase)
	5.3	55 (52% of α phase + 3% of γ phase)
isotactic α PP (α-iPP-f)	0	55 (α phase)
smectic isotactic PP (smectic iPP)	0	smectic
	Nascent forms of PP	
isotactic α PP (α-iPP)	0	68 (α phase)
isotactic γ PP (γ-iPP)	0	10 (γ phase)
	0	20 (γ phase)
	0	35 (γ phase)
	0	70 (γ phase)
amorphous PP (amorphous PP)	0	0
isotactic polybutene (iPB)	0	~ 50

As recapitulated in Table 1, the studied random copolymers of propylene (C_3H_6) with 1-octene (C_8H_{16}), denoted as CPO, contain 24.0 and 4.5 mol % of 1-octene, while the random copolymers of propylene with 1-butene (C_4H_8), denoted as CPB, contain 19.5 and 5.3 mol % of 1-butene. The CPO and CPB samples as well as the samples of the α and smectic modifications of isotactic PP (α-iPP-f and smectic iPP, respectively) were fabricated as films with a thickness of 1.5 mm. We also have investigated the nascent forms of the α and γ modifications of isotactic PP and isotactic polybutene as well as the fully-amorphous PP, referred to as α-iPP, γ-iPP, iPB, and amorphous PP, respectively.

The CPO and CPB copolymers and the α modification of isotactic PP were synthesized in the medium of liquid propylene over the homogeneous metallocene catalyst of C_2-symmetry, *rac*-$Me_2Si(4\text{-}Ph\text{-}2\text{-}MeInd)_2ZrCl_2$/MAO. The synthesis method and X-ray analysis data were published earlier in references [3,4]. The amorphous PP sample and the γ-iPP samples with a degree of crystallinity of 10, 20, and 35 % were synthesized in the medium of liquid propylene over the homogeneous metallocene catalyst of C_1 - symmetry C_2H_4(Flu)(5,6cyclopenta-2-MeInd)$ZrCl_2$/MAO. The corresponding synthesis methods and X-ray analysis data were described in earlier studies [16, 17].

Similar information about the γ-iPP sample with the degree of crystallinity of 70 % can be found in a different study reported earlier [18]. Smectic iPP was prepared by melting of the isotactic PP at 200°C followed by subsequent pressing and quenching in icy water. Table 1 summarizes the description of the samples under study and the values of the degree of crystallinity, which were determined by X-ray analysis [3, 4, 16-18].

To record Raman spectra from the polymer films, we used the 514.5 nm line of an Ar^+ laser source (Innova 308 series, Coherent, U.S.) for excitation. The scattered light was analyzed by a triple monochromator (T6400, Jobin Yvon, France) using a 1800-grooves-per-mm diffraction grating and a slit width of 100 μm in order to achieve a spectral resolution of approximately 2 cm^{-1}. A liquid-N_2 cooled CCD detector (CCD 3500, Jobin Yvon, France) was used for signal detection. The scattered light was collected in a micro-Raman backscattering geometry using a 50× (N.A. 0.5) long working distance microscope objective.

The Raman spectra of the nascent PP samples were recorded by using another Raman setup consisting of a Sapphire SF 532 laser (Coherent Inc., USA) emitting at a wavelength of 532 nm for excitation, a double monochromator (U1000, Jobin Yvon, France) and a Peltier-cooled photomultiplier detector operating in photon counting regime. For this setup, the Raman spectra were recorded at 90° scattering with a spectral resolution of approximately about 5 cm^{-1}. The Raman spectrum of toluene in the region 0-600 cm^{-1} was used for wavenumber calibration of the Raman spectrometers.

3. Results and Discussion

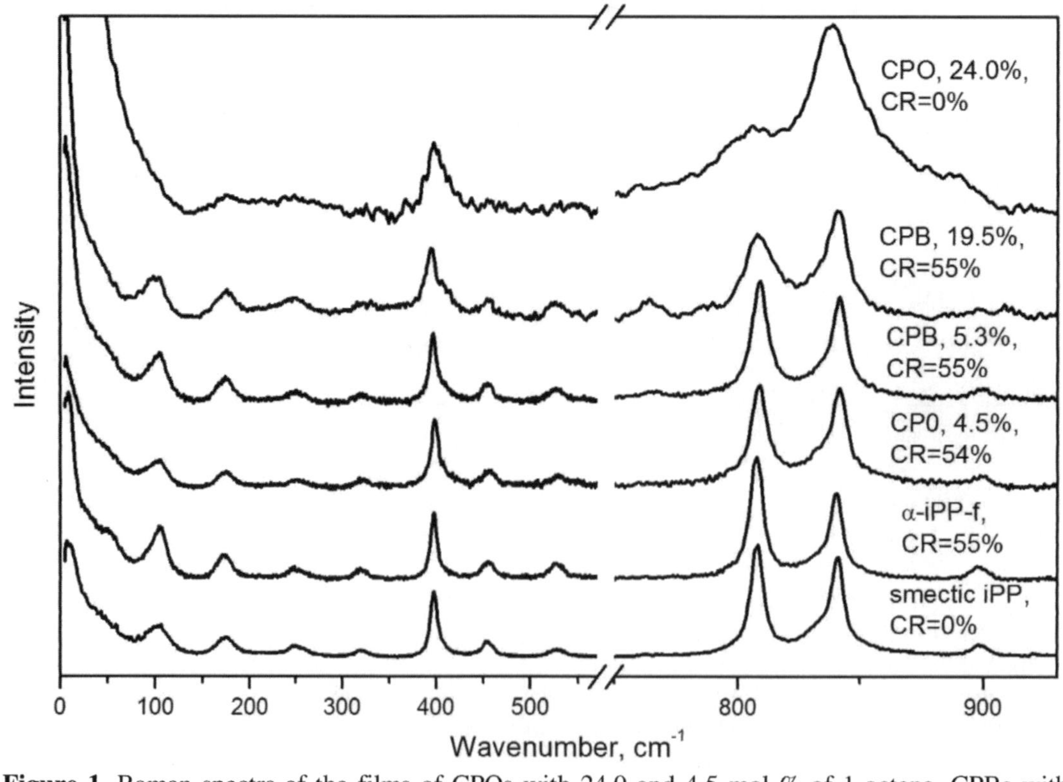

Figure 1. Raman spectra of the films of CPOs with 24.0 and 4.5 mol % of 1-octene, CPBs with 19.5 and 5.3 mol % of 1-butene, α isotactic PP (α-iPP-f), and smectic iPP in the region 0-900 cm^{-1}.

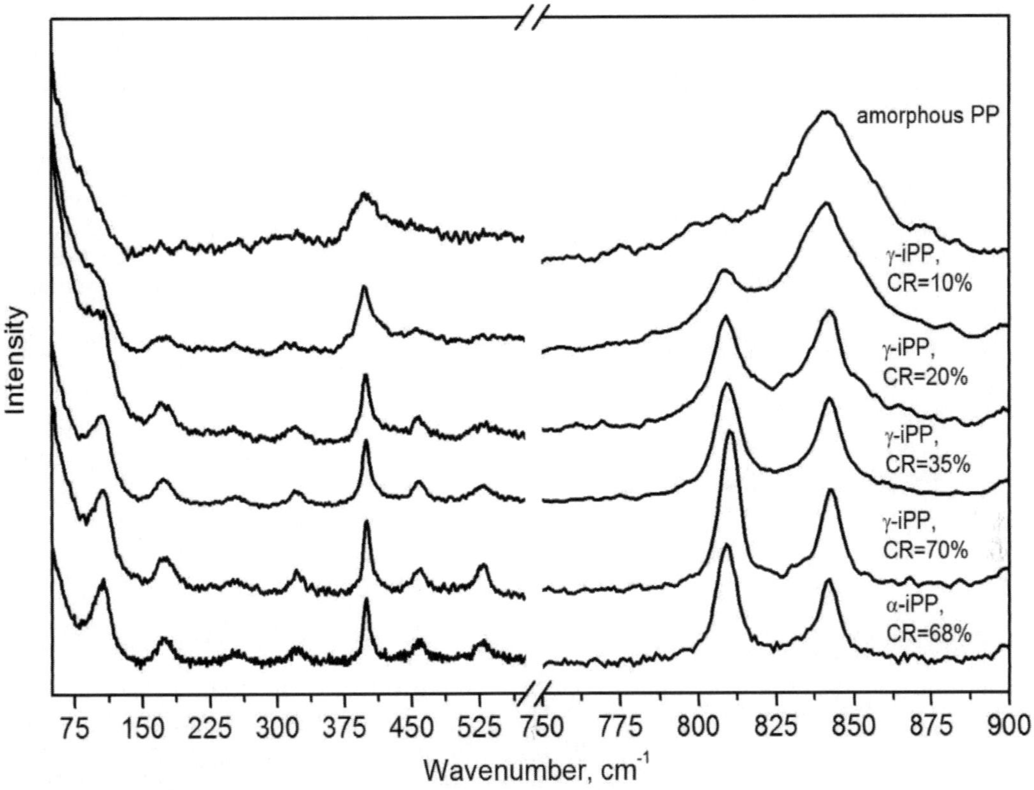

Figure 2. Raman spectra of PP nascent forms including the α- and γ-iPP samples with different degree of crystallinity CR and the amorphous PP in the region 50-900 cm^{-1}.

Figure 1 shows the Raman spectra of the polymer films including the copolymers CPOs and CPBs, the α-modification of isotactic PP, and the smectic iPP in the region 0-900 cm^{-1}. Figure 2 presents the Raman spectra of the studied α and γ modifications of PP nascent forms with different degree of crystallinity (CR), as well as that of the amorphous PP in the region 50-900 cm^{-1}. Figures 3 and 4 illustrate the evolution of the bands at 105 and 398 cm^{-1} observed in the spectra of the polymer films and the PP nascent forms when changes in their supramolecular structure occur. Figure 5 compares the Raman spectra of CPBs and iPB in the regions 30-150 cm^{-1} and 325-450 cm^{-1}. Except for the spectrum of iPB, intensities in all the Raman spectra in figures 1-5 are normalized to the peak intensity of the band at 398 cm^{-1}. Intensity in the iPB spectrum is normalized to the peak intensity of the band at 374 cm^{-1} and is scaled by factor of 3 for a clearer presentation.

The strongest changes in the spectral region 0-600 cm^{-1} due to variations in the supramolecular structure of the copolymers and PPs are observed for the 105 and 398 cm^{-1} bands (Figures 3 and 4). In contrast to this, we have not observed any shifts of the peak positions of the bands at 175, 250, 319, 455, and 530 cm^{-1} in the Raman spectra of all the copolymers and PPs, although these bands show a weak dependence of their peak intensities on the type (α, γ, or smectic) of modification (Figure 4). For instance, as can be seen in Figures 3 and 4, the ratio of the peak intensities of the bands at 530 and 455 cm^{-1} for smectic iPP is slightly lower compared with that for the α-modification of isotactic PP with CR = 55% (the sample α-iPP-f). Also, this ratio decreases more clearly with the reduction of CR of the neat γ-modification samples. Thus, we consider that there are no significant fingerprints in the spectra in the region 0-600 cm^{-1}, which allow for discrimination of the types (the α, γ or smectic) of the isotactic PP modification.

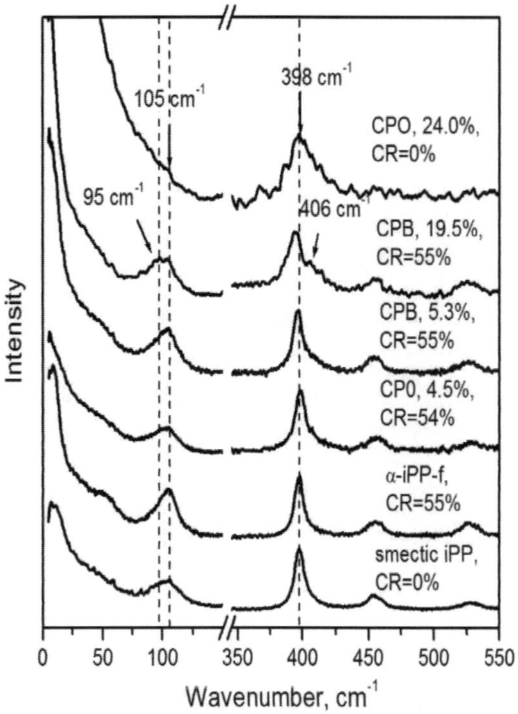

Figure 3. Raman spectra of the films of CPOs with 24.0 and 4.5 mol % of 1-octene, CPBs with 19.5 and 5.3 mol % of 1-butene, isotactic α PP (α-iPP-f, film), and smectic iPP in the region 0-550 cm^{-1}.

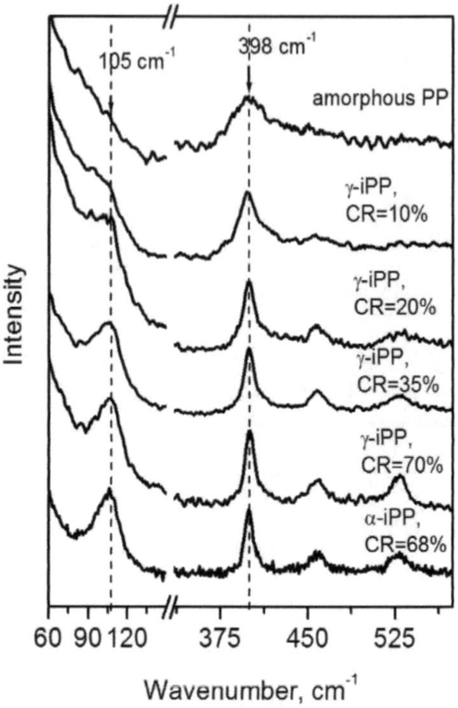

Figure 4. Raman spectra of the PP nascent forms including the α- and γ-iPP samples with different degree of crystallinity CR and the amorphous PP in the region 50-550 cm^{-1}.

For all the samples, except CPBs, the intensity of the band at 105 cm^{-1} diminishes and the band at 398 cm^{-1} is broadening together with the appearance of a high-wavenumber shoulder, when the degree of crystallinity of the sample decreases. Such an evolution of these bands well correlates with the changes of the spectral features in the region 800-850 cm^{-1} (Figures 1 and 2).

If the smectic phase is absent, then the Raman bands at 809 and 841 cm^{-1} should be assigned to the vibrations of helical macromolecules, localized in the crystalline and amorphous phases, respectively [6, 8, 9]. These bands are overlapping with the band at approximately 830 cm^{-1}, which is assigned to the vibrations of macromolecules in non-helical conformation [6, 8, 9].

According to the X-ray analysis data, the smectic phase is absent in the samples of the CPOs, α- and γ – iPPs. This means that the decrease in the intensity of the Raman band at 809 cm^{-1} and the increase and broadening of the band at 841 cm^{-1} are explained by the reduction of degree of crystallinity CR and the growth of conformational disorder in the amorphous phase [19]. Thus, the band at 105 cm^{-1} (the torsion vibrations of the CH_2-$CHCH_3$ groups in the chain backbone) can be assigned to helical macromolecules in the α and γ crystallites and the smectic phase in the Raman spectra of the CPOs and isotactic PP.

Taking into account the evolution of the Raman bands at 809 and 841 cm^{-1}, we assume that the broadening of the band at 398 cm^{-1} with the appearance of a high-wavenumber shoulder is due to the increase in the content of macromolecules in non-helical conformations. This assumption is in a good agreement with the result of theoretical treatment, which was presented by V.M. Hallmark *et al.* [13].

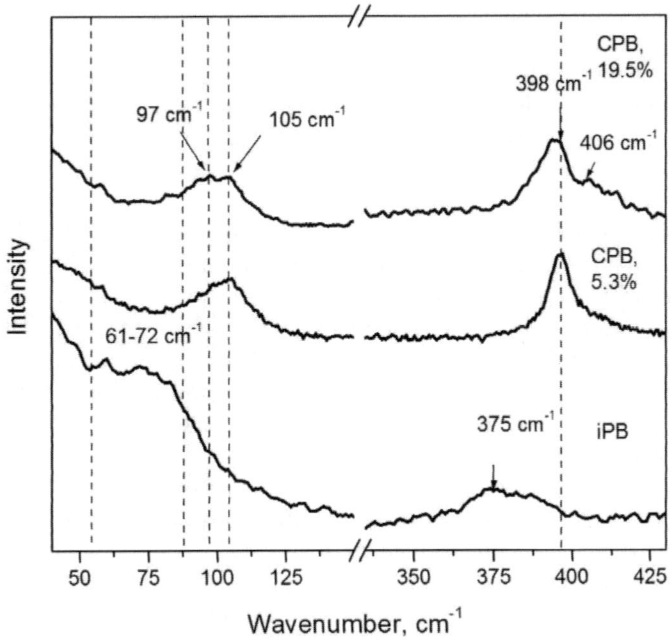

Figure 5. Raman spectra of iPB and CPBs with 19.5 and 5.3 mol % content of 1-butene in the region 25-425 cm^{-1}.

The behavior of the bands at 105 and 398 cm^{-1} in the Raman spectra of CPBs differs from the evolution of the same bands in the Raman spectra of CPOs and different PP modifications. The band observed at 398 cm^{-1} in the spectrum of CPB with 5.3 mol % of 1-butene exhibits a shift toward lower wavenumbers in the spectrum of CPB with 19.5 mol % of 1-butene, being observed in this case at 394 cm^{-1}. Simultaneously, additional lines at approximately 97 and 406 cm^{-1} appear in the spectrum of CPB with 19.5 mol % of 1-butene (Figures 3 and 5). However, the degree of crystallinity of CPBs with 5.3 and 19.5 mol % contents of 1-butene is the same (CR=55 %). The Raman spectrum of iPB in the region 25-425 cm^{-1} contains two bands at 61-75 cm^{-1} and 375 cm^{-1} (Figure 5). The broad band at 61-75 cm^{-1}, which appears as a shoulder band on the tail of Rayleigh scattering, partially overlaps with the PP Raman band at 105 cm^{-1}. It would be possible that the additional scattering at 97 cm^{-1} in the CPB spectrum is due to the sufficiently long iPB segments, which are formed in CPBs at a high content of 1-butene [3]. However, the iPB Raman band at 375 cm^{-1} extends up to 400 cm^{-1}, therefore, the Raman band at 406 cm^{-1} in the CPB spectrum cannot be due to vibrations of the 1-butene segments. Note that similar effects are not observed in the spectra of CPOs and the samples of the γ-modification of PP.

Earlier X-ray studies showed that macromolecules in the studied samples of CPB form a crystalline lattice similar to that of the α-modification of isotactic PP, but with enlarged unit-cell dimensions and an area of the cross-section of helical macromolecules in comparison with neat α-PP [3]. In other words, the growth of 1-butene content results in the expansion of the unit cell. Therefore, the expansion of the unit cell due to the incorporation of 1-butene in the PP chains allows to explain the appearance of both additional lines at approximately 97 and 406 cm^{-1}. The unit cell of the α modification of isotactic PP consists of four macromolecules. The point group symmetry of an infinitely-long helical macromolecule of isotactic PP is C_3, and the vibrational spectrum consists of the A- and the doubly-degenerated E-symmetry vibrations [2]. The splitting of the band at 105 cm^{-1} (A+E mixed symmetry) into two lines was observed at cryogenic temperatures [2, 5]. Theoretical considerations [20] of the vibrational spectrum of isotactic PP suggest two bands at approximately 400 cm^{-1}, namely, at 398 cm^{-1} (A-symmetry, ω (CH_2) + δ (CH)) and at 408 cm^{-1} (E-symmetry, ω (CH_2) + δ (CH)) + rocking (CH_2)). The changes in the unit cell parameters or in the cross-section of the PP helical chain affect the inter- and intramolecular interaction, and as, a consequence, the frequencies of the molecular vibrations are also changed.

Probably, this explains the splitting of the band at 398 cm^{-1} or an increase in the scattering cross-section of the band at 408 cm^{-1}. Thus, in principle, the reasons of the change of the band shape at 398 cm^{-1} can be different for neat isotactic PP, CPOs and CPBs.

4. Conclusions

We have studied low-frequency Raman spectra of different polypropylene (PP) polymorphs (α, γ, and smectic modifications), as well as those of a fully amorphous PP and of the random propylene/1-butene and propylene/1-octene copolymers with various comonomer content.

We have found that the band at 105 cm^{-1} (the torsion vibrations of the CH_2-$CHCH_3$ groups in the chain backbone) can be assigned to the helical macromolecules in the α and γ crystallites and the smectic phase in the Raman spectra of the propylene/1-octene copolymers and neat isotactic PP. If the smectic phase is absent, the intensity of the band at 105 cm^{-1} can be assigned to the vibrations of macromolecules in the crystallites.

The band at 398 cm^{-1} shows a broadening toward the high-wavenumber side when the degree of crystallinity of PP decreases or when the content of the incorporated monomer (1-butene or 1-octene) increases. We suggest two possible explanations of this effect: namely, the growth of the degree of conformational disorder (for the neat PP and the copolymers of propylene with 1-octene) and the increase in the dimensions of the unit cell or in the cross-section of helixes, which can appear due to the incorporation of 1-butene. As a matter of fact, the growth of the unit cell dimensions and of the cross-section of the copolymer helical chains should be taken into account for the interpretation of the Raman spectra of propylene copolymers. Additional studies are planned to clarify the behavior of the band at 398 cm^{-1}.

Acknowledgments

This work is supported by the Russian Foundation for Basic Research (project code 16-02-00808-a). The authors are very grateful to Dr. O.I. Kudinova (Semenov Institute of Chemical Physics, Russian Academy of Sciences, Moscow, Russia) for providing with the sample of isotactic polybutene and to Dr. B.F. Shklyaruk (Topchiev Institute of Petrochemical Synthesis, Russian Academy of Sciences, Moscow, Russia) for providing with the sample of the γ modification of isotactic PP with the degree of crystallinity of 70% as well as the X-ray analysis data for all the samples under study.

References

[1] White J L and Choi D D 2004 *Polyolefins: Processing, Structure Development, and Properties* (Munich: Carl Hanser Verlag)
[2] Fraser G V, Hendra P J, Watson D S, Gall M J, Willis H A and Cudby M E A 1973 *Spectrochim. Acta* A **29** 1525-33
[3] Nedorezova P M, Chapurina A V, Koval'chuk A A, Klyamkina A N, Aladyshev A M, Baranov A O and Shklyaruk B F 2012 *Polym. Sci.* B **54** 1-14
[4] Nedorezova P M, Chapurina A V, Koval'chuk A A, Klyamkina A N, Aladyshev A M, Optov V A and Shklyaruk B F 2010 *Polym. Sci.* B **52** 15-25
[5] Chalmers J M, Edwards H G M, Lees J S, Long D A, Mackenzie M W and Willis H A, 1991 *J. Raman Spectrosc.* **22** 613-8
[6] Nielsen A S, Batchelder D N and Pyrz R 2002 *Polymer* **43** 2671-6
[7] Vasko P D and Koenig J L 1970 *Macromolecules* **3** 597-601
[8] Gen D E, Prokhorov K A, Nikolaeva G Yu, Sagitova E A, Pashinin P P, Shklyaruk B F and Antipov E M 2011 *Laser Phys.* **21** 125-9
[9] Prokhorov K A, Nikolaeva G Yu, Sagitova E A, Pashinin P P, Nedorezova P M and Klyamkina A N 2016 *Vib. Spectrosc.* **85** 22-8
[10] Minardi A, Boudeulle M, Duval E and Etienne S 1996 Polymer **38** 3957-64

[11] Hsu S L and Krimm S 1976 *J. Appl. Phys.* **47** 4265-70
[12] Khafagy R M 2006 *J. Polym. Sci.* B **44** 2173-82
[13] Hallmark V M, Bohan S P, Strauss H L, Snyder R G 1991 *Macromolecules* **24** 4025-32
[14] Hahn T, Suen W, Kang S, Hsu L S, Stidham H D and Siedle A R 2001 *Polymer* **42** 5813-22
[15] Ibrahim E E, Chipara D G, Thapa R, Lozano K and Chipara M 2012 *Journal of Nanomaterials* **2012** article ID 793084
[16] Nedorezova P M, *et al.* 2005 *Russ. Chem. Bull.* **54** 400-13
[17] Nedorezova P M, Veksler E N, Aladyshev A M, Tsvetkova V I, Optov V A, Brusova G P and Lemenovskii D A 2006 *Kinet. Catal.* **47** 245-50
[18] Shklyaruk B F, Dreval' V E, Bocharova V A, Kulichikhin V G, Kuleznev V N and Antipov E M 2004 *Polym. Sci.* A **46** 605-12
[19] Gen D E, *et al.* 2010 *Laser Phys.* **20** 1354-67
[20] Snyder R G and Schachtschneider J H 1964 *Spectrochim. Acta* **20** 853-69

High-order harmonic generation by polyatomic molecules

S Odžak[1], **E Hasović**[1] **and D B Milošević**[1,2,3]

[1] Faculty of Science, University of Sarajevo, Zmaja od Bosne 35, 71000 Sarajevo, Bosnia and Herzegovina

[2] Max-Born Institut, Max-Born Strasse 2a, 12489 Berlin, Germany

[3] Academy of Sciences and Arts of Bosnia and Herzegovina, Bistrik 7, 71000 Sarajevo, Bosnia and Herzegovina

E-mail: senad.odzak@pmf.unsa.ba

Abstract. We present a theory of high-order harmonic generation by arbitrary polyatomic molecules based on the molecular strong-field approximation (MSFA) in the framework of the S-matrix theory. A polyatomic molecule is modeled by an $(N+1)$-particle system, which consists of N heavy atomic (ionic) centers and an electron. We derived various versions (with or without the dressing of the initial and/or final molecular state) of the MSFA. The general expression for the T-matrix element takes a simple form for neutral polyatomic molecules. We show the existence of the interference minima in the harmonic spectrum and explain these minima as a multiple-slit type of interference. This is illustrated by numerical examples for the nitrous oxide (N_2O) molecule exposed to strong linearly polarized laser field.

1. Introduction

With the advent of high-power lasers, many interesting nonlinear phenomena have been observed when exposing atoms and molecules to strong laser fields (see review articles [1, 2] and references therein). High-order harmonic generation (HHG) is a unique highly nonlinear process which has attracted attention of experimental and theoretical physicists over the past three decades. As a source of coherent extreme ultraviolet and soft x-ray radiation, this process has important applications in time-resolved spectroscopy and attoscience [3, 4, 5, 6, 7]. HHG may be described by the so-called three-step model [8]. In the first step, the considered molecular target absorbs more photons from the laser field than is necessary for ionization. Influenced by the laser field, the released electron may return to the parent molecular ion (the second step). In the third step, the electron recombines with the parent ion and one high-energy photon is emitted.

HHG carries information about the atomic or molecular targets, like internuclear distance of diatomic molecules or symmetry and the shape of the highest occupied molecular orbital (HOMO) of simple linear molecules [9, 10]. One powerful method to probe the structure of diatomic molecular targets is to analyze characteristic minima in the HHG spectra caused by the two-center interference of returning electron wave packet. The interference minima are first theoretically investigated for the simplest diatomic molecules H_2^+ and H_2 [11, 12, 13]. The experimental observation of the minima was first achieved for aligned CO_2 molecules [14, 15]. While the position of the minima can be determined by a simple interference formula for diatomic molecules, the case of polyatomic molecules is considerably more complex. In [16, 17] a general interference minima condition, which is valid for arbitrary homonuclear diatomic molecules, is derived and tested.

The HHG process for polyatomic molecules was analyzed experimentally in [18, 19, 20, 21, 22, 23, 24, 25, 26]. From a theoretical point of view, numerical solution of the time-dependent Schrödinger equation is not practical for polyatomic molecules. Therefore, most calculations of HHG spectra for polyatomic molecules are performed using approximate methods like strong-field approximation [27] or quantitative rescattering theory [24]. In this paper we present the molecular strong-field approximation (MSFA) and apply it to calculate and analyze HHG spectra of nitrous oxide molecule (N_2O).

2. Theory of molecular HHG

The MSFA theory is based on our previous work [27]. A polyatomic molecule is considered as a system of N atomic (ionic) centers and an electron which can be freed under the action of the laser field with the electric field vector $\mathbf{E}(t)$ in the molecular HHG process. The Hamiltonian of this system, in the dipole approximation and the length gauge, can be written in the form

$$H = \sum_{k=1}^{N} \frac{\mathbf{p}_k^2}{2m_k} + \frac{\mathbf{p}_e^2}{2m_e} + V(\mathbf{r}_e, \mathbf{r}_1, \ldots, \mathbf{r}_N) - \mathbf{E}(t) \cdot \left(e\mathbf{r}_e + \sum_{k=1}^{N} e_k \mathbf{r}_k \right), \tag{1}$$

where $\mathbf{p}_k$, $\mathbf{r}_k$, m_k, and e_k, are, respectively, the momenta, coordinates, masses, and charges of atoms (ions), while $\mathbf{p}_e$, $\mathbf{r}_e$, m_e, and e are the momentum, coordinate, mass, and charge of the electron ($m_e = 1$ and $e = -1$ in the atomic units). It is convenient to introduce the Jacobi coordinates and to replace the system of $N + 1$ coordinates $\mathbf{r}_k$ ($k = 1, \ldots, N$), $\mathbf{r}_e$, with the system of N relative coordinates $\mathbf{R}_k$, $k = 1, \ldots, N$, and the center-of-mass coordinate $\mathbf{R}_{\mathrm{c.m.}}$. In the new coordinates, the Hamiltonian of our system can be written in the form

$$H = \frac{\mathbf{P}_{\mathrm{c.m.}}^2}{2M} + \sum_{k=1}^{N} \left[\frac{\mathbf{P}_k^2}{2\mu_k} + V_k(\mathbf{r}_e - \mathbf{r}_k) \right] + V_{\mathrm{C}}(\{\mathbf{R}\}) - \mathbf{E}(t) \cdot \left(\tilde{e}_{\mathrm{c.m.}} \mathbf{R}_{\mathrm{c.m.}} + \sum_{k=1}^{N} \tilde{e}_k \mathbf{R}_k \right), \tag{2}$$

where the laser-free interaction $V(\mathbf{r}_e, \mathbf{r}_1, \ldots, \mathbf{r}_N)$ is written as the sum of all electron-kth molecular center potentials $V_k(\mathbf{r}_e - \mathbf{r}_k)$ and the potential energy of these centers $V_{\mathrm{C}}(\{\mathbf{R}\})$. The set of $(N-1)$ relative coordinates of atomic (ionic) centers is denoted by $\{\mathbf{R}\} \equiv (\mathbf{R}_1, \ldots, \mathbf{R}_{N-1})$. The reduced masses are determined by the following expressions

$$\mu_j = \frac{m_{j+1} M_j}{M_{j+1}} = \lambda_j m_{j+1} = \kappa_{j+1} M_j, \quad j = 1, \ldots, N, \tag{3}$$

where $M_j = \sum_{k=1}^{j} m_k$, $\lambda_j = M_j / M_{j+1}$ and $\kappa_j = m_j / M_j$ while $\mathbf{P}_{\mathrm{c.m.}} = \sum_{k=1}^{N+1} \mathbf{p}_k$ and $\tilde{e}_{\mathrm{c.m.}} = \sum_{k=1}^{N+1} e_k$ are the momentum and charge of the center of mass (we introduced the notation $\mathbf{p}_{N+1} = \mathbf{p}_e$ and $e_{N+1} = e$). The relative momenta and charges are given by ($j = 1, \ldots, N$)

$$\mathbf{P}_j = \lambda_j \mathbf{p}_{j+1} - \kappa_{j+1} \sum_{k=1}^{j} \mathbf{p}_k, \qquad \tilde{e}_j = \lambda_j e_{j+1} - \kappa_{j+1} \sum_{k=1}^{j} e_k. \tag{4}$$

For a $T = 2\pi/\omega$–periodic laser field the rate of emission of a harmonic photon having the frequency $\omega_{\mathbf{K}} = n\omega + E_{\mathrm{ei}}(\{\mathbf{R}\}) - E_{\mathrm{ef}}(\{\mathbf{R}\})$, wave vector $\mathbf{K}$, and the polarization $\hat{\mathbf{e}}_{\mathbf{K}}$, into a solid angle $d\Omega_{\hat{\mathbf{K}}}$, is given by [in atomic units (au)] [28]

$$w_n^{q'q} = \frac{1}{2\pi} \left(\frac{n\omega}{c} \right)^3 \left| T_n^{q'q} \right|^2, \tag{5}$$

where the T-matrix element is

$$\mathbf{T}_n^{qq'} = \int_0^T \frac{dt}{T} e^{in\omega t} \mathbf{d}_{\text{fi}}^{qq'}(t) = T_n^{qq'} \hat{\mathbf{e}}_{\mathbf{K}} = \sum_{\xi=x,y,z} T_{n\xi}^{qq'} \hat{\mathbf{e}}_{\xi}. \tag{6}$$

The time-dependent dipole matrix element $\mathbf{d}_{\text{fi}}^{qq'}(t)$ can be calculated within the strong-field approximation, fixed-nuclei approximation and solving the integral over the intermediate electron momentum $\mathbf{k}$ by the saddle-point method, as it was done in [27], with the final result

$$\begin{aligned}\mathbf{d}_{\text{fi}}^{qq'}(t) &= -i\left(\frac{2\pi}{i}\right)^{3/2} \int_0^\infty \frac{d\tau}{\tau^{3/2}} e^{i[S_{\mathbf{k}_{\text{st}}}(t') - S_{\mathbf{k}_{\text{st}}}(t) + E_{\text{ei}}\tau]} \\ &\times \sum_{l=1}^{N} \sum_{a} c_{la} \eta_q(\boldsymbol{\rho}_l, t) \langle \psi_a | \tilde{e}_N \mathbf{r} + \boldsymbol{\rho}_{\text{C}_l} | \mathbf{k}_{\text{st}} - \tilde{e}_N \mathbf{A}(t) \rangle \\ &\times \sum_{j=1}^{N} \sum_{a'} c_{ja'} \eta_{q'}^*(\boldsymbol{\rho}_j, t') \langle \mathbf{k}_{\text{st}} - \tilde{e}_N \mathbf{A}(t') | (\tilde{e}_N \mathbf{r} + \boldsymbol{\rho}_{\text{C}_j}) \cdot \mathbf{E}(t') | \psi_{a'} \rangle.\end{aligned} \tag{7}$$

Integration in the previous equation is performed over the travel time $\tau = t - t'$, while the quasiclassical action is given by $S_{\mathbf{k}}(t) = \frac{1}{2}\int_{-\infty}^{t} dt'[\mathbf{k} - \tilde{e}_N \mathbf{A}(t')]^2$ and is calculated at the stationary momentum $\mathbf{k}_{\text{st}} = \tilde{e}_N \int_{t-\tau}^{t} \mathbf{A}(t')dt'/\tau$ where $\mathbf{A}(t) = -\int_{-\infty}^{t} \mathbf{E}(t')dt'$. The ionization (recombination) matrix element is obtained by expanding the molecular orbitals as a linear combination of atomic orbitals $|\psi_a\rangle$ with expansion coefficients $c_{ja'}$ (c_{la}). Furthermore we write $\boldsymbol{\rho}_{\text{C}_l} = \tilde{e}_N \boldsymbol{\rho}_l + \sum_{j=1}^{N-1} \tilde{e}_{j_l} \mathbf{R}_j$, $\boldsymbol{\rho}_k(\{\mathbf{R}\}) = \lambda_{k-1}\mathbf{R}_{k-1} - \sum_{j=k}^{N-1} \kappa_{j+1}\mathbf{R}_j$ and we use dressing of the initial wave function with factors $\eta_{\text{d}}(\boldsymbol{\rho}_j, t) = \exp\{i\mathbf{k}_{\text{st}} \cdot \boldsymbol{\rho}_j\}$, $\eta_{\text{u}}(\boldsymbol{\rho}_j, t) = \exp\{i\,[\mathbf{k}_{\text{st}} - \tilde{e}_N \mathbf{A}(t)] \cdot \boldsymbol{\rho}_j\}$. The index l on $\tilde{e}_j$ means that the charge distribution is such that the electron is ionized from (recombined with) the center l. For neutral polyatomic molecules we have $\tilde{e}_N = -1$ and $\boldsymbol{\rho}_{\text{C}l} = \mathbf{0}$, $l = 1, \ldots, N$, so that the expression for the time-dependent dipole simplifies.

3. Numerical results

Now we are going to demonstrate the application of the presented theory using a simple example. We consider the HHG process of the nitrous oxide (N_2O) molecule. Nitrous oxide, commonly known as laughing gas, is a colorless, odorless non-flammable gas. The N_2O molecule is a linear molecule with N-N (N-O) internuclear distances of 1.128Å(1.184Å) at the equilibrium geometry. Structure of the N_2O molecule belongs to the $C_{\infty v}$ point group and its double degenerate HOMO, shown in Fig. 1, has the π symmetry. Vertical ionization energy of the N_2O molecule is $I_{\text{p}} = 12.89$ eV.

Now, we present the calculated harmonic emission rate of the HHG process in a linearly polarized laser field, defined as $\mathbf{E}(t) = E_0 \sin(\omega t)\hat{\mathbf{e}}_{\text{L}}$ and having the intensity 4×10^{14} W/cm^2 and the wavelength 800 nm. The molecule is oriented along z axis and $\hat{\mathbf{e}}_{\text{L}} \equiv (\theta_{\text{L}}, \varphi_{\text{L}})$ in spherical coordinates. We fix the laser field in the xz plane ($\varphi_{\text{L}} = 0$) and change the polar angle $\theta_{\text{L}} \in [0°, 180°]$. In the upper panel of Fig. 2 we show in false colors the logarithm of the harmonic emission rate, Eq. (5), as a function of the angle θ_{L} and the harmonic order n.

In the remaining panels we show the rates obtained using only the x (middle panel) and only the z (bottom panel) component of the T-matrix element, Eq. (6). One can see that the contributions of both components are comparable.

The HHG spectra exhibit a pronounced minimum for the angle $\theta_{\text{L}} = 90°$. This minimum is caused by the nodal plane of the N_2O molecule. Furthermore, there are minima for the

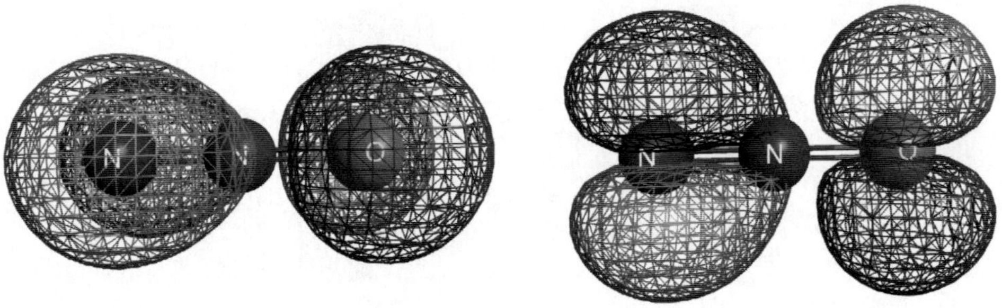

Figure 1. Coordinate-space wave function for the HOMOs of N_2O molecule.

angles of $\theta_{\rm L} \approx 45^\circ$ and $\theta_{\rm L} \approx 135^\circ$. These type of minima can be related to the interference of three contributions of the recombination matrix element, i.e. the interference of the recombining electronic wave packets. The three-point interference can be explained by the presented MSFA theory. It is possible to extract an interference factor [27] which becomes small for some specific values of n, $\theta_{\rm L}$, and $\varphi_{\rm L}$.

4. Conclusions

We have used a strong-field-approximation-based theory of high-order harmonic generation by polyatomic molecules to calculate the harmonic emission rates of the nitrous oxide N_2O molecules. In our model, a polyatomic molecule is considered as a system of $N+1$ particles: N atomic (ionic) centers and one valence electron, emitted by strong-field ionization. In the Born-Oppenheimer and the fixed-nuclei approximation we derived an expression for the harmonic emission T-matrix element. Our theory is formulated in such a way that the strong-field approximation with the laser-field dressed or undressed initial and final states can be treated on the same footing.

We have shown that the harmonic emission rate of the (N_2O) molecules exhibits minima for particular values of the orientation angle $\theta_{\rm L}$. The appearance of these minima is related to the position of the nodal planes of the molecular HOMO as well as to the destructive interference of the recombining electronic wave packets. In general, for neutral polyatomic molecules the T-matrix element can be simplified and its recombination part can be extracted as a factor. This factor is responsible for a multicenter interference of the recombining electronic wave packets.

Acknowledgments

We gratefully acknowledge support by the Federal Ministry of Education and Science, Bosnia and Herzegovina.

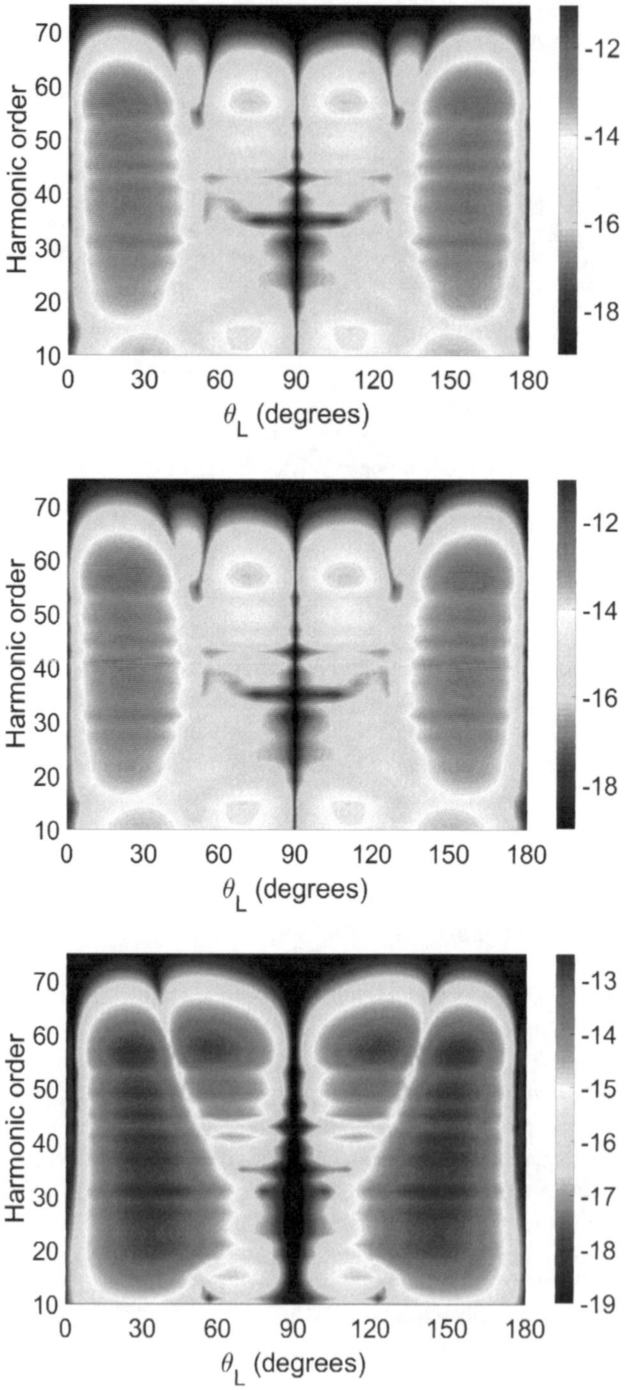

Figure 2. Logarithm of the HHG rate of the N_2O molecule presented in false colors as a function of the orientation angle θ_L and the harmonic order n. The results are obtained using both (top panel), only the x (middle panel), and only the z (bottom panel) component of the T-matrix element. Laser field is linearly polarized having the intensity 4×10^{14} W/cm^2 and the wavelength 800 nm. The initial state is dressed ($q' = \mathrm{d}$), while the final state is undressed ($q = \mathrm{u}$).

References

[1] Becker W, Grasbon F, Kopold F, Milošević D B, Paulus G G and Walther H 2002 *Adv. At. Mol. Opt. Phys.* **48** 35
[2] Winterfeldt C, Spielmann C and Gerber G 2008 *Rev. Mod. Phys.* **80** 117
[3] Agostini P and DiMauro L F 2004 *Rep. Prog. Phys.* **67** 813
[4] Krausz F and Ivanov M 2009 *Rev. Mod. Phys.* **81** 163
[5] Gaarde M B, Tate J L and Schafer K J 2008 *J. Phys.* B **41** 132001
[6] Midorikawa K, Nabekawa Y and Suda A 2008 *Prog. Quant. Electron.* **32** 43
[7] Pfeifer T, Abel M J, Nagel P M, Jullien A, Loh Z H, Bell M J, Neumark D M and Leone S R 2008 *Chem. Phys. Lett.* **463** 11
[8] Corkum P B 1993 *Phys. Rev. Lett.* **71** 1994
[9] Lein M, Hay N, Velotta R, Marangos J P and Knight P L 2002 *Phys. Rev. Lett.* **88** 183903
[10] Itatani J, Levesque J, Zeidler D, Niikura H, Pepen H, Kiefer J C, Corkum P B and Villeneuve D M 2004 *Nature* **432** 867
[11] Kopold R, Becker W and Kleber M 1998 *Phys. Rev.* A **58** 4022
[12] Lein M, Hay N, Velotta R, Marangos J P and Knight P L 2002 *Phys. Rev.* A **66** 023805
[13] Lagmago Kamta G and Bandrauk A D 2004 *Phys. Rev.* A **70** 011404(R)
[14] Kanai T, Minemoto M and Sakai H 2005 *Nature* **435** 470
[15] Vozzi C *et al* 2005 *Phys. Rev. Lett.* **95** 153902
[16] Odžak S and Milošević D B 2009 *Phys. Rev.* A **79** 023414
[17] Odžak S and Milošević D B 2009 *J. Phys.* B **42** 071001
[18] Wagner N L, Wüest A, Christov I P, Popmintchev T, Zhou X, Murnane M M and Kapteyn H C 2006 *Proc. Natl Acad. Sci. USA* **103** 13279
[19] Torres R *et al* 2007 *Phys. Rev. Lett.* **98** 203007
[20] Li W, Zhou X, Lock R, Patchkovskii S, Stolow A, Kapteyn H C and Murnane M M 2008 *Science* **322** 1207
[21] Wong M C H, Brichta J P and Bhardwaj V R 2010 *Phys. Rev.* A **81** 061402(R)
[22] Wong M C H, Brichta J P, Spanner M, Patchkovskii S and Bhardwaj V R 2011 *Phys. Rev.* A **84** 051403
[23] Wong M C H, Le A T, Alharbi A F, Boguslavskiy A E, Lucchese R R, Brichta J P, Lin C D and Bhardwaj V R 2013 *Phys. Rev. Lett.* **110** 033006
[24] Le A T, Lucchese R R and Lin C D 2013 *Phys. Rev.* A **87** 063406
[25] Ferré A *et al* 2015 *Nat. Commun.* **6** 6592
[26] Kraus P M *et al* 2015 *Nat. Commun.* **6** 8039
[27] Odžak S, Hasović E and Milošević D B 2016 *Phys. Rev.* A **93** 043413
[28] Milošević D B and Ehlotzky F 2003 *Adv. At. Mol. Opt. Phys.* **49** 373

Above-threshold ionization in a bicircular field: quantum orbits unfolding in a plane

W Becker[1,2] and D B Milošević[1,3,4]

[1] Max-Born-Institut, Max-Born-Str. 2a, 12489 Berlin, Germany
[2] National Research Nuclear University MEPhI, Kashirskoe Shosse 31, 115409, Moscow, Russia
[3] Faculty of Science, University of Sarajevo, Zmaja od Bosne 35, 71000 Sarajevo, Bosnia and Herzegovina
[4] Academy of Sciences and Arts of Bosnia and Herzegovina, Bistrik 7, 71000 Sarajevo, Bosnia and Herzegovina

E-mail: `wbecker@mbi-berlin.de,milo@bih.net.ba`

Abstract. Above-threshold ionization (ATI) of atoms by a strong bicircular laser field is investigated using the strong-field approximation and the quantum-orbit theory. The bicircular field consists of two coplanar counterrotating circularly polarized fields with a frequency ratio of 2:1. The velocity map of the angle-resolved ATI spectra, both for direct and rescattered electrons, reflects the shape of a parametric plot of the bicircular field and its symmetries. It is shown that the main characteristics of the ATI spectra can be explained using only a few quantum orbits having short travel times. We also analyze a recently discovered [Phys. Rev. A **93**, 052402(R) (2016)] bicircular-field-induced spin asymmetry of the ATI electrons and show that the momentum dependence of the spin-asymmetry parameter is stronger for longer wavelengths.

1. Introduction

The interaction of intense bicircular laser fields with atoms has received a great deal of recent attention [1, 2, 3, 4, 5, 6, 7, 8, 9, 10, 11, 12, 13, 14, 15, 16, 17, 18, 19, 20, 21, 22, 23, 24, 25, 26, 27, 28, 29, 30, 31, 32, 33, 34, 35, 36, 37, 38, 39, 40, 41, 42, 43]. A bicircular field consists of the superposition of two circularly polarized fields that rotate in the same plane, usually in opposite directions (counterrotating), with different frequencies. The case of a frequency ratio of 2:1 is the standard situation, but other frequency ratios and incommensurate frequencies are also considered. Counterrotating bicircular fields are known to generate very intense circularly polarized high-order harmonics that obey certain characteristic selection rules. The velocity map of angle-resolved above-threshold ionization (ATI) induced by such fields reflects the shape of the parametric plot of the bicircular field and its symmetries [33]. The strong-field approximation (with or without rescattering taken into account) well accounts for the experimental observations [32].

If the pertinent temporal integrations are carried out by the saddle-point method, quantum orbits result. They allow for a space-time visualization of the electronic trajectories subsequent to tunneling and before and after the rescattering event. Unlike the standard case of a linearly polarized driving field, the quantum orbits fill the entire polarization plane, with the electrons recolliding with their parent ions at arbitrary angles. It is important to discriminate between backward and forward scattering orbits. The former shape the spectrum for comparably high energies, while the latter cause involved low-energy structures. Identifying the contributions of specific quantum orbits to the velocity map, it is possible to attribute certain features to certain quantum orbits. If the ionic dynamics are such that fine-structure splitting is important, then the rescattered electrons exhibit significant spin asymmetry. Since rescattered electrons are emitted on the attosecond time scale, the possibility opens up for the electron spin to enter attoscience as an observable [34].

Our theory, which is based on the S-matrix formalism introduced in [33] and [34], is briefly presented in Sec. 2. The numerical results and discussions are presented in Sec. 3 and, finally, the conclusions are summarized in Sec. 4. The atomic system of units ($\hbar = e = m_e = 4\pi\varepsilon_0 = 1$) is used throughout the paper.

2. Theory

Using the S-matrix theory and the strong-field approximation (SFA) in [33] we derived the following expression for the averaged differential ionization rate,

$$\overline{w}_{\mathbf{p}i}(n) = \frac{2\pi p N_e}{2l+1} \sum_{m=-l}^{l} |T_{\mathbf{p}lm}(n)|^2, \tag{1}$$

where N_e is the number of equivalent electrons in the ionizing shell and $T_{\mathbf{p}lm}(n)$ is the T-matrix element, which can be written as

$$T_{\mathbf{p}lm}(n) = \int_0^T \frac{dt}{T} e^{i[\mathbf{p}\cdot\boldsymbol{\alpha}(t)+\mathcal{U}_1(t)+n\omega t]} \left[\langle \mathbf{p}+\mathbf{A}(t)|\mathbf{r}\cdot\mathbf{E}(t)|\psi_{lm}\rangle + \mathcal{T}^{\mathrm{R}}_{\mathbf{p}lm}(t)\right]. \tag{2}$$

Here $T = 2\pi/\omega$ is the optical period and n is the number of photons exchanged with the laser field, which is related to the energy-conserving condition $n\omega = E_{\mathbf{p}} + I_{\mathrm{p}} + U_{\mathrm{p}}$, where $E_{\mathbf{p}} = \mathbf{p}^2/2$ is the final electron kinetic energy in the continuum state $|\mathbf{p}\rangle$ [$\mathbf{p}$ is the electron's drift momentum, which equals the momentum outside the field and $\langle \mathbf{r}|\mathbf{p}\rangle = (2\pi)^{-3/2}\exp(i\mathbf{p}\cdot\mathbf{r})$], I_{p} is the ionization potential of the atom in the bound state $|\psi_{lm}\rangle e^{iI_{\mathrm{p}}t}$, $\boldsymbol{\alpha}(t) = \int^t \mathbf{A}(t')dt'$, $\mathbf{A}(t) = -\int^t \mathbf{E}(t')dt'$ [$\mathbf{E}(t)$ is the electric field vector], and $\mathcal{U}_1(t) = \int^t \mathbf{A}^2(t')dt'/2 - U_{\mathrm{p}}t$ with $U_{\mathrm{p}} = \int_0^T \mathbf{A}^2(t)dt/(2T)$ the ponderomotive energy. The first term of the integrand on the right-hand side of Eq. (2) corresponds to direct ionization, while the rescattering part of the transition amplitude is given by

$$\mathcal{T}^{\mathrm{R}}_{\mathbf{p}lm}(t) = -i\int_0^{\tau_{\max}} d\tau \left(\frac{2\pi}{i\tau}\right)^{3/2} \langle \mathbf{p}|V|\mathbf{k}_s\rangle\langle \mathbf{k}_s+\mathbf{A}(t-\tau)|\mathbf{r}\cdot\mathbf{E}(t-\tau)|\psi_{lm}\rangle e^{i[S_{\mathbf{k}_s}(t-\tau)-S_{\mathbf{k}_s}(t)-I_{\mathrm{p}}\tau]}, \tag{3}$$

where $\tau_{\max} \to \infty$, $\mathbf{k}_s \equiv -\int_{t-\tau}^t dt'\mathbf{A}(t')/\tau$ is the intermediate electron stationary momentum and $S_{\mathbf{k}}(t) = \int^t dt'[\mathbf{k}+\mathbf{A}(t')]^2/2$. In Sec. 3 we will analyze the influence of finite values of the parameter $\tau_{\max}$. The ground-state atomic wave function is modeled by a linear combination of Slater-type orbitals obtained by using the Hartree-Fock-Roothaan method [44]. For atoms having a p ground state ($l = 1$) the magnetic quantum number is $m = 0, \pm 1$.

Within the quantum-orbit theory, the integrals in Eq. (3) are approximated by the sum over the saddle-point solutions of the equations

$$\frac{1}{2}\left[\mathbf{k}_s + \mathbf{A}(t_0)\right]^2 = -I_{\mathrm{p}}, \;\; \frac{1}{2}\left[\mathbf{k}_s + \mathbf{A}(t)\right]^2 = \frac{1}{2}\left[\mathbf{p} + \mathbf{A}(t)\right]^2, \tag{4}$$

which express energy conservation at the ionization time $t_0 = t - \tau$ and the rescattering time t. Not going into details, we only mention that in [33] these solutions were classified by the multiindex (α, β, m) for backward-scattering solutions and (ν, ρ, μ) for forward-scattering solutions. More details about quantum orbits can be found in references [45, 46, 47, 48, 49, 50, 51, 52].

We use a bichromatic circularly polarized field with the coplanar counterrotating components having the angular frequencies ω and 2ω. The corresponding electric-field vector in the xy plane, determined by the unit vectors $\hat{\mathbf{e}}_x$ and $\hat{\mathbf{e}}_y$, is defined by

$$\mathbf{E}(t) = \frac{i}{2\sqrt{2}} \left[E_1(\hat{\mathbf{e}}_x + i\hat{\mathbf{e}}_y)e^{-i\omega t} + E_2(\hat{\mathbf{e}}_x - i\hat{\mathbf{e}}_y)e^{-i2\omega t}\right] + \text{c.c.}, \tag{5}$$

where c.c. denotes the complex conjugate of the first term on the right-hand side and $I_1 = E_1^2$ and $I_1 = E_1^2$ are the component intensities. The bicircular field $\mathbf{E}(t)$ and its vector potential $\mathbf{A}(t)$ obey the following dynamical symmetry: a rotation by the angle $\alpha_j = -2\pi j/3$ about the z axis is equivalent to a translation in time by $\tau_j = jT/3$, i.e. $R_z(\alpha_j)\mathbf{E}(t) = \mathbf{E}(t + \tau_j)$ where j is an integer. The diagonal matrix elements of the rotation matrix $R_z(\alpha_j)$ are $\cos\alpha_j$, while the off diagonal elements are $\pm\sin\alpha_j$.

In ATI of xenon atoms there are two continua corresponding to the two ground states of the Xe^+ ion: ${}^2P_{3/2}$ ($I_\text{p}{}^{3/2} = 12.14$ eV) and ${}^2P_{1/2}$ ($I_\text{p}{}^{1/2} = 13.44$ eV). Because of this fine-structure splitting of 1.31 eV, it is possible to observe different momentum distributions of electrons having different spin in the case where the ATI rate shows a strong $m = \pm 1$ asymmetry. This is the case for ATI by a circularly polarized field. This was predicted in [53] and recently observed [54]. For ATI by a bicircular field this effect was considered in [34], where the normalized spin-asymmetry parameter was defined with respect to the maximal summed rate by the relation

$$\tilde{A}_\mathbf{p} = (W_{\mathbf{p}\uparrow} - W_{\mathbf{p}\downarrow})/\max_\mathbf{p}(W_{\mathbf{p}\uparrow} + W_{\mathbf{p}\downarrow}). \tag{6}$$

Here $W_{\mathbf{p}\uparrow}$ ($W_{\mathbf{p}\downarrow}$) denotes the rate for electrons having spin up (down). The method of calculation of these rates is presented in [34].

3. Numerical results

In Figs. 1 and 2 we present numerical results for ATI of Ne atoms by a bicircular ω–2ω laser field having the fundamental wavelength of 800 nm and equal intensity of both components, $I_1 = I_2 = 2 \times 10^{14}$ W/cm^2. The T-matrix element (2) is evaluated on a grid of points that satisfy the energy-conservation condition $n\omega = E_\mathbf{p} + I_\text{p} + U_\text{p}$ and the resulting discrete distribution is smoothed. This is why no ATI rings are visible in the subsequent figures. In Fig. 1 the logarithm of the differential ionization rate of Ne atoms is presented as a function of the electron momentum. In the upper left panel we show the spectra of the direct electrons [electrons that reach the detector without interacting with the parent ion after ionization, as described by the first term in Eq. (2)]. Inspecting the lower left panel of Fig. 2 we conclude that the direct electrons are emitted predominantly in the direction opposite to the vector potential $\mathbf{A}(t)$ at the time of ionization (i.e. in the directions determined by the emission angles $\theta = 60°, 180°$ and $300°$). The remaining panels of Fig. 1 present the differential ionization rate obtained taking into account only the rescattered electrons. The results shown in the left panels are obtained by two-dimensional numerical integration [rescattering term in Eqs. (2) and (3)], using four $2p$ Slater-type orbitals to represent the ground state of the Ne atom, for different values of the parameter τ_max: $0.66T$ (bottom panel), $1T$ (second panel from the bottom) and $5T$ (third panel from the bottom). For $\tau_\text{max} > 5T$ the results do not change, which means that the contribution of electrons having travel times larger than five optical cycles can be neglected. It is interesting that the results obtained with $\tau_\text{max} = 2T/3$ qualitatively reproduce the exact ATI spectra. This means that the short quantum orbits give the main contribution to the ATI rate.

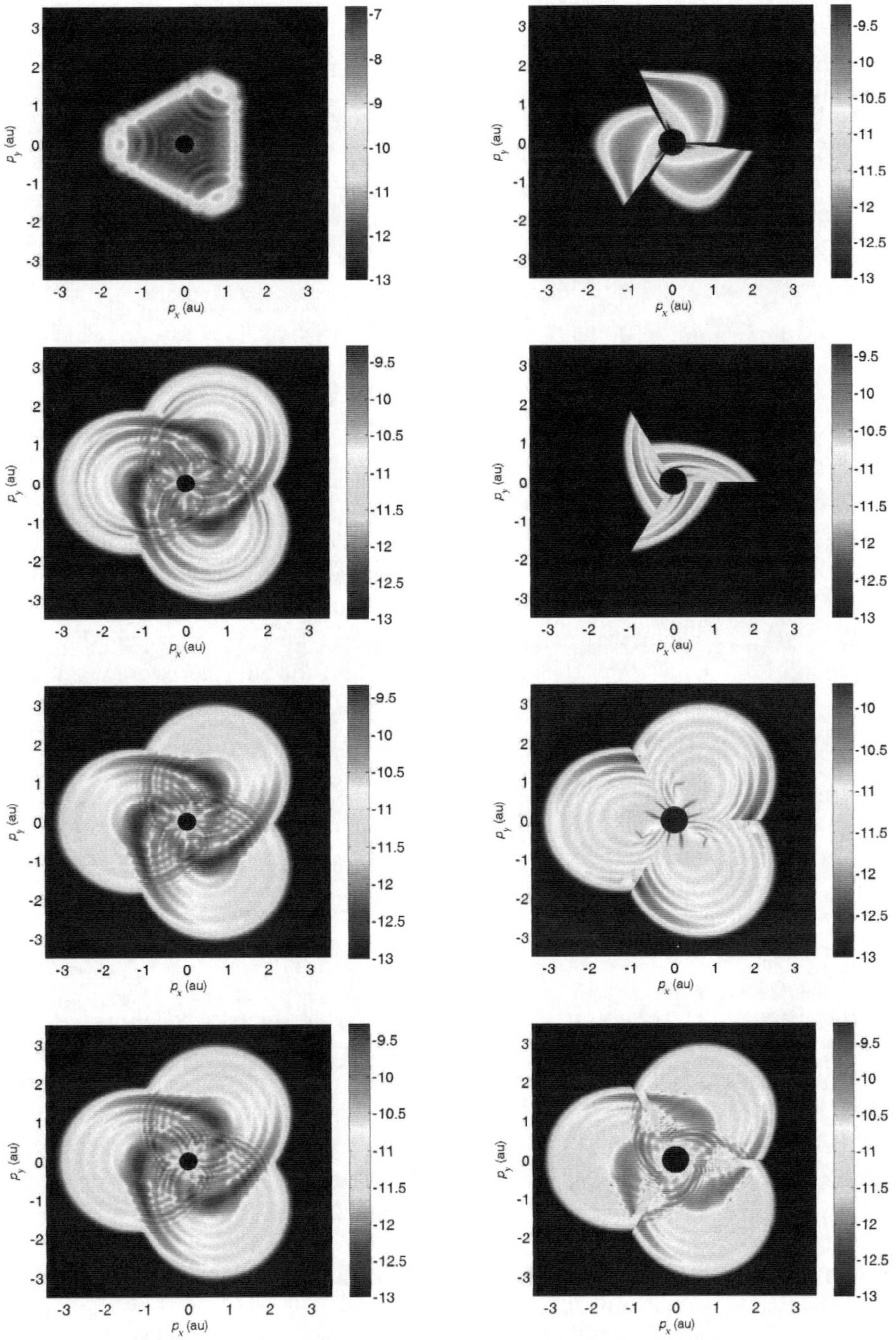

Figure 1. (Color online) The logarithm of the differential ionization rate (in au) of Ne atoms presented in false colors in the electron-momentum plane for ionization by a bicircular ω–2ω laser field with equal intensity of both components, $I_1 = I_2 = 2 \times 10^{14}$ W/cm^2 and the fundamental wavelength of 800 nm. The false-color scale covers four orders of magnitude. Left panels: results obtained by numerical integration. The subpanels from top to bottom correspond to the direct SFA (top) and rescattering improved SFA with the upper limit of integration over the travel time equal to 5, 1 and 0.66 optical cycles. Right panels: results obtained using the quantum-orbit formalism with the following quantum-orbit solutions (from top to bottom): $(\nu, \rho, \mu) = (1, 0, 0)$, $(\nu, \rho, \mu) = (\pm 1, 2, 0)$, $(\alpha, \beta, m) = (\pm 1, 1, 0)$ and the coherent sum of all these five contributions.

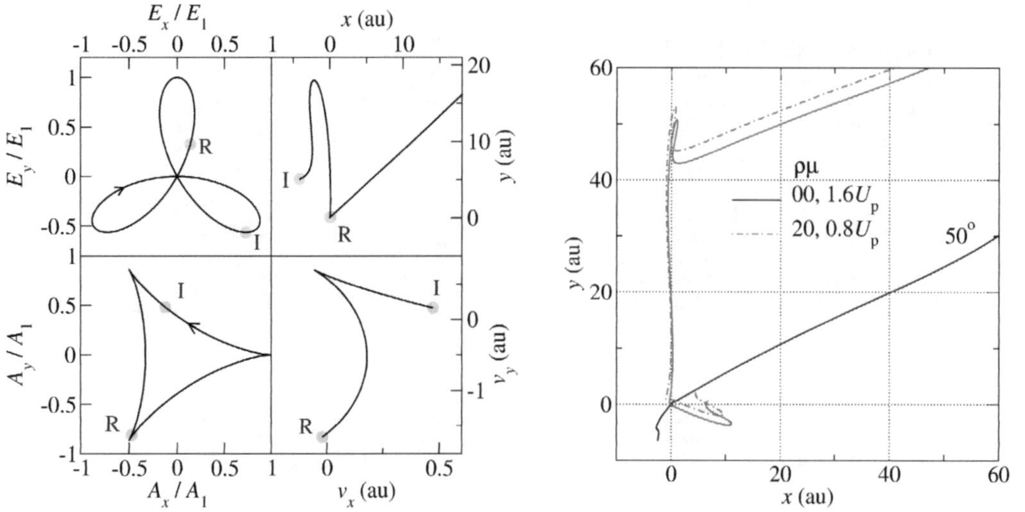

Figure 2. (Color online) Left: The ω–2ω bicircular electric-field vector (upper left panel), vector potential (lower left), electron trajectory (upper right), and velocity (lower right) between the ionization (I) and rescattering (R) times that correspond to the orbit $(\alpha, \beta, m) = (1, 1, 0)$ and the energy $8U_\mathrm{p}$. Right: Electron trajectories for three short forward-scattering-like quantum orbits and the energies as denoted in the legend. Laser and atomic parameters are as in the previous figure and the electron is emitted in the direction $\theta = 50°$.

For a closer analysis, in the right-hand panels of Fig. 1 we present the results obtained using only some particular short quantum orbits. The top right panel displays the results obtained using only the shortest forward-scattering quantum orbit, which is denoted by the multiindex $(\nu, \rho, \mu) = (1, 0, 0)$ (notation of [33] is used). The second panel shows the contribution of the next forward-scattering quantum orbits $(\nu, \rho, \mu) = (\pm 1, 2, 0)$. These orbits come in pairs ($\nu = \pm 1$), which leads to interference structures. The third panel corresponds to the pair of the shortest backward-scattering quantum orbits $(\alpha, \beta, m) = (\pm 1, 1, 0)$. These orbits dominate the high-energy part of the spectrum. Interference of the two solutions of the pair is again clearly visible. The coherent sum of the contributions of all five quantum orbits mentioned yields the spectrum shown in the bottom right panel. By comparison with the panels shown on the left-hand side we see that these five shortest quantum orbits qualitatively reproduce the exact spectrum.

In order to better understand the physics of the bicircular-field-induced ATI process, in Fig. 2 we analyze the relevant quantum orbits in more detail. In the left-hand panel we show for $0 \leq t \leq T$ the parametric plot of the electric field vector $(E_x(t), E_y(t))$ (upper left subpanel) and the corresponding vector potential $(A_x(t), A_y(t))$ (lower left subpanel). The field and the vector potential at the ionization time t_0 and the rescattering time t, which correspond to the shortest backward-scattering quantum orbit $(\alpha, \beta, m) = (1, 1, 0)$ are denoted by the filled circles and the letters I and R, respectively. We see that the field between ionization and rescattering approximately follows a line oriented by the angle $120°$ with respect to the positive x axis. As a consequence, the corresponding electron trajectory between $\mathrm{Re}\, t_0$ and $\mathrm{Re}\, t$, which is obtained projecting the quantum orbit onto the real xy plane for real times, is almost parallel to the y axis (see the upper right subpanel of the left panel in Fig. 2). The electron is "born at the tunnel exit", a few atomic units away from the origin, moves in the direction of the positive y axis and, when the x component of the field changes its sign, turns around and moves back to the parent ion, approximately along the negative y-axis direction. Thereafter, the electron rescatters at the origin and moves towards the detector, which is situated at the angle $\theta = 50^\circ$ with respect to the x axis. In the lower right subpanel of the left panel of Fig. 2 we display the electron velocity. This velocity at the ionization time is different from zero, which allows the electron to return to the ion at the rescattering time t. The force in the negative x direction [$-E_x(t)$; the field is shown in the upper left subpanel] is largely compensated by the initial momentum in this direction. After ionization the electron is first accelerated to positive momenta, then decelerated and bent around such that it recollides with substantial momentum. The vector potential is maximum at this return time t.

In the right-hand panel of Fig. 2 we show the electron trajectory for the three short forward-scattering-like quantum orbits $(\nu, \rho, \mu) = (1, 0, 0)$ (maroon line) and $(\nu, \rho, \mu) = (\pm 1, 2, 0)$ (red lines). For the maroon line the electron moves directly from the tunnel exit to the nucleus where it forward scatters and leaves in the direction determined by the angle $\theta = 50^\circ$ having the final energy $E_{\mathbf{p}} = 1.6U_{\mathrm{p}}$. Since the trajectory is short the probability of this process is higher. Between ionization and recombination the electron is close to the nucleus so that this probability is further enhanced by the influence of the Coulomb potential. The contribution of this solution is responsible for the experimentally observed blobs in the electron-momentum distribution [26]. The trajectories for the solutions $(\nu, \rho, \mu) = (\pm 1, 2, 0)$ (red solid line and red dot-dashed line) correspond to the energy $E_{\mathbf{p}} = 0.8U_{\mathrm{p}}$. The electron forward scatters under the angle 60°, then moves in the direction of the positive y axis and then, finally, in the direction towards the detector.

Let us now present results for direct ATI of Xe atoms. We use the bicircular ω–2ω laser field with $I_1 = I_2 = 1\times 10^{14}$ W/cm^2 and the fundamental wavelength of 1300 nm. Similarly as in [34], we exhibit in Fig. 3 the partial differential-ionization rate for initial states having the magnetic quantum number $m = -1$ (upper left panel), $m = +1$ (upper right panel) and the rate summed over $m = \pm 1$ (lower right panel; the contribution of the $m = 0$ term is zero [34]). It is obvious that we have a large asymmetry between the $m = -1$ and $m = +1$ case. As a consequence, the normalized spin-asymmetry parameter $\tilde{A}_{\mathbf{p}}$, shown in the lower right panel, takes on large values. The results presented exhibit symmetry with respect to rotation by the angle 120° about the z axis and reflection symmetry with respect to axes at the angles 60°, 180°, and 300° with respect to the x axis. These symmetries are explained in [33]. Near the reflection axes the rate exhibits maxima, which are at different positions for $m = \pm 1$. The spin-asymmetry parameter and, therefore, the spin of the emitted electron, considerably changes with the electron energy for a fixed emission angle (look, for example, at $\theta = 60^\circ$). This effect was explained in detail in [34]. In the present paper we repeated the calculations of [34] for the fundamental wavelength of 1300 nm (800 nm was used in [34]). As a consequence of the longer wavelength, the oscillations of the rate are much larger and the observed structures change more rapidly than in the 800 nm case.

4. Conclusions

The theory of ATI by a bicircular field was presented in [33], while the corresponding spin-asymmetry effect was reported in [34]. In the present paper we confirmed the results of these references and analyzed some aspects of the bicircular-field-induced ATI in more detail. In particular, we investigated numerically the influence of the electron travel time on the rescattering process and found that the main contributions to the electron spectra come from short travel times (shorter than 2/3 of the optical cycle). These numerical results agree with the results obtained using the quantum-orbit theory. We have analyzed how the relevant quantum orbits unfold in the bicircular-laser-field polarization plane.

Another important result of the present paper concerns the spin asymmetry of ATI by a bicircular field with a longer fundamental wavelength than considered in [34]. We found that the oscillations of the rate with the electron energy are much stronger and more rapid for the

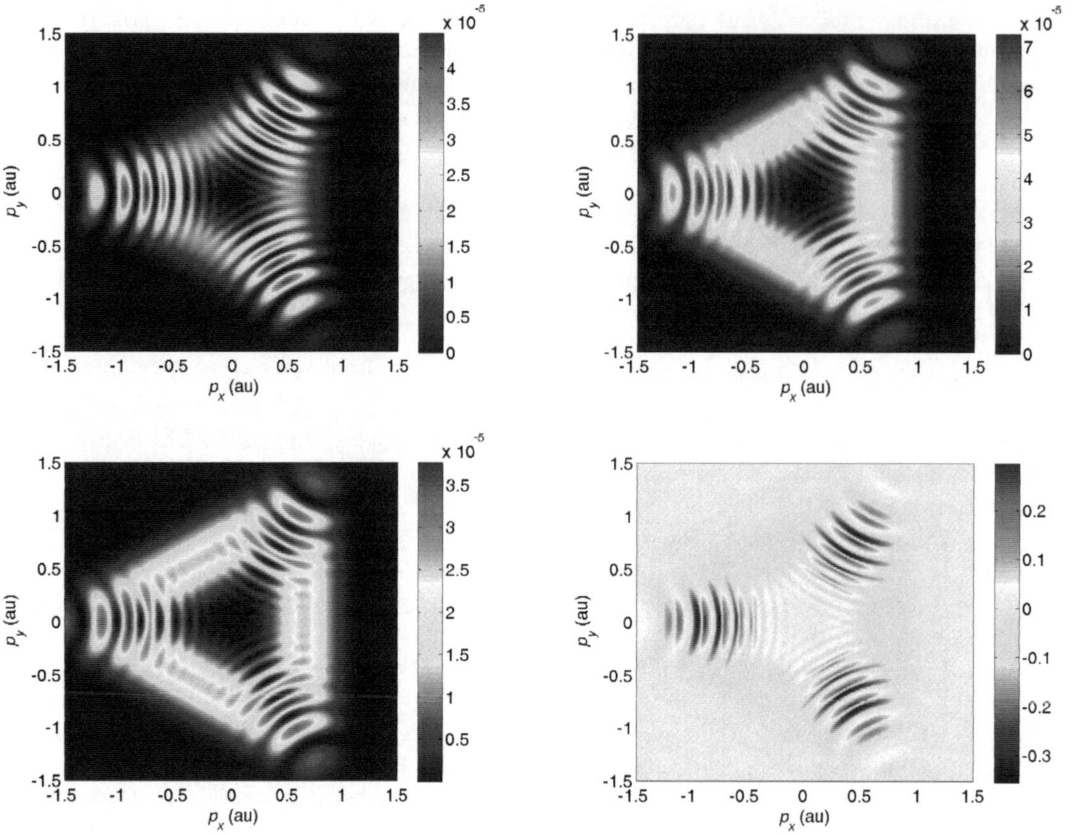

Figure 3. (Color online) The differential ionization rate (in au) of Xe atoms, presented in false colors in the electron-momentum plane for ionization by a bicircular ω–2ω laser field with equal intensity of both counterrotating components, $I_1 = I_2 = 1 \times 10^{14}$ W/cm^2, and the fundamental wavelength of 1300 nm. Upper panels: the rates for the initial states having the magnetic quantum number $m = -1$ (left) and $m = +1$ (right). Lower left panel: the rate summed over m. Lower right panel: normalized asymmetry parameter $\tilde{A}_{\mathbf{p}}$.

longer wavelength of 1300 nm than for 800 nm used in [34]. This is an important result since spin-polarized electrons have many practical applications (for example, to study magnetic materials [55] and chiral molecules [56, 57]).

References

[1] Eichmann H, Egbert A, Nolte S, Momma C, Wellegehausen B, Becker W, Long S and McIver J K 1995 *Phys. Rev. A* **51** R3414(R)
[2] Long S, Becker W and McIver J K 1995 *Phys. Rev. A* **52** 2262
[3] Zuo T and Bandrauk A D 1995 *Nonlinear Opt. Phys. Mat.* **04** 533–546
[4] Alon O E, Averbukh V and Moiseyev N 1998 *Phys. Rev. Lett.* **80** 3743–3746
[5] Becker W, Chichkov B N and Wellegehausen B 1999 *Phys. Rev. A* **60** 1721–1722
[6] Milošević D B, Becker W and Kopold R 2000 *Phys. Rev. A* **61** 063403
[7] Milošević D B and Sandner W 2000 *Opt. Lett.* **25** 1532
[8] Milošević D B and Becker W 2000 *Phys. Rev. A* **62** 011403(R)
[9] Milošević D B, Becker W, Kopold R and Sandner W 2001 *Laser Phys.* **11** 165
[10] Milošević D B, Becker W and Kopold R 2001 *Atoms, Molecules and Quantum Dots in Laser Fields: Fundamental Processes* (Conference Proceedings vol 71) ed N Bloembergen, N Rahman and A Rizzo (Bologna: Societá Italiana di Fisica) pp 239–252

[11] Ceccherini F, Bauer D and Cornolti F 2001 *J. Phys. B* **34** 5017–5029
[12] Nilsen H M, Madsen L B and Hansen J P 2002 *J. Phys. B* **35** L403–L408
[13] Bandrauk A D and Lu H Z 2003 *Phys. Rev. A* **68** 043408
[14] Ceccherini F, Bauer D and Cornolti F 2003 *Phys. Rev. A* **68** 053402
[15] Milošević D B and Becker W 2005 *J. Mod. Opt.* **52** 233
[16] Hasović E, Milošević D B and Becker W 2006 *Laser Phys. Lett.* **3** 200–204
[17] Kramo A, Hasović E, Milošević D B and Becker W 2007 *Laser Phys. Lett.* **4** 279
[18] Hasović E, Kramo A and Milošević D B 2008 *Eur. Phys. J. Spec. Top.* **160** 205
[19] Fleischer A, Kfir O, Diskin T, Sidorenko P and Cohen O 2014 *Nature Photon.* **8** 543
[20] Pisanty E, Sukiasyan S and Ivanov M 2014 *Phys. Rev. A* **90** 043829
[21] Kfir O *et al* 2015 *Nature Photon.* **9** 99
[22] Milošević D B 2015 *Opt. Lett.* **40** 2381
[23] Milošević D B 2015 *J. Phys. B* **48** 171001
[24] Medišauskas L, Wragg J, van der Hart H and Ivanov M Yu 2015 *Phys. Rev. Lett.* **115** 153001
[25] Milošević D B 2015 *Phys. Rev. A* **92** 043827
[26] Mancuso C A *et al* 2015 *Phys. Rev. A* **91** 031402(R)
[27] Odžak S and Milošević D B 2015 *Phys. Rev. A* **92** 053416
[28] Fan T *et al* 2015 *Proc. Natl. Acad. Sci. USA* **112** 14206–14211
[29] Chen C *et al* 2016 *Science Advances* **2** e1501333
[30] Hasović E, Becker W and Milošević D B 2016 2016 *Optics Express* **24** 6413–6424
[31] Reich D M and Madsen L B 2016 *Phys. Rev. A* **93** 043411
[32] Mancuso C A *et al* 2016 *Phys. Rev. A* **93** 053406
[33] Milošević D B and Becker W 2016 *Phys. Rev. A* **93** 063418
[34] Milošević D B 2016 *Phys. Rev. A* **93** 051402(R)
[35] Chaloupka J L and Hickstein D D 2016 *Phys. Rev. Lett.* **116** 143005
[36] Mancuso C A *et al* 2016 *Phys. Rev. Lett.* **117** 133201
[37] Eckart S *et al* 2016 *Phys. Rev. Lett.* **117** 133202
[38] Baykusheva D, Ahsan M S, Lin N and Wörner H J 2016 *Phys. Rev. Lett.* **116** 123001
[39] Mauger F, Bandrauk A D and Uzer T 2016 *J. Phys. B* **49** 10LT01
[40] Du H, Zhang J, Ben S, Zhong H-Y, Xu T-T, Guo J and Liu X-S 2016 *Chin. Phys. B* **25** 043202
[41] Reich D M and Madsen L B 2016 *Phys. Rev. Lett.* **117** 133902
[42] Liu X, Zhu X, Li L, Li Y, Zhang Q, Lan P and Lu P 2016 *Phys. Rev. A* **94** 033410
[43] Odžak S, Hasović E and Milošević D B 2016 *Phys. Rev. A* **94** 033419
[44] Radzig A A and Smirnov B M 1985 *Reference Data on Atoms, Molecules and Ions* (Berlin: Springer)
[45] Kopold R, Becker W and Kleber M 2000 *Opt. Commun.* **179** 39
[46] Milošević D B 2000 *J. Phys. B* **33** 2479
[47] Kopold R, Milošević D B and Becker W 2000 *Phys. Rev. Lett.* **84** 3831
[48] Salières P, Carré B, Le Déroff L, Grasbon F, Paulus G G, Walther H, Kopold R, Becker W, Milošević D B, Sanpera A and Lewenstein M 2001 *Science* **292** 902
[49] Milošević D B and Becker W 2002 *Phys. Rev. A* **66** 063417
[50] Becker W, Grasbon F, Kopold R, Milošević D B, Paulus G G and Walther H 2002 *Adv. At. Mol. Opt. Phys.* **48** 35
[51] Milošević D B, Bauer D and Becker W 2006 *J. Mod. Opt.* **53** 125–134
[52] D. B. Milošević D B 2014 *Phys. Rev. A* **90** 063414
[53] Barth I and Smirnova O 2013 *Phys. Rev. A* **88** 013401
[54] Hartung A, Morales F, Kunitski M, Henrichs K, Laucke A, Richter M, Jahnke T, Kalinin A, Schöffler M, Schmidt L Ph H, Ivanov M, Smirnova O and Dörner R 2016 *Nature Photon.* **10** 526–528
[55] Stöhr J, Wu Y, Hermsmeier B D, Samant M G, Harp G R, Koranda S, Dunham D and Tonner B P 1993 *Science* **259** 658
[56] Janssen M H M and Powis I 2014 *Phys. Chem. Chem. Phys.* **16** 856
[57] N Berova, P N Polavarapu, K Nakanishi and R W Woody (eds) 2012 *Comprehensive Chiroptical Spectroscopy* Vol 1 (New York: Wiley)

Electron Rescattering in a Bicircular Laser Field

E Hasović[1], W. Becker[2] and D B Milošević[1,2,3]

[1] Faculty of Science, University of Sarajevo, Zmaja od Bosne 35, 71000 Sarajevo, Bosnia and Herzegovina

[2] Max-Born Institut, Max-Born Strasse 2a, 12489 Berlin, Germany

[3] Academy of Sciences and Arts of Bosnia and Herzegovina, Bistrik 7, 71000 Sarajevo, Bosnia and Herzegovina

E-mail: ahasovic1@yahoo.com

Abstract. We investigate high-order above-threshold ionization (HATI) of krypton atoms by a bicircular laser field, which consists of two coplanar co- or counter-rotating circularly polarized fields of frequencies $r\omega$ and $s\omega$. We show that the photoelectron spectra in the HATI process, presented in the momentum plane, exhibit the same discrete rotational symmetry as the driving field. We also analyze HATI spectra for various combinations of the intensities of two field components for co- and counter-rotating fields. We find that the appearance of high-energy plateau for the counter-rotating case is vary sensitive to the laser intensity ratio, while the plateau is always absent for the co-rotating bicircular field.

1. Introduction

High-order harmonics generation (HHG) and high-order above threshold ionization (HATI) by the so-called bicircular field are very interesting and useful tools in the laser-atom/molecule interaction. The bicircular field consists of two coplanar counter- or co-rotating circularly polarized fields having different angular frequencies. For linear polarization, HHG and HATI processes have been observed with high precision and theoretically analyzed with various methods [1, 2, 3, 4, 5]. Both processes can be explained within the three-step model [6]: in the first step the electron is ionized, then it is driven back to the origin by the laser field, and, finally, in the third step, the electron recombines to the ground state (HHG) or scatters elastically of the parent ion (HATI). For circular polarization, there is no rescattering because when a classical electron is released in the laser-field environment with zero initial velocity, it will never return to the site of its release. For a bicircular field, however, the harmonic-generation efficiency is surprisingly high, which indicates that the absence of rescattering for one circularly polarized field does not extend to the superposition of two such fields.

The first experimental and theoretical studies of HHG by bicircular fields are dating back to the mid and late nineties. HHG spectra of argon atoms in bichromatic linearly and circularly polarized fields, were investigated for the first time experimentally in [7]. In fact, it was observed that strong high-harmonic emission can be achieved using a counter-rotating bicircular field having the frequencies ω and 2ω. Calculations based on a zero-range potential model have confirmed this [8]. The influence of a strong magnetic field on HHG spectra induced by two color laser fields was studied theoretically by solving time-dependent Schrödinger equation (TDSE) for H_2^+ molecular ion [9]. Schemes for generation of circularly polarized high-order harmonics by an $\omega - 2\omega$ circularly polarised laser field were proposed in [10]. In the following time, a series of theoretical studies of HHG in two color fields were reported. Namely, the HHG process in a bicircular field was explained using a semiclassical three-step model based on the quantum-orbit formalism [11]. It was found that the harmonic-emission efficiency is very high and the harmonics are circularly polarized with alternating helicities [12]. Besides that, the two-dimensional electron trajectories responsible for HHG process were clearly identified. Experimental confirmation of the circular nature of harmonics, generated in a bicircular field, was reported in 2014 [13]. Modern theoretical and experimental studies of HHG in bicircular laser fields can be found in [14, 15, 16, 17, 18, 19, 20, 21, 22, 23].

The (H)ATI process by a bicircular field was considered in Ref. [24], where a superposition of two counter-rotating circularly polarized pulses, one long and one few-cycle, having the same frequency, was considered. It was shown that high-energy electrons, generated in ionization by such a combination of pulses, are emitted in a direction correlated with the carrier-envelope phase. Angle-resolved electron energy spectra in strong-field ionization by a bicircular field of arbitrary frequencies $r\omega$ and $s\omega$ were analyzed in [25, 26]. For the case where $r = 1$ and $s = 2$, the predicted three-lobed shape of these spectra was recently confirmed experimentally [27]. Results on high-energy spectra obtained theoretically and experimentally have been reported recently [28, 29]. A detailed analysis of the HATI process, based on quantum-orbit theory, is given in [30].

Other atomic processes such as laser-assisted recombination [31], non-sequential double ionization [32, 33, 34], the possibility of introducing spin into attoscience [35], etc. were also investigated in strong bicircular fields.

2. Strong-Field Approximation Theory

The Strong-Field Approximation (SFA) theory of (H)ATI process in a bicircular field is described in detail in [30]. The observable quantity is the averaged differential ionization rate given by

$$\overline{w}_{\mathbf{p}i}(n) = \frac{N_e}{2l+1}\sum_{m=-l}^{l} w_{\mathbf{p}ilm}(n), \tag{1}$$

where N_e is the number of equivalent electrons in the ionizing shell and $w_{\mathbf{p}ilm}(n) = 2\pi p|T_{\mathbf{p}i}(n)|^2$, while n is the number of absorbed photons such that $n\omega = E_{\mathbf{p}} + U_{\mathrm{p}} + I_{\mathrm{p}}$, with I_{p} the ionization potential and U_{p} the ponderomotive energy. The T-matrix element $T_{\mathbf{p}i}(n)$ can be written as

$$T_{\mathbf{p}i}(n) = \int_0^T \frac{dt}{T} e^{i[\mathbf{p}\cdot\boldsymbol{\alpha}(t)+U_1(t)]}\left[\mathcal{T}_{\mathbf{p}}^{(0)}(t) + \mathcal{T}_{\mathbf{p}}^{(1)}(t)\right]e^{in\omega t}, \tag{2}$$

where

$$\mathcal{T}_{\mathbf{p}i}^{(0)}(t) = \langle \mathbf{p} + \mathbf{A}(t)|\mathbf{r}\cdot\mathbf{E}(t)|\psi_0\rangle \tag{3}$$

is the part of the transition amplitude describing the ATI process, while

$$\mathcal{T}_{\mathbf{p}i}^{(1)}(t) = -i\int_0^\infty d\tau \left(\frac{2\pi}{i\tau}\right)^{3/2}\langle\mathbf{p}|V|\mathbf{k}_s\rangle\langle\mathbf{k}_s+\mathbf{A}(t-\tau)|\mathbf{r}\cdot\mathbf{E}(t-\tau)|\psi_i\rangle e^{i[S_{\mathbf{k}_s i}(t-\tau)-S_{\mathbf{k}_s i}(t)]}. \tag{4}$$

allows for one additional interaction of the liberated electron with its parent ion (HATI process). The phase factor in the previous equation is given by $S_{\mathbf{k}_s i}(t) = \int^t dt'\{[\mathbf{k}_s + \mathbf{A}(t')]^2/2 + I_{\mathrm{p}}\} = (\mathbf{k}_s^2/2 + U_{\mathrm{p}} + I_{\mathrm{p}})t + \mathbf{k}_s\cdot\boldsymbol{\alpha}(t) + U_1(t)$, with $U_1(t) = \int^t \mathbf{A}^2(t')dt'/2 - U_{\mathrm{p}}t$ and $\boldsymbol{\alpha}(t) = \int^t \mathbf{A}(t')dt'$.

IOP Conf. Series: Journal of Physics: Conf. Series **826** (2017) 012009 doi:10.1088/1742-6596/826/1/012009

3. Numerical results

We are going to present numerical results of (H)ATI spectra and momentum distributions for krypton atoms exposed to a strong bicircular laser field. A bicircular laser field is a superposition of two coplanar circularly polarized counter-rotating or co-rotating fields having the intensities $I_1 = E_1^2$ and $I_2 = E_2^2$ and the angular frequencies $r\omega$ and $s\omega$, which are integer multiples of the same fundamental frequency ω. It is defined by

$$E_x(t) = \left[E_1 \sin(r\omega t + \phi_1) + E_2 \sin(s\omega t + \phi_2)\right]/\sqrt{2}, \tag{5}$$

$$E_y(t) = \left[-E_1 \cos(r\omega + \phi_1) \pm E_2 \cos(s\omega t + \phi_2)\right]/\sqrt{2}, \tag{6}$$

where the counter-rotating (co-rotating) fields correspond to the sign +1 (-1). A change of the phase ϕ_1, for a fixed value of ϕ_2, corresponds to a rotation of the field around the z axis by the angle $\alpha = s\phi_1/(r+s)$. In the present work we choose $\phi_1 = \phi_2 = 0$. Denoting $A_1 = E_1/(r\omega)$ and $A_2 = E_2/(s\omega)$, we obtain for the ponderomotive energy $U_\mathrm{p} = (A_1^2 + A_2^2)/4$. Normalized electric-field vector $\mathbf{E}(t)$ and the corresponding normalized vector potential $\mathbf{A}(t)$ for various combinations of $(r, s) \in (1,2), (1,3), (1,4)$, equal intensities of the field components and counter-rotating case are presented in the upper panels of Fig. 1.

In the bottom panels of Fig. 1 we present momentum distributions of HATI electrons obtained by ionising Kr atoms by a bicircular field having equal components intensities $I_1 = I_2 = 2\times 10^{14}$ W/cm^2 for the combinations $(r, s) = (1,2), (1,3), (1,4)$. The fundamental wavelength is 800 nm. The counter-rotating bicircular fields obeys dynamical symmetry, with the result that the rotation by the angle $\alpha_j = -r\omega\tau_j$ about the z axis is equivalent to a translation in time by $\tau_j = jT/(r+s)$. It has been shown [30] that both, the direct and the rescattering T-matrix elements, are invariant with respect to a rotation by the angle $\alpha_j = 2\pi jr/(r+s)$ up to the same phase factor. Momentum distributions, presented in the bottom panels of Fig. 1, clearly obeys $(r+s)$-fold rotational symmetry of the differential ionization rate which confirms previous conclusions. If we compare polar diagrams of the vector potentials, shown by the blue dashed curves in top panels of Fig. 1, with the corresponding differential ionization rate, shown in the bottom panels of Fig. 1, we see that the rate is maximal for $\mathbf{p} = -\mathbf{A}(\tau)$, i.e. the electrons are predominantly emitted opposite to the direction of the vector potential at the ionization time τ.

One way to enhance the process of electron-ion recombination and the harmonic emission rate is to vary the relative intensity between the two components of counter-rotating bicircular field [29]. Here we present the results for ATI spectra obtained by ionizing Kr atoms using a $\omega - 3\omega$ bicircular field. In Fig. 2 we plot the electron yield as function of the electron kinetic energy for various combinations of the relative intensity $I_{3\omega}/I_\omega$. The spectra of direct electrons, calculated using T-matrix element given by (3), are presented by black solid curves, while the overall spectra, which include direct and rescattered electrons are shown by red dashed lines. It is clear that the photoelectron distributions in this case are extremely sensitive to the intensity ratio of the driving laser fields. For low values of the ratio $I_{3\omega}/I_\omega$ the total spectra are comparable to the spectra of the direct electrons. The absence of a plateau in the spectra indicates that the rescattering does not occur. As the $I_{3\omega}/I_\omega$ ratio is increased to the values of 1, 3 or 9, the shape of the spectra is changing and a plateau begins to form, which is an indication of the rescattering of the released electrons. The cutoff energy in the rescattering plateau takes its maximal value for the ratio $I_{3\omega}/I_\omega = 9$. In this case, the ponderomotive energies of the two components, $U_{\mathrm{p}1} = A_1^2/4$ and $U_{\mathrm{p}2} = A_2^2/4$, have the same value.

In order to consider the ionization by a co-rotating bicircular field, we present photoelectron spectra for various values of the intensity ratio $I_{3\omega}/I_\omega$. Similar to the counter-rotating case, we plot on the same panels of Fig. 2 the electron spectra including only the direct electrons (green

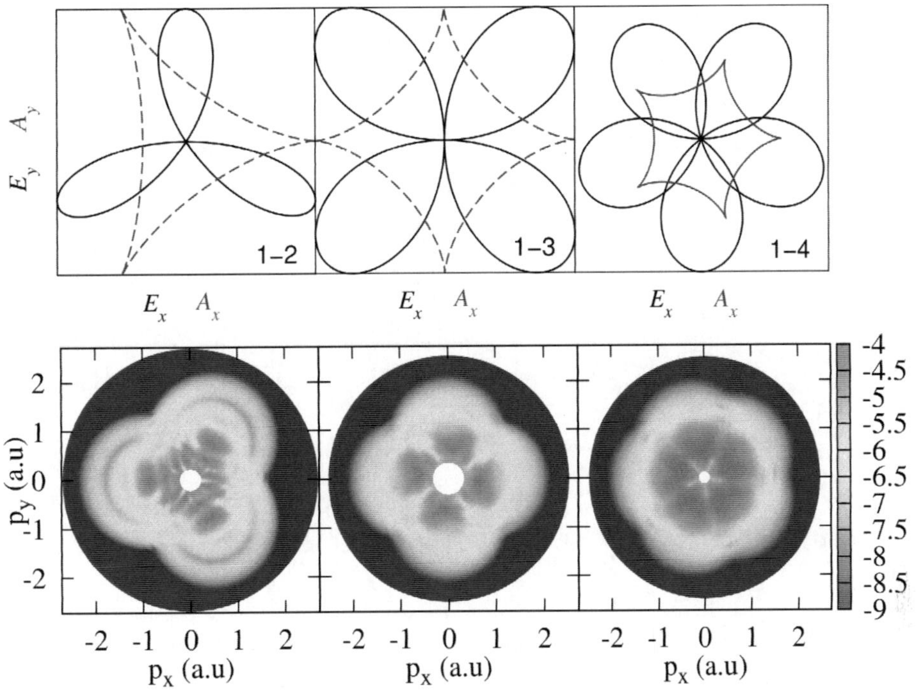

Figure 1. Upper panels: normalized electric-field vector $\mathbf{E}(t)$ (black solid lines) and vector potential $\mathbf{A}(t)$ (blue dashed lines) of the $r\omega - s\omega$ counter-rotating bicircular laser field for equal laser field component intensities and $(r, s) \in (1, 2), (1, 3), (1, 4)$, plotted for $0 \leq t \leq T$, $T = 2\pi/\omega$. Lower panels: the logarithm of the differential ionization rate of Kr atoms presented in false colors in the electron momentum plane for ionization by a counter-rotating bicircular laser field with $(r, s) \in (1, 2), (1, 3), (1, 4)$ and equal intensity of both components $I_1 = I_2 = 2 \times 10^{14}$ W/cm^2. The fundamental wavelength is 800 nm.

dotted lines) and the overall spectra consisting of the direct as well as rescattered electrons (blue long-dashed lines). It is clear that the electron spectra do not change significantly with the change of the intensity ratio $I_{3\omega}/I_\omega$. The spectra are dominated by the direct electrons for all intensity ratios, which indicates the absence of the rescattering event in the co-rotating bicircular field and confirms previous results [29, 30].

4. Conclusions

We presented momentum distributions of the electrons, freed in ionization by a strong bicircular laser field of frequencies $r\omega$ and $s\omega$. The presented spectra, obtained by applying SFA theory, exhibit characteristic symmetry properties, which are a consequence of the dynamical symmetry of the counter-rotating bicircular field. Furthermore, we calculated photoelectron momentum distributions for a broad range of intensity ratios of the two field components for the $(r, s) = (1, 3)$ case. We found that the photoelectron spectra for the counter-rotating bicircular field change significantly with the change of the intensity ratio $I_{3\omega}/I_\omega$ and optimize for the value of $I_{3\omega}/I_\omega = 9$. For lower values of the ratio $I_{3\omega}/I_\omega$ the spectra are dominated by direct electrons, while for values higher than 1 a plateau begins to form. The photoelectron spectra obtained in a co-rotating bicircular field are dominated by the direct electrons for all intensity ratios. This behaviour indicates the absence of the rescattering event in the co-rotating bicircular field.

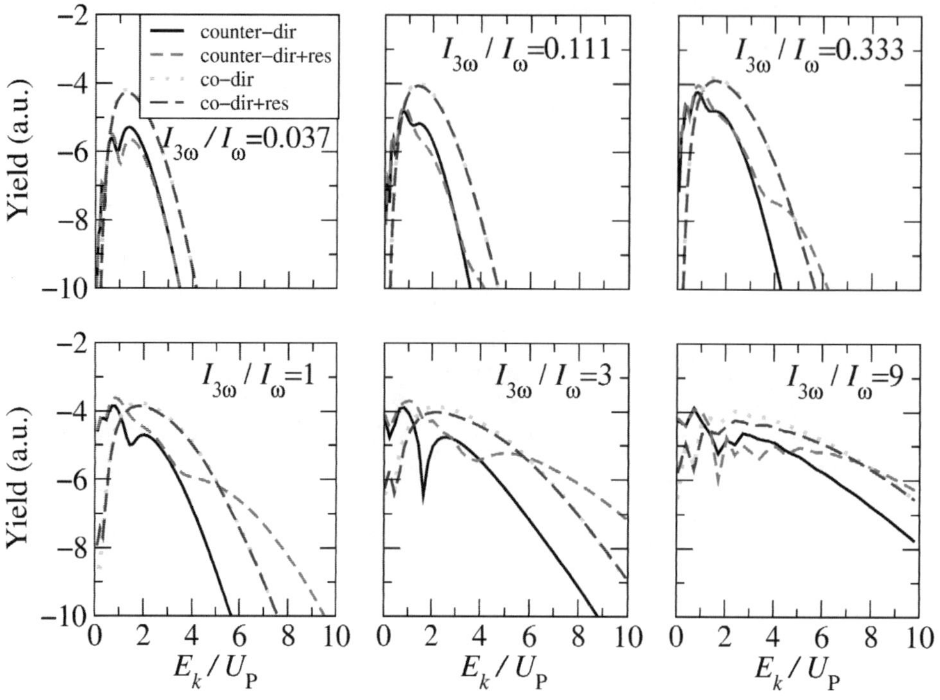

Figure 2. Photoelectron distributions of Kr atoms ionized by co- and counter-rotating $\omega - 3\omega$ bicircular fields, obtained for various intensity ratios of the field components. We present spectra of direct electrons for co-rotating (counter-rotating) field by green dotted (black solid) lines as well as the overall spectra consisting of direct and rescattered electrons by blue, long-dashed (red dashed) lines. The laser field parameters are the same as in Fig. 1.

References

[1] Becker W, Grasbon F, Kopold F, Milošević D B, Paulus G G and Walther H 2002 *Adv. At., Mol., Opt. Phys.* **48** 35
[2] Agostini P and DiMauro L F 2004 *Rep. Prog. Phys.* **67** 813
[3] Scrinzi A, Ivanov M Y, Kienberger R and Villeneuve D M 2006 *J. Phys.* B **39** R1
[4] Milošević D B, Paulus G G, Bauer D and Becker W 2006 *J. Phys.* B **39** R203
[5] Krausz F and Ivanov M 2009 *Rev. Mod. Phys.* **81** 163
[6] Corkum P B 1993 *Phys. Rev. Lett.* **71** 1994
[7] Eichmann H, Egbert A, Nolte S, Momma C, Wellegehausen B, Becker W, Long S and McIver J K 1995 *Phys. Rev.* A **51** R3414
[8] Long S, Becker W and McIver J K 1995 *Phys. Rev.* A **52** 2262
[9] Zuo T and Bandrauk A D 1995 *J. Nonlinear Opt. Phys. Mater.* **04** 533
[10] Becker W, Chichkov B N and Wellegehausen B 1999 *Phys. Rev.* A **60** 1721
[11] Milošević D B, Becker W and Kopold R 2000 *Phys. Rev.* A **61** 063403
[12] Milošević D B and Sandner W 2000 *Opt. Lett.* **25** 1532
[13] Fleischer A, Kfir O, Diskin T, Sidorenko P and Cohen O 2014 *Nat. Photon.* **8** 543
[14] Pisanty E, Sukiasyan S and Ivanov M 2014 *Phys. Rev.* A **90** 043829
[15] Milošević D B 2015 *J. Phys.* B **48** 171001
[16] Milošević D B 2015 *Opt. Lett.* **40** 2381
[17] Medišauskas L, Wragg J, van der Hart H and Ivanov M. Y. 2015 *Phys. Rev. Lett.* **115** 153001
[18] Milošević D B 2015 *Phys. Rev.* A **92** 043827
[19] Kfir O *et al.* 2015 *Nat. Photonics* **9** 99
[20] Fan T *et al.* 2015 *Proc. Natl. Acad. Sci. USA* **112** 14206
[21] Chen C *et al.* 2016 *Sci. Adv.* **2** e1501333

[22] Reich D M and Madsen L B 2016 *Phys. Rev.* A **93** 043411
[23] Odžak S, Hasović E and Milošević D B 2016 *Phys. Rev.* A **93** 043413
[24] Hasović E, Milošević D B and Becker W 2006 *Laser Phys. Lett.* **3** 200
[25] Kramo A, Hasović E, Milošević D B and Becker W 2006 *Laser Phys. Lett.* **4** 279
[26] Hasović E, Kramo A and Milošević D B 2008 *Eur. Phys. J. Spec. Top.* **160** 205
[27] Mancuso C A *et al.* 2015 *Phys. Rev. A* **91** 031402(R)
[28] Hasović E, Becker W and Milošević D B 2016 *Opt. Express* **24** 6413
[29] Mancuso C A *et al.* 2016 *Phys. Rev.* A **93** 053406
[30] Milošević D B and Becker W 2016 *Phys. Rev.* A **93** 063418
[31] Odžak S and Milošević D B 2015 *Phys. Rev.* A **92** 053416
[32] Chaloupka J L and Hickstein D D 2016 *Phys. Rev. Lett.* **116** 143005
[33] Mancuso C A *et al.* 2016 *Phys. Rev. Lett.* **117** 133201
[34] Eckart S *et al.* 2016 *Phys. Rev. Lett.* **117** 133202
[35] Milošević D B 2016 *Phys. Rev.* A **93** 051402(R)

Ionization without interference

F. Cajiao Vélez, J. Z. Kamiński, and K. Krajewska

Institute of Theoretical Physics, Faculty of Physics, University of Warsaw, Pasteura 5, 02-093 Warszawa, Poland

E-mail: Felipe.Cajiao-Velez@fuw.edu.pl

Abstract. Ionization of hydrogen-like ions by intense, circularly polarized laser pulses is analyzed under the scope of the relativistic strong-field approximation. We show that, for specific parameters of the laser field, the energy spectra of photoelectrons present a broad region without interference (supercontinuum) which can be controlled by modifying the laser field intensity. The physical interpretation of the process is developed according to the Keldysh theory, emphasizing the importance of the complex-time saddle point contributions to the total probability of photoionization. The corresponding polar-angle distributions present an asymmetry attributed to radiation pressure effects.

1. Introduction

In nonlinear optics, the supercontinuum generation refers to the process for which a narrow-band laser field presents a considerable spectral broadening, resulting in radiation frequencies with a large bandwidth while evincing temporal and/or spatial coherence [1, 2]. In a similar way, the concept of supercontinuum can be applied to describe the energy spectrum of electrons emitted by photoionization of atoms or ions by strong laser fields. When the energy spectra of photoelectrons exhibit broad structures without important modulations in the scale of tenths (or even hundreds) of laser photon energies, it is considered to present a supercontinuum.

The ionization by strong laser fields is typically dominated by interference effects. The latter are demonstrated, for instance, when the spectrum of photoelectrons consists of series of equally separated peaks. This is also true when the driving pulse comprises just few oscillations. When the monochromatic plane-wave approximation is considered, those peaks are separated by the laser carrier frequency and are commonly recognized as multiphoton peaks. In the pioneering work by Keldysh [3] (see the review by Popov [4]), it was shown that the total amplitude of photoionization by strong fields contains the contribution of multiple factors arising from different complex-time saddle points. Those contributions lead to pronounced interference effects and to the formation of the aforementioned peaks in the energy spectrum of photoelectrons.

According to the Keldysh theory, if just one complex-time saddle point contributes predominantly to the total probability of photoionization, the interference effects are expected to be suppressed (ionization without interference) leading to the creation of a supercontinuum. On the contrary, if two or more saddle points present similar contributions, interference dominates the process with the consequent formation of series of peaks in the spectra of photoelectrons.

The aim of this paper is to investigate the interference-free ionization process for the relativistically intense and short laser pulses.

In this paper we set $\hbar = 1$. In our numerical calculations we use relativistic units with $\hbar = m_e = c = 1$, where m_e is the electron rest mass and c is the speed of light. Moreover, the relativistic scalar product between two arbitrary four-vectors a and b is written as $a \cdot b = a^\mu b_\mu = a^0b^0 - a^1b^1 - a^2b^2 - a^3b^3$ and the Feynman notation $\not{a} = \gamma \cdot a = \gamma^\mu a_\mu$, where γ^μ are the Dirac gamma matrices, is used. As usual, $\bar{u} = u^\dagger\gamma^0$. When necessary, we employ the so-called light-cone variables, i.e., for an arbitrary unit vector $\boldsymbol{n}$ and an arbitrary four-vector a, we define $a^\parallel = \boldsymbol{n} \cdot \boldsymbol{a}$, $a^- = a^0 - a^\parallel$, $a^+ = (a^0 + a^\parallel)/2$, and $\boldsymbol{a}^\perp = \boldsymbol{a} - a^\parallel \boldsymbol{n}$. Thus, $a \cdot b = a^+b^- + a^-b^+ - \boldsymbol{a}^\perp \cdot \boldsymbol{b}^\perp$ and $d^4x = dx^+dx^-d^2x^\perp$.

2. Theory

Let us consider the interaction between a relativistically-intense laser field and a hydrogen-like ion. The exact probability amplitude of ionization is (see, Ref. [5])

$$\mathcal{A}_{\rm fi} = -{\rm i} \int d^4x {\rm e}^{-{\rm i}(E_0/c)x^0} \bar{\Psi}_{\rm f}(x) e \not{A}_{\rm R}(x) \Psi_{\rm i}(\boldsymbol{x}), \tag{1}$$

where the bispinor $\Psi_{\rm i}(\boldsymbol{x})$ describes the electron bound state of energy E_0, $\Psi_{\rm f}(x)$ represents the exact scattering state, $A^\nu_{\rm R}(x)$ is the four-vector potential describing the laser pulse, and $e < 0$ is the electron charge.

While the ground state wavefunction for hydrogen-like systems is known exactly (see, e.g., Ref. [6]), the scattering state $\Psi_{\rm f}(x)$ has to be approximated, specially when high intensity laser fields are involved. Assuming that the kinetic energy of the photoelectron characterized by the asymptotic momentum $\boldsymbol{p}$ is much larger than the ionization potential, i.e., $E_{\boldsymbol{p}} = \sqrt{(m_e c^2)^2 + (c\boldsymbol{p})^2} - m_e c^2 \gg m_e c^2 - E_0$, the scattering state can be approximated by means of the Born expansion. In the zeroth order, such approximation consists in replacing the exact state $\Psi_{\rm f}(x)$ in Eq. (1) by the solution of the Dirac equation in the laser field $\Psi^{(0)}_{\boldsymbol{p}\lambda}(x)$, which does not account for the interaction with the atomic potential,

$$({\rm i}\not{\partial} - e\not{A}_{\rm R}(x) - m_e c)\Psi^{(0)}_{\boldsymbol{p}\lambda}(x) = 0. \tag{2}$$

Here, the subscript $\lambda = \pm$ stands for the electron spin polarization. The solutions of Eq. (2) are known as the Volkov states and can be derived exactly for laser fields in the plane-wave front approximation. This is the essence of the relativistic strong-field approximation (RSFA). Therefore, the probability amplitude of ionization under the RSFA [now denoted as $\mathcal{A}(\boldsymbol{p}, \lambda; \lambda_{\rm i})$], takes the form

$$\mathcal{A}(\boldsymbol{p}, \lambda; \lambda_{\rm i}) = -{\rm i} \int \frac{d^3q}{(2\pi)^3} \int d^4x \, {\rm e}^{-{\rm i}q\cdot x} \bar{\Psi}^{(0)}_{\boldsymbol{p}\lambda}(x) e \not{A}_{\rm R}(x) \tilde{\Psi}_{\rm i}(\boldsymbol{q}), \tag{3}$$

where $\tilde{\Psi}_{\rm i}(\boldsymbol{q})$ represents the Fourier transform of the atomic bound state $\Psi_{\rm i}(\boldsymbol{x})$. In Eq. (3) we have introduced $q = (q^0, \boldsymbol{q}) = (E_0/c, \boldsymbol{q})$, which is not a four-vector as it does not transform properly under the relativistic Lorentz transformations. Nevertheless, this notation helps us to simplify the formulas presented below. Note that the probability amplitude of ionization $\mathcal{A}(\boldsymbol{p}, \lambda; \lambda_{\rm i})$ depends on the initial and final spin states $\lambda_{\rm i}$ and λ, respectively, which are denoted as '+' for a spin up and '−' for a spin down.

Up to now, our considerations have been very general. In the remaining part of this paper the calculations are going to be carried out in the velocity gauge. In order to proceed, we model the electromagnetic potential describing the laser field using the plane-wave front approximation,

$$A_{\rm R}(x) \equiv A(\phi) = A_0[\varepsilon_1 f_1(\phi) + \varepsilon_2 f_2(\phi)], \tag{4}$$

where $\phi = k \cdot x = k^0 x^-$, $k = k^0 n = k^0(1, \boldsymbol{n})$, $k^0 = \omega/c$, and $\omega = 2\pi/T_\mathrm{p}$. In our notation ω represents the fundamental frequency of the pulse and T_p corresponds to its duration. The polarization of the laser field is determined by two real and normalized four-vectors, $\varepsilon_j \equiv (0, \boldsymbol{\varepsilon}_j)$, which are perpendicular to the pulse propagation direction (i.e., $k \cdot \varepsilon_j = -\boldsymbol{k} \cdot \boldsymbol{\varepsilon}_j = 0$). The two shape functions, $f_j(\phi)$, are real, with continuous second derivatives, and vanish for $\phi < 0$ and $\phi > 2\pi$. The unitary vector $\boldsymbol{n}$ represents the direction of propagation of the laser pulse.

The Volkov solution for the vector potential (4) is given by [5, 7]

$$\psi^{(+)}_{\boldsymbol{p}\lambda}(x) = \sqrt{\frac{m_\mathrm{e}c^2}{VE_{\boldsymbol{p}}}}\Big(1 + \frac{m_\mathrm{e}c\mu}{2p\cdot k}\big[f_1(k\cdot x)\not{\varepsilon}_1\not{k} + f_2(k\cdot x)\not{\varepsilon}_2\not{k}\big]\Big)\mathrm{e}^{-\mathrm{i}S^{(+)}_p(x)}u^{(+)}_{\boldsymbol{p}\lambda}, \tag{5}$$

where

$$S^{(+)}_p(x) = p\cdot x + \int_0^{k\cdot x} d\phi\Big[-\frac{m_\mathrm{e}c\mu}{p\cdot k}(\varepsilon_1\cdot p f_1(\phi) + \varepsilon_2\cdot p f_2(\phi)) + \frac{(m_\mathrm{e}c\mu)^2}{2p\cdot k}(f_1^2(\phi) + f_2^2(\phi))\Big]. \tag{6}$$

In Eqs. (5) and (6), the superscript (+) indicates that $\psi^{(+)}_{\boldsymbol{p}\lambda}(x)$ is a positive-energy solution of the Dirac equation in the laser field [Eq. (2)], V represents the quantization volume, and $p = (p^0, \boldsymbol{p}) = (E_{\boldsymbol{p}}/c, \boldsymbol{p})$ is the on-mass-shell four-vector. In addition, the Dirac free particle bispinors $u^{(+)}_{\boldsymbol{p}\lambda}$ are normalized such that $\bar{u}^{(+)}_{\boldsymbol{p}\lambda}u^{(+)}_{\boldsymbol{p}\lambda'} = \delta_{\lambda\lambda'}$. Note that we have introduced the dimensionless relativistically invariant parameter μ, defined as

$$\mu = \frac{|e|A_0}{m_\mathrm{e}c}. \tag{7}$$

This parameter is related to the relativistic character of ionization. If $\mu \ll 1$ then the quantum mechanical evolution of the field-particle interaction can be analyzed according to the Schrödinger equation. In contrast, if $\mu \approx 1$ or $\mu > 1$, which is the case studied in this paper, relativistic effects need to be accounted for and the time-evolution of the system needs to be studied according to the Dirac equation.

As it is shown in Ref. [5], the probability amplitude of ionization [Eq. (3)] for a laser pulse described by Eq. (4) can be written, in the velocity gauge, as

$$\mathcal{A}(\boldsymbol{p}, \lambda; \lambda_\mathrm{i}) = \int \frac{d^3q}{(2\pi)^3}\int d^4x\,\mathrm{e}^{\mathrm{i}S^{(+)}_p(x) - \mathrm{i}q\cdot x}M_{\lambda,\lambda_\mathrm{i}}(k\cdot x), \tag{8}$$

where

$$\begin{aligned} M_{\lambda,\lambda_\mathrm{i}}(k\cdot x) = \mathrm{i}m_\mathrm{e}c\mu\sqrt{\frac{m_\mathrm{e}c^2}{VE_{\boldsymbol{p}}}}\big[&f_1(k\cdot x)B^{(1,0)}_{\boldsymbol{p}\lambda;\lambda_\mathrm{i}}(\boldsymbol{q}) + f_2(k\cdot x)B^{(0,1)}_{\boldsymbol{p}\lambda;\lambda_\mathrm{i}}(\boldsymbol{q}) \\ &- \frac{m_\mathrm{e}c\mu}{2p\cdot n}([f_1(k\cdot x)]^2 + [f_2(k\cdot x)]^2)B^{(0,0)}_{\boldsymbol{p}\lambda;\lambda_\mathrm{i}}(\boldsymbol{q})\big]. \end{aligned} \tag{9}$$

Here, in order to simplify the notation, we have introduced the following functions related to the Fourier transform of the ground-state wavefunction $\Psi_\mathrm{i}(\boldsymbol{x})$ and the Dirac free particle bispinors,

$$B^{(0,0)}_{\boldsymbol{p}\lambda;\lambda_\mathrm{i}}(\boldsymbol{q}) = \bar{u}^{(+)}_{\boldsymbol{p}\lambda}\not{n}\tilde{\Psi}_\mathrm{i}(\boldsymbol{q}),\quad B^{(1,0)}_{\boldsymbol{p}\lambda;\lambda_\mathrm{i}}(\boldsymbol{q}) = \bar{u}^{(+)}_{\boldsymbol{p}\lambda}\not{\varepsilon}_1\tilde{\Psi}_\mathrm{i}(\boldsymbol{q}),\quad \text{and}\quad B^{(0,1)}_{\boldsymbol{p}\lambda;\lambda_\mathrm{i}}(\boldsymbol{q}) = \bar{u}^{(+)}_{\boldsymbol{p}\lambda}\not{\varepsilon}_2\tilde{\Psi}_\mathrm{i}(\boldsymbol{q}). \tag{10}$$

Now, defining the laser-dressed momentum $\bar{p}$ as [5, 8, 9, 10]

$$\bar{p} = p - \frac{m_\mathrm{e}c\mu}{p\cdot k}(\varepsilon_1\cdot p\langle f_1\rangle + \varepsilon_2\cdot p\langle f_2\rangle)k + \frac{(m_\mathrm{e}c\mu)^2}{2p\cdot k}(\langle f_1^2\rangle + \langle f_2^2\rangle)k, \tag{11}$$

the function $S_p^{(+)}(x)$ in Eq. (8) is rewritten as

$$S_p^{(+)}(x) = \bar{p}^+ x^- + p^- x^+ - \boldsymbol{p}^\perp \cdot \boldsymbol{x}^\perp + G_p(k^0 x^-), \tag{12}$$

where

$$G_p(\phi) = \int_0^\phi d\phi' \Big[-\frac{m_e c\mu}{p\cdot k}(\varepsilon_1 \cdot p(f_1(\phi') - \langle f_1 \rangle) + \varepsilon_2 \cdot p(f_2(\phi') - \langle f_2 \rangle)) + \frac{(m_e c\mu)^2}{2p\cdot k}(f_1^2(\phi') - \langle f_1^2 \rangle + f_2^2(\phi') - \langle f_2^2 \rangle)\Big]. \tag{13}$$

Note that in Eqs. (11) and (13) the average of a function $F(\phi)$, which vanishes for $\phi < 0$ and $\phi > 2\pi$, is given by

$$\langle F \rangle = \frac{1}{2\pi} \int_0^{2\pi} d\phi F(\phi). \tag{14}$$

According to Ref. [5], the multidimensional integral in Eq. (8) can be treated analytically with the help of the following Fourier transforms defined for $j = 1, 2$ and $0 \leqslant \phi \leqslant 2\pi$,

$$[f_1(\phi)]^j \exp[\mathrm{i}G_p(\phi)] = \sum_{N=-\infty}^{\infty} G_N^{(j,0)} \mathrm{e}^{-\mathrm{i}N\phi}, \tag{15}$$

$$[f_2(\phi)]^j \exp[\mathrm{i}G_p(\phi)] = \sum_{N=-\infty}^{\infty} G_N^{(0,j)} \mathrm{e}^{-\mathrm{i}N\phi}. \tag{16}$$

Hence, the probability amplitude of ionization $\mathcal{A}(\boldsymbol{p}, \lambda; \lambda_\mathrm{i})$ can be represented as

$$\mathcal{A}(\boldsymbol{p}, \lambda; \lambda_\mathrm{i}) = \mathrm{i} m_e c\mu \sqrt{\frac{m_e c^2}{V E_{\boldsymbol{p}}}} \mathcal{D}(\boldsymbol{p}, \lambda; \lambda_\mathrm{i}), \tag{17}$$

were $\mathcal{D}(\boldsymbol{p}, \lambda; \lambda_\mathrm{i})$ involves an infinite sum,

$$\mathcal{D}(\boldsymbol{p}, \lambda; \lambda_\mathrm{i}) = \sum_{N=-\infty}^{\infty} \frac{\mathrm{e}^{2\pi\mathrm{i}(\bar{p}^+ - q^+ - Nk^0)/k^0} - 1}{\mathrm{i}(\bar{p}^+ - q^+ - Nk^0)} \Big[G_N^{(1,0)} B_{\boldsymbol{p}\lambda;\lambda_\mathrm{i}}^{(1,0)}(\boldsymbol{Q}) + G_N^{(0,1)} B_{\boldsymbol{p}\lambda;\lambda_\mathrm{i}}^{(0,1)}(\boldsymbol{Q}) - \frac{m_e c\mu}{2p\cdot n}[G_N^{(2,0)} + G_N^{(0,2)}] B_{\boldsymbol{p}\lambda;\lambda_\mathrm{i}}^{(0,0)}(\boldsymbol{Q}) \Big], \tag{18}$$

and

$$\boldsymbol{Q} = \boldsymbol{p} + (q^0 - p^0)\boldsymbol{n}. \tag{19}$$

Taking into account Eq. (17) and keeping in mind that the final density of electron states, according to our current normalization conventions, is equal to $V d^3 p/(2\pi)^3$, the spin-dependent probability of ionization is given by

$$P(\lambda; \lambda_\mathrm{i}) = \mu^2 \frac{(m_e c)^3}{(2\pi)^3} \int \frac{d^3 p}{p^0} |\mathcal{D}(\boldsymbol{p}, \lambda; \lambda_\mathrm{i})|^2. \tag{20}$$

On the other hand, the initial-spin-averaged triply-differential probability distribution, which is obtained by averaging the values corresponding to the initial spin states and summing up the values corresponding to the final spin states, in atomic units, takes the form

$$\mathcal{P}(\boldsymbol{p}) = \frac{\alpha^2 m_e c^2}{2} \sum_{\lambda,\lambda_\mathrm{i}=\pm} \frac{d^3 P(\boldsymbol{p}, \lambda; \lambda_\mathrm{i})}{dE_{\boldsymbol{p}} d^2\Omega_{\boldsymbol{p}}} \equiv \frac{\alpha^2 \mu^2}{2} \frac{(m_e c)^4}{(2\pi)^3} \sum_{\lambda,\lambda_\mathrm{i}=\pm} |\boldsymbol{p}| \cdot |\mathcal{D}(\boldsymbol{p}, \lambda; \lambda_\mathrm{i})|^2, \tag{21}$$

where $\alpha = e^2/(4\pi\varepsilon_0 c)$ is the fine-structure constant.

Up to now we have calculated the probability distribution of photoionization of hydrogen-like systems in the RSFA framework and the velocity gauge. The fact that short laser pulses were considered assures that the electron ground-state wavefunction is well-defined and unperturbed before the interaction begins, even for arbitrarily intense laser fields. In contrast, when the infinite plane-wave approximation is used (see, e.g., Refs. [11, 12, 13]), the analysis is restricted to the interaction of highly charged positive ions with fields of moderate intensity. This is in order to guarantee that the ground-state wavefunction is not heavily distorted by the action of the oscillating laser field.

Even though Eqs. (17) and (18) allow us to calculate the energy and angular probability amplitude of ionization for given initial and final spin states, they do not offer a simple insight into the physics behind the process. For this reason, in the next Section, we introduce the saddle-point analysis of the integrals in Eq. (8) with the sole purpose of interpreting our numerical results.

3. Saddle-point approximation

To perform the saddle-point analysis of the integrals in the probability amplitude of ionization, we rewrite Eq. (8) in terms of the light-cone variables,

$$\mathcal{A}(\boldsymbol{p},\lambda;\lambda_{\rm i}) = \frac{1}{k^0}\int_0^{2\pi} d\phi \int \frac{d^3q}{(2\pi)^3}\int dx^+ d^2x^\perp \mathrm{e}^{\mathrm{i}(p^- - q^-)x^+ - \mathrm{i}(\boldsymbol{p}^\perp - \boldsymbol{q}^\perp)\cdot\boldsymbol{x}^\perp}\mathrm{e}^{\mathrm{i}G(\phi)} M_{\lambda,\lambda_{\rm i}}(\phi), \tag{22}$$

where $M_{\lambda,\lambda_{\rm i}}(\phi)$ is defined by (9) and

$$G(\phi) \equiv G(g_0, g_1, g_2, h; \phi) = \int_0^\phi d\phi' \big[g_0 + g_1 f_1(\phi') + g_2 f_2(\phi') + h\big(f_1^2(\phi') + f_2^2(\phi')\big)\big]. \tag{23}$$

Here we have introduced the functions

$$g_0 = \frac{p^+ - q^+}{k^0}, \quad h = \frac{(m_{\rm e}c\mu)^2}{2k\cdot p}, \quad \text{and} \quad g_j = -m_{\rm e}c\mu\frac{\varepsilon_j\cdot p}{k\cdot p}, \tag{24}$$

for $j = 1, 2$. The integration over $dx^+d^2x^\perp$ leads to the conservation relations

$$p^- = q^- \quad \text{and} \quad \boldsymbol{p}^\perp = \boldsymbol{q}^\perp, \tag{25}$$

and allows us to perform the integration over d^3q. Finally, the probability amplitude of ionization takes the form

$$\mathcal{A}(\boldsymbol{p},\lambda;\lambda_{\rm i}) = \frac{1}{k^0}\int_0^{2\pi} d\phi\, \mathrm{e}^{\mathrm{i}G(\phi)}\big[M_{\lambda,\lambda_{\rm i}}(\phi)\big]_{\boldsymbol{q}=\boldsymbol{Q}}, \tag{26}$$

where $\boldsymbol{Q}$ is defined in Eq. (19). Note that, according to the relations (25), $g_0 = (p^0 - q^0)/k^0$ and it only depends on the energy of the initial and final states.

As the function $\mathrm{e}^{\mathrm{i}G(\phi)}$ is considered to be fast oscillating compared to the remaining parts of the integrand in Eq. (26), the standard saddle-point method can be used to approximate this expression. The saddle points are obtained by solving the equation

$$\frac{\mathrm{d}G(\phi)}{\mathrm{d}\phi} = 0, \tag{27}$$

which, in general, has complex solutions. The only saddle points that contribute to the integral, denoted as ϕ_s, are those which satisfy the relation $\mathrm{Im}\, G(\phi_s) > 0$. With that in mind, the probability amplitude of ionization is approximated as

$$\mathcal{A}(\boldsymbol{p},\lambda;\lambda_{\rm i}) = \frac{1}{k^0}\sum_s \mathrm{e}^{\mathrm{i}G(\phi_s)}\sqrt{\frac{2\pi\mathrm{i}}{G''(\phi_s)}}\big[M_{\lambda,\lambda_{\rm i}}(\phi_s)\big]_{\boldsymbol{q}=\boldsymbol{Q}}. \tag{28}$$

By analyzing the previous expression, it is clear that the interference pattern in photoionization arises when two or more saddle points contribute importantly to the probability amplitude $\mathcal{A}(\boldsymbol{p}, \lambda; \lambda_\text{i})$. In contrast, if just one of them is dominant over a range of photon energies and emission angles [i.e., if the $\operatorname{Im} G(\phi_s)$ is considerable small compared to the corresponding value of the remaining saddle points], then no strong interference effects are expected. Therefore, we anticipate that the supercontinuum should appear in the energy range for which just one saddle point contributes the most to Eq. (28). This was demonstrated for the case of fixed initial and final spin states in Ref. [5].

4. Numerical calculations

We consider the photoionization of He^+ ions (with atomic number $Z = 2$) by a relativistically-intense and circularly polarized laser pulse. The latter is characterized by the electric field shape functions with a $\sin^2$ envelope defined as

$$F_j(\phi) = F_0(\phi, \delta_j, \chi)\cos(\delta + \delta_j), \tag{29}$$

with

$$F_0(\phi, \delta_j, \chi) = N_0 \sin^2\Big(\frac{\phi}{2}\Big)\sin(N_\text{osc}\phi + \delta_j + \chi) \tag{30}$$

for $0 < \phi < 2\pi$ and 0 otherwise. Here, N_osc represents the number of field oscillations within the pulse, χ is the carrier-envelope phase, δ and δ_j determine the polarization properties of the pulse, and N_0 is a normalization constant chosen such that the average intensity of the field is independent of the number of cycles (see, Ref. [5]). The function F_0, explicitly written as a function of time, is

$$F_0(\boldsymbol{r}, t, \delta_j, \chi) = N_0 \sin^2\Big(\frac{1}{2N_\text{osc}}\omega_\text{L}(t - \boldsymbol{n}\cdot\boldsymbol{r}/c)\Big)\sin(\omega_\text{L}(t - \boldsymbol{n}\cdot\boldsymbol{r}/c) + \delta_j + \chi), \tag{31}$$

for $0 < t - \boldsymbol{n}\cdot\boldsymbol{r}/c < T_\text{p}$ and it is 0 otherwise. Here we have introduced the carrier frequency of the laser field, $\omega_\text{L} = N_\text{osc}\omega$.

For the numerical calculations presented below we have chosen a circularly polarized laser pulse ($\delta_1 = 0$, $\delta_2 = \pi/2$, and $\delta = \pi/4$), propagating along the z-axis ($\boldsymbol{n} = \boldsymbol{e}_z$), and comprising four field oscillations within the $\sin^2$ envelope ($N_\text{osc} = 4$). The polarization vectors are $\boldsymbol{\varepsilon}_1 = \boldsymbol{e}_x$ and $\boldsymbol{\varepsilon}_2 = \boldsymbol{e}_y$. The carrier-envelope phase is $\chi = \pi/2$ and the carrier frequency is $\omega_\text{L} = 20\text{eV}$.

As the electric field is related to the vector potential by the relation $\boldsymbol{\mathcal{E}}(\phi) = -\partial_t \boldsymbol{A}(\phi)$, the shape functions $f_1(\phi)$ and $f_2(\phi)$ in Eq. (4) are calculated as

$$f_j(\phi) = -\int_0^\phi d\phi' F_j(\phi'), \tag{32}$$

with $j = 1, 2$.

In Fig. 1, we present the time evolution of the tips of the vector potential (left panel) and electric field (right panel) in the xy-plane (which is perpendicular to the laser field propagation direction) for the pulse described above with an average intensity of $I = 4 \times 10^{20}\ \text{W/cm}^2$. Both curves start at the origin of coordinates and evolve counterclockwise during the ramp up (blue color) and ramp down (red color), and are presented in relativistic units. The parameter $A_\text{S} = m_\text{e}c/|e|$ and the ratio $A(\phi)/A_\text{S}$ is given by

$$\frac{A(\phi)}{A_\text{S}} = \mu[\varepsilon_1 f_1(\phi) + \varepsilon_2 f_2(\phi)]. \tag{33}$$

Note that, for this particular intensity, $|A(\phi)|/A_\text{S}$ reaches values larger than the unity, so one needs to consider the problem in a fully relativistic way. On the other hand, the parameter

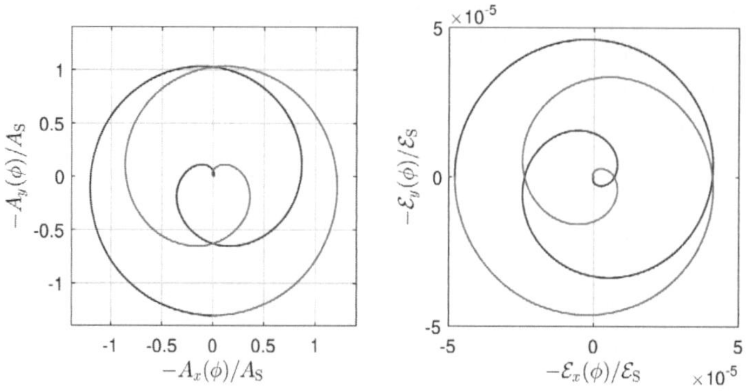

Figure 1. (Color online) Trajectories of the tips of the electromagnetic vector potential $\boldsymbol{A}(\phi)$ (left panel) and the electric field vector $\boldsymbol{\mathcal{E}}(\phi)$ (right panel), in relativistic units, for the laser pulse discussed below. All trajectories start from and end up at the origin $(0,0)$. In order to show the direction of the time-evolution, we mark with colors blue and red the ramp up and ramp down parts of the laser pulse, respectively. We observe the azimuthal symmetry of the electromagnetic potential, $\varphi \to \pi - \varphi \bmod 2\pi$ or $(x,y) \to (-x,y)$, and of the electric field, $\varphi \to -\varphi \bmod 2\pi$ or $(x,y) \to (x,-y)$. The time-averaged intensity of the laser pulse is $I = 4 \times 10^{20}$ W/cm^2, the carrier laser frequency is $\omega_{\rm L} = 20$ eV, and $N_{\rm osc} = 4$. Note that for these parameters, $\mu > 1$.

$\mathcal{E}_{\rm S} = m_{\rm e}^2 c^3/|e|$, known as the Sauter-Schwinger critical electric field (see, Refs. [5, 14] and references therein), is related to the probability of electron-positron pair production. As the ratio $|\boldsymbol{\mathcal{E}}(\phi)|/\mathcal{E}_{\rm S} \ll 1$, such effect can be ignored in our calculations.

4.1. Energy and polar-angle spectra of photoelectrons
In the upper panels of Fig. 2 we present the energy spectra of photoelectrons calculated according to Eq. (21). The results are plotted for the electron asymptotic momentum $\boldsymbol{p}$ with azimuthal angle $\varphi_{\boldsymbol{p}} = 0$ and polar angles $\theta_{\boldsymbol{p}} = 0.5\pi$ (left panel) and $\theta_{\boldsymbol{p}} = 0.48\pi$ (right panel). While the averaged intensity is $I = 2 \times 10^{20}$ W/cm^2, the remaining parameters are the same as in Fig. 1. We observe that the energy spectra of photoelectrons present a single and very broad structure which ranges from 3 keV up to 20 keV (corresponding to hundreds of single-photons energy), without the distinctive signatures of interference; the supercontinuum. Moreover, the distributions exhibit a maximum located at positions depending on the polar angle. For $\theta_{\boldsymbol{p}} = 0.5\pi$, the spectral maximum appears at electron kinetic energies close to 10.96 keV, whereas for $\theta_{\boldsymbol{p}} = 0.48\pi$ it is shifted to 11.17 keV.

It follows from our analysis in Sec. 3 and the results shown in Ref. [5] that the presence of a supercontinuum in the energy distributions can be related to absence of interference effects, i.e., to the existence of a single saddle point which contributes the most to the probability amplitude of ionization (28). In the lower panels of Fig. 2 we present the energy dependence of $\operatorname{Im} G(\phi_s)$ for all relevant saddle points. It can be seen that, for the parameters chosen in our calculations, there is just one of such points for which $\operatorname{Im} G(\phi_s)$ is much smaller than the others (magenta curves in the lower panels). Note that its minimum is located around the region for which the probability distribution is maximal.

In Fig. 3, we present the same as in Fig. 2 but for $I = 4 \times 10^{20}$ W/cm^2 and kinetic energies ranging from 10 keV up to 35 keV. One can see that, for those parameters, supercontinua are also formed. This time the maxima are located at 20.78 keV for $\theta_{\boldsymbol{p}} = 0.5\pi$ and 21.83 keV for $\theta_{\boldsymbol{p}} = 0.48\pi$. Therefore, we conclude that the position of the maximum in the energy spectra of photoelectrons scales linearly with the averaged intensity of the laser pulse (see,

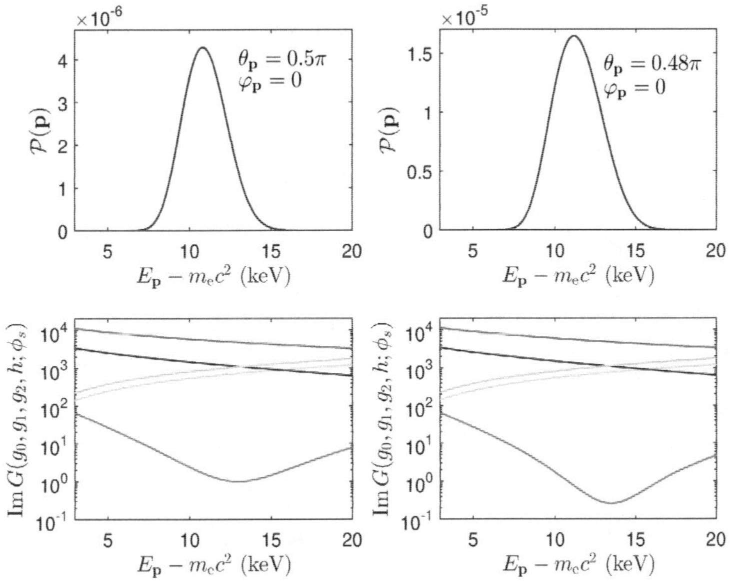

Figure 2. (Color online) Initial-spin-averaged energy spectra of photoelectrons (upper row) for the time-averaged intensity $I = 2 \times 10^{20}$ W/cm^2 and for the emission angles indicated in the figure. In the lower row, the appropriate plots of $\mathrm{Im}\, G(\phi_s)$ for all five relevant saddle points are shown.

the results corresponding to $I = 10^{20}$ W/cm^2 presented in Ref. [5]). Note that just one saddle point contributes the most to the probability amplitude of ionization and $\mathrm{Im}\, G(\phi_s)$, related to this particular saddle point, has a minimum around the region where the distribution is maximum (magenta curves in the lower panels). Moreover, comparing Figs. 2 and 3, one can see a considerable reduction of the maximum probability distribution (from one to two orders of magnitude, depending on the polar angle), which can be attributed to stabilization against ionization (see, e.g., Refs. [15, 16, 17, 18, 19, 20, 21]).

In Fig. 4, the polar-angle spectra of photoelectrons at fixed kinetic energy are presented. The constant azimuthal angle has been chosen to be $\varphi_p = 0$ and the averaged intensities are $I = 2 \times 10^{20}$ W/cm^2 and $I = 4 \times 10^{20}$ W/cm^2 (upper left and right panels, respectively). One can see that, in both cases, a broad structure is formed. Additionally, as it was discussed in Refs. [5, 19, 22], the non-relativistic SFA predicts a maximum value of the distribution at $\theta_p = 0.5\pi$ for circularly polarized laser fields, which is not the case in our numerical calculations. One can see from the upper left panel in Fig. 4 that the actual maximum is located at $\theta_p = 0.467\pi$, i.e., it is shifted towards the direction of propagation of the laser field. Moreover, when the intensity is increased (upper right panel), the shifting is more pronounced and the maximum appears at $\theta_p = 0.455\pi$. Such effect has been attributed to the radiation pressure exerted by the laser field on the emitted photoelectrons [5, 10, 19, 22]. On the other hand, one can clearly see that the maximum value of the distribution depends on the laser field intensity. More precisely, the maximum scales as the inverse of the averaged intensity squared, which is another indication of stabilization against ionization. Finally, in the lower panels of Fig. 4 we present the plots of $\mathrm{Im}\, G(\phi_s)$ for all relevant saddle points. As expected, just one of them contributes importantly to the probability amplitude of ionization (magenta curves in the lower panels). In this case, their minima appear near to the angular regions for which the distribution acquires maximum values.

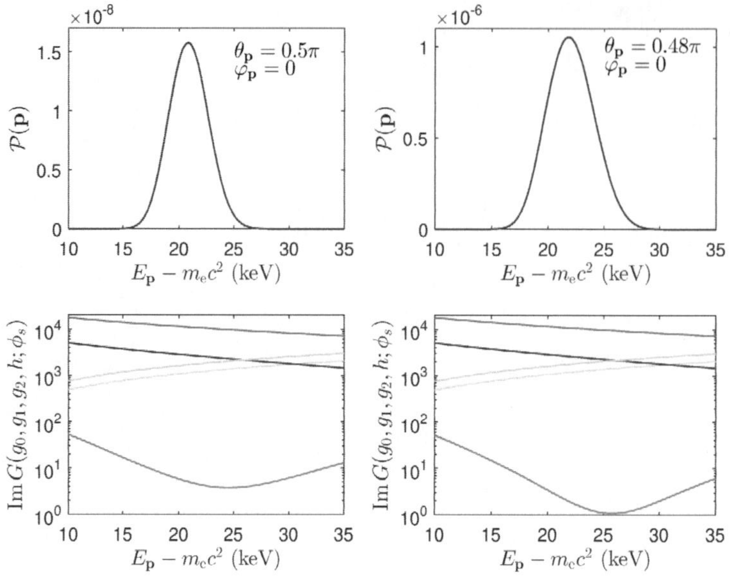

Figure 3. (Color online) The same as Fig. 2 but for the time-averaged intensity $I = 4 \times 10^{20}$ W/cm^2.

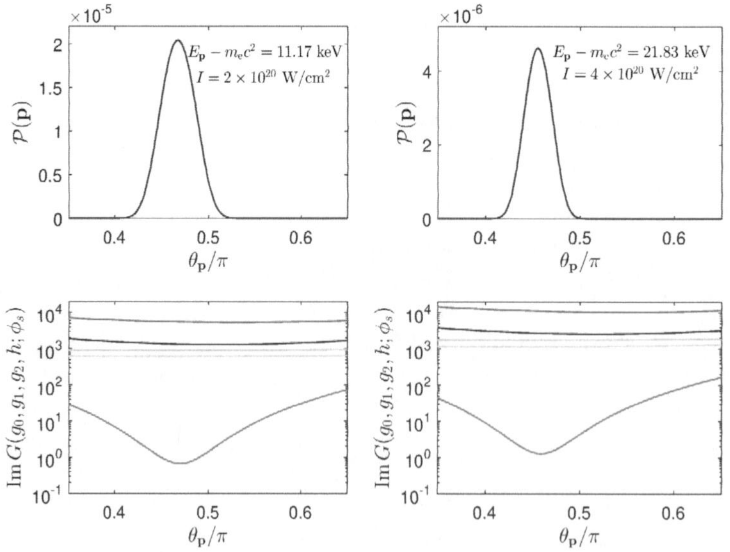

Figure 4. (Color online) Initial-spin-averaged polar-angle distributions (upper row) for $\varphi_{\boldsymbol{p}} = 0$ and for two time-averaged intensities: $I = 2 \times 10^{20}$ W/cm^2 (left column, maximum for $\theta_{\boldsymbol{p}} = 0.467\pi$) and $I = 4 \times 10^{20}$ W/cm^2 (right column, maximum for $\theta_{\boldsymbol{p}} = 0.455\pi$) with energies indicated in the figure. The maximum of these distributions scales as the inverse of intensity squared. In the lower row, we present the plots corresponding to $\mathrm{Im}\, G(\phi_s)$ for all five relevant saddle points.

5. Conclusions

We have analyzed the probability distribution of photoelectrons obtained from the interaction of relativistically-intense and short laser pulses with hydrogen-like ions under the RSFA framework. Our treatment is applicable even to very light ions, as it was illustrated for He^+, due to the fact that the ground-state wavefunction is well-defined before the interaction with the pulse. Furthermore, we have demonstrated that, by adjusting the parameters of the driving laser field, the energy spectrum of photoelectrons can exhibit a supercontinuum. Using the saddle-point approximation, we have related such broad structure to energy regions without interference (i.e., regions for which just one saddle point contributes importantly to the probability amplitude of ionization). Contrary to the results presented in Ref. [5], we have considered the initial-spin-averaged probability distributions, without restricting ourselves to fixed initial and final spin states.

In our numerical calculations we have shown that the position of the maximum of the energy spectra of photoelectrons increases linearly with the averaged intensity of the driving field. As well, its maximum value decreases with intensity due to stabilization against ionization. The polar-angle distribution presents a maximum at $\theta_{\boldsymbol{p}} < 0.5\pi$ due to the radiation pressure exerted by the laser field. Moreover, an increase of the averaged intensity of the pulse leads to a maximum located at smaller polar angles.

6. Acknowledgements

This work is supported by the National Science Centre (Poland) under Grant No. 2014/15/B/ST2/02203.

References

[1] Dudley J M, Genty G and Coen S 2006 *Rev. Mod. Phys.* **78** 1135
[2] Krajewska K, Twardy M and Kamiński J Z 2014 *Phys. Rev.* A **89** 032125
[3] Keldysh L V 1964 *Zh. Eksp. Teor. Fiz.* **47** 1945
[4] Popov V S 2004 *Phys. Usp.* **47** 855
[5] Krajewska K and Kamiński J Z 2016 *Phys. Rev.* A **94** 013402
[6] Bjorken J D and Drell S D 1964 *Relativistic Quantum Mechanics* (New York: McGraw-Hill)
[7] Krajewska K and Kamiński J Z 2010 *Phys. Rev.* A **82** 013420
[8] Krajewska K and Kamiński J Z 2012 *Phys. Rev.* A **85** 062102
[9] Krajewska K and Kamiński J Z 2012 *Phys. Rev.* A **86** 052104
[10] Krajewska K and Kamiński J Z 2015 *Phys. Rev.* A **92** 043419
[11] Klaiber M, Yakaboylu E and Hatsagortsyan K Z 2013 *Phys. Rev.* A **87** 023418
[12] Klaiber M and Hatsagortsyan K Z 2014 *Phys. Rev.* A **90** 063416
[13] Yakaboylu E, Klaiber M and Hatsagortsyan K Z 2015 *Phys. Rev.* A **91** 063407
[14] Salamin Y I, Hu S X, Hatsagortsyan K Z and Keitel C H 2006 *Phys. Rep.* **427** 41
[15] Gavrila M and Kamiński J Z 1984 *Phys. Rev. Lett.* **52** 613
[16] Offerhaus M J, Kamiński J Z and Gavrila M 1985 *Phys. Lett.* A **112** 151
[17] Fedorov M V and Movsesian A M 1988 *J. Phys.* B **21** L155
[18] Pont M and Gavrila M 1990 *Phys. Rev. Lett.* **65** 2362
[19] Crawford D P and Reiss H R 1994 *Phys. Rev.* A **50** 1844
[20] de Boer M P, Hoogenraad J H, Vrijen R B, Constantinescu R C, Noordam L D and Muller H G 1994 *Phys. Rev.* A **50** 4085
[21] Popov A M, Tikhonova O V and Volkova E A 2003 *J. Phys.* B **36** R125
[22] Reiss H R 1990 *J. Opt. Soc. Am.* B **7** 574

IOP Conf. Series: Journal of Physics: Conf. Series **826** (2017) 012011 doi:10.1088/1742-6596/826/1/012011

Quantum-Plasma Model of Atomic Cluster Excitations

A. El-Khawaldeh and H.-J. Kull

Institute for Theory of Statistical Physics, Laser Physics Group, RWTH Aachen University, 52062 Aachen, Germany

E-mail: amir.el@rwth-aachen.de

Abstract. The laser-light absorption of a thin foil is studied in the framework of the self-consistent time-dependent Hartree theory. It is found that asympotically for large numbers of electron states, a single-state Vlasov model is sufficient to describe the main features of the energy absorption to a very good approximation. Since the single-state model is numerically feasible, this allows one to calculate the energy absorption over a wide range of laser parameters. Studying the size-dependence of the absorption mechanism, quantum-mechanical modifications of the well-known classical Brunel theory can be observed for thin layers.

1. Introduction

This article summarizes parts of the results which were presented at the twenty fifth annual International Laser Physics Workshop (LPHYS'16) in Yerevan, Armenia. In the context of our presentation on the topic 'Quantum-Plasma Model of Atomic Cluster Excitations' we gave an overview on the results we achieved based on quantum-plasma calculations on spherical atomic clusters and thin plasma foils. This paper refers to the results we obtained for thin plasma foils.

The statistical properties of ideal quantum plasmas are commonly described in the framework of the self-consistent quantum Vlasov theory which determines the time evolution of the single-particle density operator. This method corresponds to a quantum-mechanical multi-stream model in which the many-particle system can be described by a finite set of single-particle wave functions interacting via a self-consistent potential [1]. Another effective many-body theory closely related to the Vlasov theory is the time-dependent Hartree theory.

In this work, both models are applied to a thin plasma foil interacting with an external laser field which is considered to be obliquely incident on the foil surfaces. A reduced capacitor model [2] effectively leads to a spatially one-dimensional (1d) description of the electron dynamics. The classical Brunel theory [3] describes the collisionless energy absorption of a plane surface by electrons which are pulled into the vacuum and accelerated in the presence of the laser field. This mechanism dominates the absorption if the solid has a sharp surface and an electron density well above the critical density where the light frequency equals the plasma frequency. In the present work, we focus attention on this non-resonant collisionless absorption. Brunel's theory predicts a characteristic scaling $E_{\text{abs}} \sim \mathcal{E}_0^3$ of the absorbed energy with the electric field amplitude $\mathcal{E}_0$ of the laser. The absorption mechanism suggested by Brunel has been analyzed classically for thin foils with two surfaces in the framework of a 1d capacitor model [4]. However, if the foil thickness approaches the nanoscale, quantum-size effects of the absorption which have not been treated so far are expected to become relevant. In this regime, the so-called spill-out effect, which describes the occurrence of a non-vanishing electron density outside the foil, is a non-thermal quantum feature. In this work, we analyze how Brunel's scaling law is affected by quantum effects in thin plasma foils.

2. Multi-Stream Model

In the present model, the background ions are treated in the jellium approximation as a homogeneously charged foil with thickness l and charge density en_0, where e denotes the elementary charge and $n_{ion} = n_0/Z$ is the density of ions with average charge state Z. The following dimensionless quantities are used for a plasma with plasma frequency $\omega_p = \sqrt{4\pi n_0 e^2/m_e}$ (CGS units),

$$\boldsymbol{r} = \sqrt{\frac{m_e\omega_p}{\hbar}}\boldsymbol{r}^\star, \quad t = \omega_p t^\star, \quad n = \frac{n^\star}{n_0}, \quad E = \frac{E^\star}{\hbar\omega_p}, \\ q = \frac{q^\star}{e}, \quad \varphi = \varphi^\star\left(\frac{\hbar}{m_e\omega_p}\right)^{\frac{1}{2}}, \tag{1}$$

where $\hbar$ is the reduced Planck constant. Dimensional quantities are denoted by a star. The prefactors in (1) define the units used throughout this work if not otherwise stated. In some figures, we use the length scale $L = \sqrt{\hbar/(m_e\omega_p)}$.

For the description of the electron system, we consider two different approaches. On the one hand, we consider the well-known time-dependent Hartree theory which describes the electron system in terms of N_s single-particle states $\varphi_k(x,t)$ interacting via individual self-consistent potentials $\phi_k(x,t)$,

$$i\partial_t\,\varphi_k = \left[-\frac{1}{2}\partial_x^2 - \phi_k - \phi_L\right]\varphi_k, \tag{2a}$$

$$\partial_x^2\phi_k = \frac{1}{N_s}\sum_{k'\neq k}|\varphi_{k'}|^2 - \rho_{ion}, \tag{2b}$$

where ϕ_L corresponds to the external laser potential in the dipole approximation and $\rho_{ion}(x) = \theta(l/2 - |x|)$ is the ionic charge density. The Hartree theory excludes self-interaction in the calculation of the electronic charge density in (2b). It should be noted that in the dimensionless representation (2) the single-particle states are normalized to the length of the foil, $\langle\varphi_k|\varphi_k\rangle = l$. On the other hand, we consider the simple single-state Vlasov model

$$i\partial_t\,\varphi = \left[-\frac{1}{2}\partial_x^2 - \phi - \phi_L\right]\varphi, \tag{3a}$$

$$\partial_x^2\phi = |\varphi|^2 - \rho_{ion}, \tag{3b}$$

where the electron system is described by a single electron state $\varphi(x,t)$ interacting with a self-consistent potential $\phi(x,t)$. The single state model (3) corresponds to a reduced version of the quantum Vlasov theory which describes the electron system by a statistical ensemble of N_s single-particle wave functions $\varphi_k(x,t)$ which we call representative electron states [5]. The Vlasov theory reduces to the above single-state model (3) when taking into account only a single representative state, $N_s = 1$. It is noted that the main difference between the Hartree and the Vlasov model arises from the fact that self-energy interaction terms are not included in the Poisson's equations of the Hartree theory.

For the numerical evaluation of the systems (2) respectively (3), a Crank-Nicolson scheme [6] is chosen for the Schrödinger equations in combination with standard finite-difference techniques for Poisson's equation.

3. Results

In this section, we summarize some results obtained for the equilibrium configurations, the single-state Vlasov approximation and finally for the scaling law of the energy absorption.

3.1. Equilibrium States

In the absence of the laser field, $\phi_L = 0$, solutions of the systems (2) and (3) with a time-independent electron density can be written in the form

$$\varphi_k(x,t) = \varphi_k^{(0)}(x)e^{-i\epsilon_k t}. \tag{4}$$

Then, the equilibrium density is given by

$$n^{(0)} = \frac{1}{N_s}\sum_k |\varphi_k^{(0)}|^2. \tag{5}$$

To determine the equilibrium states $\varphi_k^{(0)}$, we choose a relaxation method [7]. Figure 1 shows the equilibrium densities for a foil with thickness $l = 50$. Comparison is made between the single-state model and the Hartree model for different N_s. Furthermore, results for the self-consistent potentials $\phi_k^{(0)}$ in the equilibrium state are shown.

As depicted, the electron density within the Hartree model becomes homogeneous inside the foil as the number of representative states is chosen sufficiently large. The same behavior of the electron density inside the foil can be observed for the single-state Vlasov model.

However, the individual equilibrium states $\varphi_k^{(0)}$ look significantly different from the equilibrium wave function of the single-state Vlasov model. The single-state equilibrium wave function has the same structure as the density. In the case of the Hartree model, each peak of the density is associated with a single-electron equilibrium state k which is localized around the mean position $\langle x\rangle_k$. To get a qualitative interpretation of the equilibrium states and to compare the results to corresponding classical calculations, analytical solutions for the Hartree equilibrium states are calculated asymptotically for $N_s/l \to 0$. In this limit, the overlap between the electron states is negligible and the equilibrium becomes a set of coherent states of a harmonic oscillator separated by a distance l/N_s,

$$\varphi_k(x) = \frac{1}{\sqrt[4]{\pi}}e^{-0.5(x-x_k)^2}, \qquad x_k = -\frac{l}{2}\left(1+\frac{1-2k}{N_s}\right), \qquad k = 1,2,\ldots,N_s. \tag{6}$$

This result suggests that the Hartree model (2) can be interpreted as the quantum-mechanical analogue of the classical particle model defined in [4]. Thereby, the equation of motion for the point charge k is replaced by the Schrödinger equation for the wave function φ_k. Since the electron density of a single wave function within the Hartree description is not point-like, the calculation of the electrostatic potential is not as simple as for point-particles.

The self-consistent potential of the single-state model exhibits two pronounced minima close to the foil surface which generate the binding of the electrons. The local minima of the potential occur due to the overshoot of the electron density at the foil surface. Within the Hartree model, each electron state sees an individual potential with a single local minimum. As one can see, the potential ϕ_k increases linearly far away from the region electron k is localized to. This fact is related to the 1d nature of the underlying model. According to Poisson's equation in 1d, a localized charge distribution $\rho(x)$ generates a constant electric field $|\mathcal{E}| = |q|/2$ outside the region of its localization. Thus, since a Hartree state sees a charge distribution with total charge $q = l/N_s$, there is no ionization in the present Hartree model and an outgoing electron will always be accelerated back to the foil. This feature of the 1d description is not present in

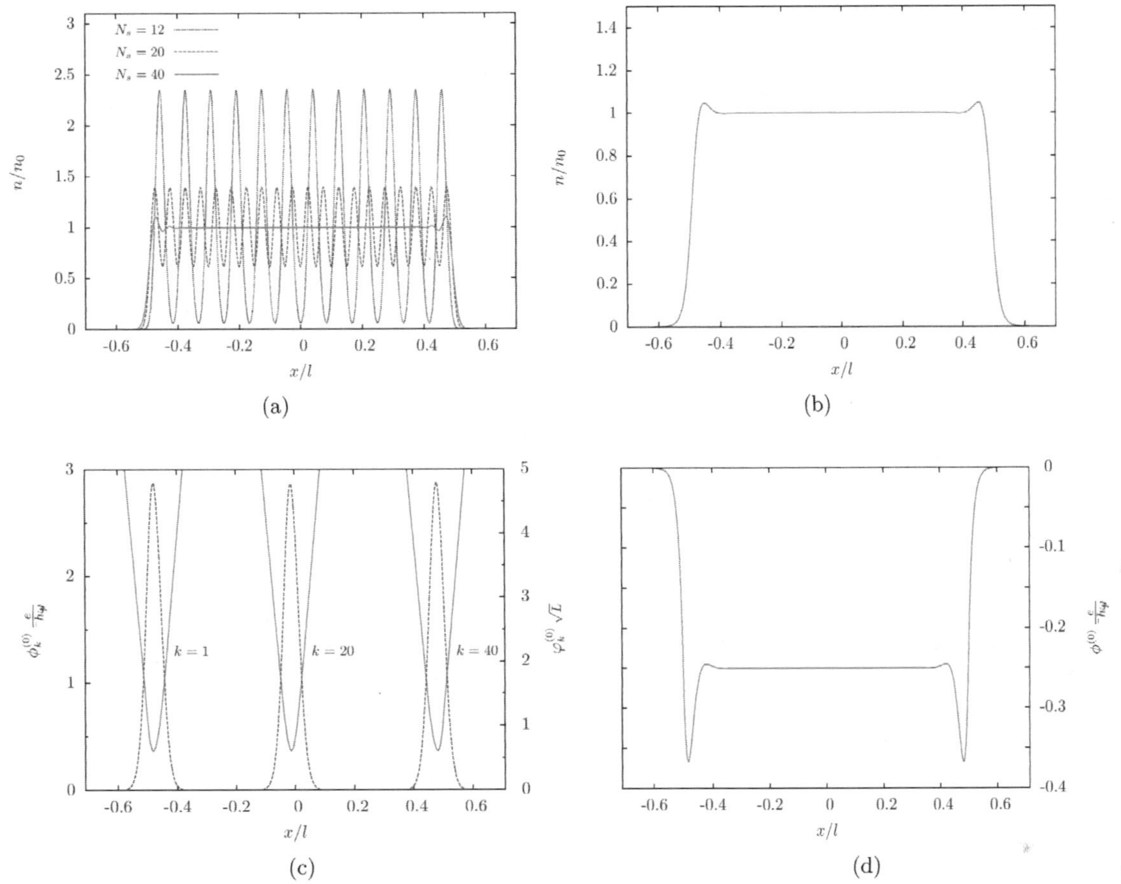

Figure 1. Equilibrium quantities for a foil with thickness $l = 50$. The figure shows the equilibrium densities for (a) the Hartree and (b) the single-state model. Furthermore, exemplary results for the self-consistent potentials in the equilibrium state are shown for (c) the Hartree model with $N_s = 40$ and (d) the single-state model. In the case of the Hartree model, the corresponding equilibrium wave functions $\varphi_k^{(0)}$ are indicated.

the single-state model since the self-consistent potential is calculated from a neutrally charged distribution.

Furthermore, figures 1a and 1b clearly show the spill-out effect of the electron density at the foil surface. To investigate the regime of l-values for which quantum effects are suggested to become significant, we analyze how the spill-out effect depends on the size of the foil. Figure 2 shows the fraction of spill-out electrons f_{out} as a function of the foil thickness in the single-state model. One calculates for $l = 5$ that 13% of the electrons are located outside the foil in the equilibrium state. For $l \gtrsim 5$, it is found that f_{out} scales with the inverse foil thickness. A fit to the data with $l \gtrsim 5$ using the model function $f_{out}(l) = \alpha/l^{\beta}$ yields the values $\alpha = 0.659$ and $\beta = 0.997$. In the following, the plasma foil is exposed to a laser field. Before the interaction with the laser at $t = 0$, the electron system is considered to be in the equilibrium state.

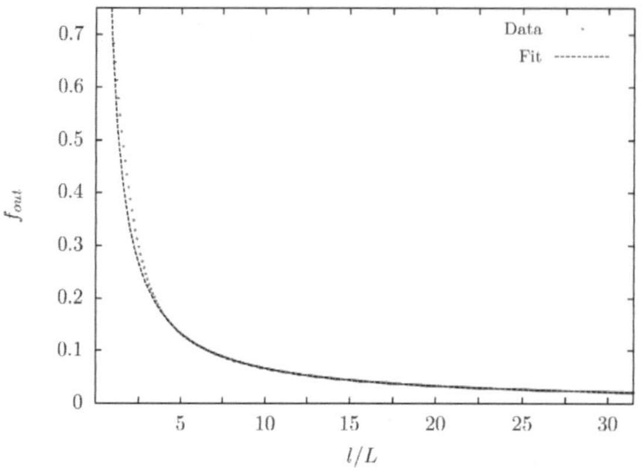

Figure 2. Fraction of spill-out electrons as a function of the foil thickness. The data points are fitted to the inverse power law $f_{out} = 0.659/l^{0.997}$.

3.2. Comparison of Hartree Model and Single-State Model

The numerical calculation of the electron dynamics in the presence of strong laser fields based on the Hartree model (2) requires on the one hand large spatial grids to account for fast electrons moving far away from the foil and on the other hand small numerical grid constants with respect to the time variable to resolve their dynamics. These requirements are already present in classical PIC calculations. Since the present quantum-mechanical description furthermore requires small spatial grid constants to resolve small de Broglie wavelengths, calculations with large N_s become time-consuming. Since we are interested in macroscopic quantities like the absorbed energy per electron, it is an instructive task to analyze two major questions. Does the dynamics of the electron system converge if the number of electron states N_s is chosen sufficiently large? If yes, is the single-state Vlasov model capable of describing the main features of the energy absorption? In particular, the last point is of great interest since calculations within the single-state Vlasov model are numerically feasible which allows one to perform calculations over a wide range of laser parameters. In order to investigate the above questions, we performed test calculations within the Hartree model for different N_s and the single-state model in the presence of a strong laser field compared to the electric field generated by the ions.

In the following, the laser field is considered in the electric dipole approximation. A $\sin^2$-pulse with n laser cylces, frequency ω and pulse duration $\tau = 2\pi n/\omega$ is chosen for the electric field of the laser,

$$\phi_L(x,t) = -x\mathcal{E}(t), \quad \mathcal{E}(t) = \mathcal{E}_0 \sin(\pi t/\tau)^2 \sin(\omega t). \tag{7}$$

The mean total energy of the foil per electron is calculated based on the expectation values of the single-particle energies,

$$\frac{E_{tot}}{N_e} = \frac{1}{N_s}\sum_{k=1}^{N_s} \frac{\langle\varphi_k|E_k|\varphi_k\rangle}{\langle\varphi_k|\varphi_k\rangle} + \frac{E_{ion}}{N_e}, \tag{8}$$

$$E_k = T + V_k, \quad T = -\frac{1}{2}\frac{\partial^2}{\partial x^2}, \quad V_k = -\phi^{(ion)} - \frac{1}{2}\phi_k^{(e)}. \tag{9}$$

where E_{ion} is the potential energy of the ion system in its own field which is a constant value within the jellium approximation. The operator T corresponds to the kinetic energy operator. The potentials $\phi^{(ion)}$ and $\phi_k^{(e)}$ specify the ion and the electron contribution to the self-consistent potential in (2b) respectively (3b), $\phi_k = \phi^{(ion)} + \phi^{(e)}$. It is noted that the single-particle energy operator E_k does not coincide with the single-particle Hamiltonian in the present self-consistent description. The prefactor 1/2 in the definition of the potential energy V_k has to be taken into account in order to avoid double-counting of the electron-electron interaction energy.

For the test calculation, a foil with $l = 10$ is considered. Since we want to perform the calculations over a wide range of N_s-values, we choose a sufficiently short single-cycle pulse ($n = 1$) with $\omega = 0.125$ and $\mathcal{E}_0/\mathcal{E}_{ion} = 40\%$ in order to make the calculation numerically affordable. Here, $\mathcal{E}_{ion} = l/2$ corresponds to the maximal value of the ionic field which is achieved at the foil boundary. Figure 3 shows the dynamics of the center of mass (COM)

$$X(t) = \frac{1}{N_s} \sum_{k=1}^{N_s} \frac{\langle \varphi_k | \hat{x} | \varphi_k \rangle}{\langle \varphi_k | \varphi_k \rangle} \tag{10}$$

as a function of time as well as the absorbed energy $E_{abs} = E_{tot}(\tau) - E_{tot}(0)$ per electron for different N_s. For the considered value of $\mathcal{E}_0$, the grid constants have to be chosen quite small, $\Delta x \lesssim 0.01$ and $\Delta t \lesssim 0.001$, in order to observe convergence of the numerical solution. To avoid boundary effects driven by outgoing electrons, one has to choose a spatial grid with about 10^5 grid points.

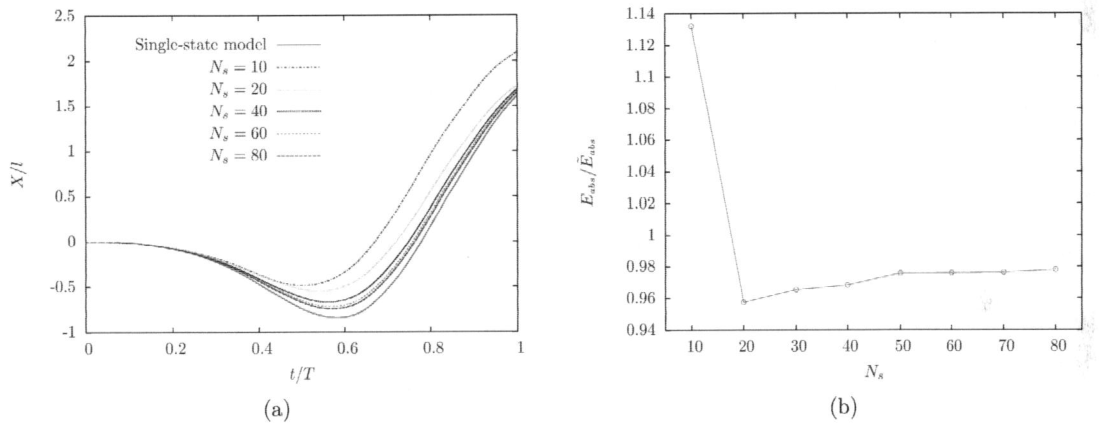

Figure 3. Motion of (a) the center of mass and (b) the absorbed energy per electron for a laser pulse with $n = 1$, $\omega = 0.125$ and $\mathcal{E}_0/\mathcal{E}_{ion} = 40\%$. The results are shown for different numbers of electron states N_s within the Hartree model and the single-state Vlasov model. In (b), the absorbed energy is measured in units of the energy $\tilde{E}_{abs}$ absorbed within the single-state Vlasov model.

Figure 3a shows that the electrons are pulled in the negative direction within the first half of the laser cycle. Thereby, a large number of electrons leave the foil. Within the second half of the cycle, electrons are accelerated back and gain enough energy to move through the foil. A large displacement of the electron system with respect to the equilibrium state can be observed at the end of the pulse.

As depicted in Figure 3b, the absorbed energy deviates from the single-state result by more than 10% for the smallest considered value of $N_s = 10$. As shown in Figure 1, also the corresponding equilibrium density looks pretty different from the single-state result. However, for $N_s > 10$ one can see that the deviation of the absorbed energy is less than 5% and varies only slightly over the range of considered N_s-values. For the largest considered value of $N_s = 80$, the Hartree result differs from the single-state result by an amount of approximately 2%. This result suggests that the single-state model is able to reproduce the absorption of the many-particle system to a good approximation if the number N_s is chosen sufficiently large.

3.3. Energy Absorption as a Function of the Laser Field Strength

In this section, we analyze the energy absorption of the foil as a function of the field strength of the laser. Brunel described the electron dynamics at a single sharp surface by an ensemble of single-particle trajectories interacting via an electrostatic potential [3]. As a major result, it was found that the absorbed energy of the surface per electron can be written in the form

$$\frac{E_{abs}}{N_e} = \frac{1}{2}\eta \frac{\mathcal{E}_0^3}{\omega^2}\frac{1}{l}, \tag{11}$$

where η is a numerical constant which depends on the pulse duration. It should be noted that the result of Brunel is given in (11) in terms of the plasma units (1), where the length dependence is due to these units. The absorption mechanism of vacuum heating for a foil with two sharp surfaces has been analyzed classically [4]. It was found that the absorbed energy can be represented by a scaling law similar to (11) but with a different numerical value for η. From a classical point of view, the characteristic scaling of the absorbed energy with the third power of $\mathcal{E}_0$ in the non-resonant case can be motivated based on the assumption that the energy absorption is mainly driven by electrons which leave the foil. Since our quantum results will be compared to the classical results of Brunel, we shortly summarize the main points. For the estimation of the absorbed energy, we approximate the COM-motion by the equation of motion without surface effects,

$$\ddot{X} + X = -\mathcal{E}(t). \tag{12}$$

In the non-resonant limit $\omega \ll 1$, the solution reads $X(t) = -\mathcal{E}(t)$ with a maximal displacement of $X_0 = \mathcal{E}_0$. Since the equilibrium density is spatially homogeneous in the classical description, the maximum number N_{out} of electrons which leave the foil over one laser cycle is given by

$$N_{out} = n_0 X_0^\star = \frac{N_e}{l}\mathcal{E}_0. \tag{13}$$

Electrons which are pulled into the vacuum are accelerated by the laser field and gain an amount of energy proportional to the mean quiver energy $U_p \sim \mathcal{E}_0^2$ of a free electron in a laser field, known as the ponderomotive potential. With this consideration, the absorbed energy exhibits a scaling of the form $E_{abs} \sim \mathcal{E}_0^3$.

In the following, we analyze the energy absorption of the foil as a function of the laser field strength. Due to the fact that the cycle-averaged absorbed energy is not expected to vary significantly during the main part of a few-cycle pulse [4], we choose for simplicity a single-cycle pulse and an overdense foil with $\omega = 0.1$. Figure 4 shows the absorbed energy per electron as a function of $\mathcal{E}_0$ for two foils with $l = 5$ and $l = 50$. On the double-logarithmic scale, one can clearly identify two regimes of $\mathcal{E}_0$-values for $l = 5$ which can be represented by scaling laws $E_{abs}/N = \alpha\,\mathcal{E}_0^\beta$. A fit yields the values $\alpha = 5.65$, $\beta = 3.33$ for $\mathcal{E}_0/\mathcal{E}_{ion} \lesssim 0.2$ and $\alpha = 5.13$, $\beta = 2.98$ for $\mathcal{E}_0/\mathcal{E}_{ion} \gtrsim 0.4$. The two fits are also indicated in Figure 4. Two scaling regimes can also be found for the thicker foil with $l = 50$. A fit yields the values $\alpha = 0.43$, $\beta = 3.15$ for $\mathcal{E}_0/\mathcal{E}_{ion} \lesssim 0.02$ and $\alpha = 0.4$, $\beta = 3.07$ for $\mathcal{E}_0/\mathcal{E}_{ion} \gtrsim 0.04$.

In summary, for sufficiently large values of $\mathcal{E}_0$, one can observe a scaling which is pretty close to the one predicted within Brunel's theory with a scaling exponent close to 3. However, for the smaller foil one can clearly see a deviation from the classical behavior over a wide range of $\mathcal{E}_0$-values. Since this effect is less pronounced for the large foil, we conclude that the deviation can be referred to quantum-size effects which become relevant for small foils. To investigate how the increased scaling exponent of E_{abs}/N_e emerges for small foils, we compare the scaling of the maximum number of free electrons (13) in Brunel's theory to a corresponding value within the present quantum-mechanical treatment. As a measure for the fraction of free electrons, we

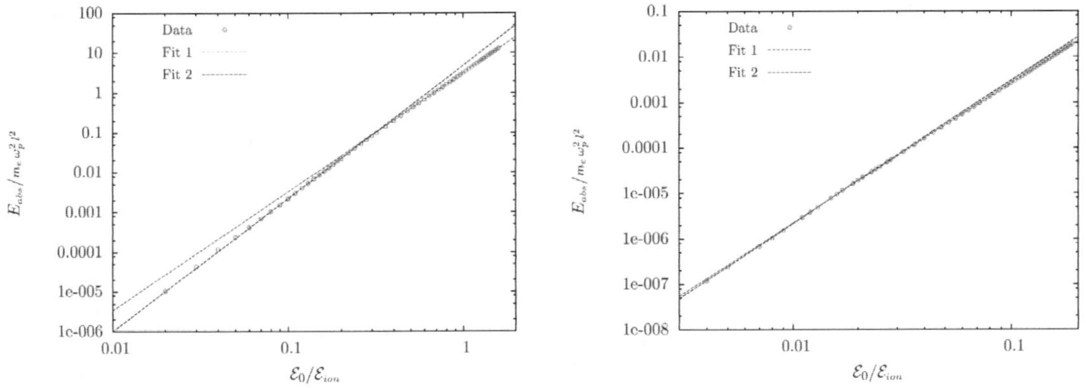

(a) $l = 5$ (b) $l = 50$

Figure 4. Absorbed energy per electron as a function of the laser field strength $\mathcal{E}_0$ for two foils with thickness $l = 5$ and $l = 50$. A laser pulse with $n = 1$ and $\omega = 0.125$ is chosen. The results are shown for different field strengths $\mathcal{E}_0$ measured with respect to the maximal value of the ionic field $\mathcal{E}_{ion}$. The data points are fitted to model functions of the form $E_{abs}/N = \alpha\,\mathcal{E}_0^{\beta}$.

calculate the time-averaged fraction of electrons $\bar{f}_{out}$ which are located outside the foil during the laser-foil interaction,

$$\bar{f}_{out} = \frac{1}{\tau}\int_0^{\tau} f_{out}(t)\,dt\,, \quad f_{out}(t) = 1 - \frac{1}{N_e}\int_{-\frac{L}{2}}^{\frac{L}{2}} n(x,t)dx. \tag{14}$$

Figure (5) shows the results for the two foils with thickness $l = 5$ and $l = 50$. One clearly

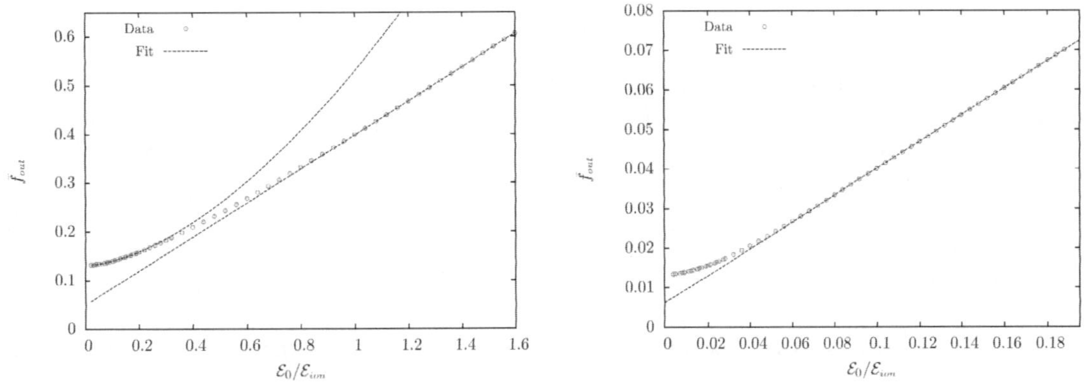

(a) $l = 5$ (b) $l = 50$

Figure 5. Time-averaged fraction of electrons which are located outside the foil during the interaction with the laser ($\omega = 0.125, n = 1$). The results are shown for different laser field strengths $\mathcal{E}_0$ measured with respect to the maximal value of the ionic field $\mathcal{E}_{ion}$. The data points for strong laser fields are fitted to a linear function.

identifies a linear dependence of $\bar{f}_{out}$ on $\mathcal{E}_0$ for sufficiently high values of $\mathcal{E}_0$. Thereby, one recognizes that the linear dependence is already fulfilled for $\mathcal{E}_0/\mathcal{E}_{ion} \gtrsim 0.06$ in the case of the large foil with $l = 50$, while it requires much stronger fields $\mathcal{E}_0/\mathcal{E}_{ion} \gtrsim 1$ compared to the ionic field to observe a linear dependence in the case of the small foil with $l = 5$. To investigate the increase of $\bar{f}_{out}$ in the regime of $\mathcal{E}_0$-values for which a faster increase of the absorbed energy can be observed, a fit is performed to the data with $\mathcal{E}_0/\mathcal{E}_{ion} \lesssim 0.2$ using the model function $\bar{f}_{out}(\mathcal{E}_0) = \alpha\,\mathcal{E}_0^\beta + \gamma$. The fit yields the values $\alpha = 0.088$, $\beta = 1.65$, $\gamma = 0.13$ and is also indicated in Figure 5a. Here, γ corresponds to the fraction of spill-out electrons in the equilibrium state. The result indicates that $\bar{f}_{out}$ grows faster as a function of $\mathcal{E}_0$ for small fields, where a faster increase of the absorbed energy can be observed compared to the Brunel-like behavior for $\mathcal{E}_0/\mathcal{E}_{ion} \gtrsim 1$.

The different behavior of $\bar{f}_{out}$ for thin foils can be understood in terms of the surface profiles of the equilibrium density. Decreasing the foil thickness, the transition region of the equilibrium density shown in Figure 1b is broadened. We consider a shift of the equilibrium distribution by an amount ξ along the x-axis, where $N_{out}(\xi)$ is the number of electrons which leave the foil due to the displacement. In the classical case, one has $N_{out} \sim \xi$ due to the homogeneity of the equilibrium density inside the foil. If the electron density is broadened, the expansion of $N_{out}(\xi)$ with respect to ξ also contains higher orders of ξ. However, if the deflection is chosen sufficiently large such that the inhomogeneous part of the equilibrium density is already located outside the foil, the density distribution becomes homogeneous at the foil boundary and a further increase $d\xi$ generates a number of outgoing electrons $dN_{out} \sim d\xi$.

The scaling law

$$\frac{E_{abs}}{N_e} = a(l)\,\mathcal{E}_0^b \tag{15}$$

we found for large values of $\mathcal{E}_0$, with values of b close to 3 and a constant a which depends on the foil thickness, allows for a comparison with the corresponding classical result calculated in [4]. The authors find a scaling law similar to the result (11) of Brunel with a numerical constant of $\eta = 8.75$ for a laser pulse with $n = 10$ laser cycles. The corresponding value for a single-cycle pulse is given by $\eta = 0.875$, where we divided the result $\eta = 8.75$ suggested in [4] by the number of laser cycles. For the comparison with our results, we evaluate the a-parameter of the classical result (11) with the laser frequency $\omega = 0.125$ we used in our calculations,

$$a = \frac{0.875}{2}\frac{1}{0.125^2}\frac{1}{l} = \frac{28}{l}. \tag{16}$$

We calculate the a-parameter based on our calculations for different values of l. The results are shown in Figure 6. A model function $a(l) = \alpha/l^\beta$ is fitted to the data which is also presented in Figure 6. We obtain the fit parameters $\alpha = 30.1$ and $b = 1.09$. One can see that the found value of α is close to the value 28 suggested by the classical result (16).

4. Conclusion

In the presented work, we have analyzed the energy absorption of a thin foil based on the Hartree model and a single-state Vlasov model. We found that the energy absorption within the Hartree model can be described to a very good approximation by the simple Vlasov model if the number of electron states is chosen sufficiently large. For a foil with thickness $l = 10$ and $N_s = 80$ electron states, we have found that the Hartree and the single-state result for the absorbed energy differ only by 2%.

At sufficiently high laser fields, we found that the energy absorption can be represented by a simple scaling law $E_{abs} \sim \mathcal{E}_0^b$, where the numerical value of b is close to the value 3 predicted by the classical Brunel theory independent of the foil thickness. For thin foils where quantum-size effects are relevant, we found that the scaling exponent is larger than the value $b = 3$ over a wide range of laser field strengths.

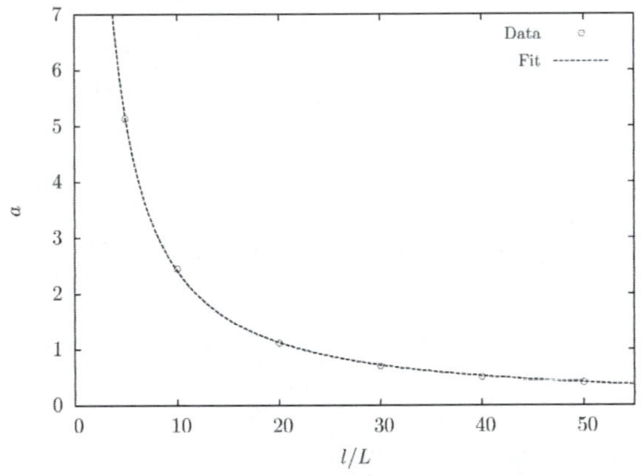

Figure 6. Scaling parameter a as a function of the layer thickness l. The data points are fitted to the inverse power law $a = 30.1/l^{1.09}$.

References

[1] Haas F, Manfredi G and Feix M 2000 *Phys. Rev. E* **62**(2) 2763–2772
[2] Forslund D W, Kindel J M, Lee K and Lindman E L 1976 *Phys. Rev. Lett.* **36**(1) 35–38
[3] Brunel F 1987 *Phys. Rev. Lett.* **59**(1) 52–55
[4] Greschik F, Dimou L and Kull H J 2000 *Laser and Particle Beams* **18** 367–373
[5] Schmidt-Bleker A, Gassen W and Kull H J 2011 *EPL (Europhysics Letters)* **95** 55003
[6] Crank J and Nicolson P 1947 *Mathematical Proceedings of the Cambridge Philosophical Society* **43**(01) 50–67
[7] Goldberg A and Schwartz J L 1967 *Journal of Computational Physics* **1** 433 – 447

IOP Conf. Series: Journal of Physics: Conf. Series **826** (2017) 012012 doi:10.1088/1742-6596/826/1/012012

Quantum Two-Stream Instability with Exchange Interaction

H.-J. Kull

Institute for Theory of Statistical Physics, Laser Physics Group, RWTH Aachen University, 52056 Aachen, Germany

E-mail: kull@ilt-extern.fraunhofer.de

Abstract. Quantum plasmas can be described in the framework of the self-consistent field approach by a set of time-dependent Schrödinger equations (TDSE) for the electrons in a self-consistent potential. This approach is also well-known as the multistream model of quantum plasmas. The self-consistent field is commonly calculated in the Vlasov approximation from the average charge density of all electrons thereby neglecting the quantum-mechanical exchange interaction. In the present work, we consider a linear Hartree-Fock model with screened Coulomb interactions to examine the role of the exchange interaction. The two-stream instability is analysed and results for the Vlasov, Hartree and Hartree-Fock approximations are compared in detail. It is found that the exchange interaction considerably modifies the stability behavior for weakly screened Coulomb interactions. In particular, crossings of the two branches of the dispersion relation lead to oscillating unstable solutions that are not present in the absence of the exchange interaction.

1. Introduction

The study of strong-field laser-atom interactions has led to increased interest in the solution of the time-dependent Schrödinger equation (TDSE) for single-electron and multi-electron atoms. For many-electron problems, on the other hand, direct solution methods become computationally very expensive, although they are presently explored in the fields of warm dense matter and high-energy density plasmas. Another standard approach to many-electron systems is the self-consistent field method. Here the electron-electron interaction is replaced by an effective interaction of each electron with the medium. Variants of this approach are the Vlasov-theory, the Hartree-theory and the density functional theory (DFT). An approximate formulation of the quantum Vlasov theory is provided by the multistream model [1–5], which is a quantum-mechanical version of the classical beam model by Dawson [6]. Here the single-particle statistical operator is approximated by an ensemble of representative quantum states. The numerical solution of the TDSE for representative states with an effective self-consistent potential is feasible in many cases. Using this method, we have recently studied the linear [7] and nonlinear dynamics [5] of quantum plasmas in the weak-coupling regime. The reduced single- and two-stream models have also attracted considerable attention in the past [3; 8–11].

In this work, we wish to discuss the role of the exchange interaction in this context, which has mostly been ignored in previous analyses. General discussions on the exchange corrections in degenerate electron gases can be found in [12] and [13]. It is well-known that the wavefunction of identical particles has to satisfy the postulate of symmetry for bosons and of antisymmetry for fermions. In the simplest case the electronic wavefunction can be chosen as a Slater determinant and by minimizing a variational principle for the TDSE, the familiar Hartree-Fock equations are obtained [14; 15]. These equations include the exchange interaction, which is not present in the Vlasov or the Hartree approximations and is only treated on the average in most DFT formulations. To explore the role of the exchange term, we study a simple analytically solvable model of the two-stream instability. It is our goal, to compare for this particular case in detail the Vlasov, the Hartree and the Hartree-Fock approaches to gain more insight in the role of the exchange interaction in multistream plasma treatments.

2. Two-Stream Model

In this section, we briefly summarize the Hartree-Fock equations of the multistream-model of quantum plasmas. We then restrict attention to the two-stream model and derive the basic equations for small amplitude perturbations. These equations will be further considered in the subsequent section to discuss the two-stream instability under various conditions.

2.1. Multi-Stream Equations

We consider an infinitely extended plasma with a homogeneous ion density. The multistream model of the electrons consists of an ensemble of N carrier-envelope waves with envelopes $A_s(\boldsymbol{r}, t)$, carrier momenta $\boldsymbol{p}_s$ and carrier energies E_s,

$$\psi_s(\boldsymbol{r}, t) = A_s(\boldsymbol{r}, t) e^{i\varphi_s}, \qquad \varphi_s = \frac{1}{\hbar}(\boldsymbol{p}_s \cdot \boldsymbol{r} - E_s t), \qquad s = 1, 2, 3, \cdots, N, \tag{1}$$

where $\hbar$ denotes the reduced Planck constant. These wavefunctions are assumed to be mutually orthonormal. The spin dependence of the streams is omitted for simplicity, since we will only treat an electron-pair with parallel spins and neglect any spin interactions. Using a Slater determinant of single-electron wavefunctions as a trial-function in the variational formulation of the Schrödinger equation, one arrives at the set of time-dependent Hartree-Fock equations,

$$i\hbar\partial_t \psi_s = -\frac{\hbar^2}{2m}\Delta\psi_s + q\phi_i\psi_s + q\sum_r (\phi_{rr}\psi_s - \phi_{rs}\psi_r), \tag{2}$$

where ϕ_i denotes the potential produced by the ions, q is the charge, m the mass of the electron, and the electron-electron interaction is described by a set of effective potentials ϕ_{rs}. The diagonal terms ϕ_{rr} describe the classical direct interaction, while the off-diagonal terms ϕ_{rs}, $r \neq s$ correspond to the quantum exchange interaction. It can be seen that the self-interaction for $r = s$ is excluded since the direct term is canceled by the exchange term. We also mention that the Hartree-Fock theory is a generalization of simpler approaches. In the Vlasov theory the exchange interaction is absent and the direct interaction commonly includes the self-interaction. In the Hartree theory the direct interaction is calculated by omitting the self-interaction. In this sense, the Hartree theory accounts for the $r = s$ exchange term, while all other exchange terms are omitted.

The effective potentials can be expressed by well-known integral representations or, equivalently, by a set of Poisson equations,

$$\Delta\phi_{rs} - k_0^2(1 - \delta_{rs})\phi_{rs} = -4\pi q n_{rs}, \qquad n_{rs} = \psi_r^*\psi_s. \tag{3}$$

For $r \neq s$, we have added to the standard Poisson equation a second term on the l.h.s, which corresponds to the screening of the Coulomb interaction in the Thomas-Fermi approximation. Screening is an important mechanism that competes with the exchange-interaction in plasmas. We can take the Thomas-Fermi wavenumber k_0 just as a parameter to control the strength of the exchange interaction. A self-consistent treatment of screening within the Hartree-Fock theory is beyond the scope of this work.

In the following, we will use dimensionless quantities. With the density n_0 of the positive elementary charges and the plasma frequency $\omega_p = \sqrt{4\pi q^2 n_0/m}$ in CGS-Gaussian units, the dimensionless variables are defined by

$$\tilde{\psi}_s = \frac{1}{\sqrt{n_0}}\psi_s, \quad \tilde{\phi}_{rs,i} = \frac{|q|}{\hbar\omega_p}\phi_{rs,i}, \quad \tilde{\boldsymbol{r}} = \sqrt{\frac{m\omega_p}{\hbar}}\boldsymbol{r}, \quad \tilde{k}_0 = k_0\sqrt{\frac{\hbar}{m\omega_p}}, \quad \tilde{t} = \omega_p t. \tag{4}$$

Omitting the tilde-symbol, the dimensionless Hartree-Fock and Poisson equations become

$$i\partial_t\psi_s = -\frac{1}{2}\Delta\psi_s - \phi_i\psi_s - \sum_r \phi_{rr}\psi_s - \phi_{rs}\psi_r, \tag{5a}$$

$$\Delta\phi_{rs} - k_0^2(1-\delta_{rs})\phi_{rs} = n_{rs}, \qquad n_{rs} = \psi_r^*\psi_s. \tag{5b}$$

In the present units, the wavefunctions are normalized by

$$\langle\psi_r|\psi_s\rangle = \int d^3r\; n_{rs} = \frac{1}{n_0}\delta_{rs}. \tag{5c}$$

2.2. Two-Stream Equilibrium

In this work, we will restrict attention to an analytically solvable idealized two-electron system. Let us consider two electron streams, occupied by just a pair of electrons of equal spin such that the wavefunction of both electrons has to be antisymmetric. The Hartree-Fock equations for the two streams then become a system of two coupled Schrödinger equations,

$$i\partial_t\psi_1 = -\frac{1}{2}\Delta\psi_1 - (\phi_i + \phi_{22})\psi_1 + \phi_{21}\psi_2, \tag{6a}$$

$$i\partial_t\psi_2 = -\frac{1}{2}\Delta\psi_2 - (\phi_i + \phi_{11})\psi_2 + \phi_{12}\psi_1. \tag{6b}$$

We first consider the equilibrium state of this system and specifically assume plane-wave solutions as given in (1) with equal and opposite momenta

$$\psi_1^{(0)} = e^{i\varphi_1}, \qquad \psi_2^{(0)} = e^{i\varphi_2}, \qquad \boldsymbol{p}_{1,2} = \pm\boldsymbol{p}. \tag{7}$$

This symmetrically counterstreaming configuration is time-reversible and as a consequence the 4th-order dispersion relation for the excitation frequency ω will become a simpler quadratic equation for ω^2. The charge density of each electron, $-|\psi_{1,2}^{(0)}|^2 = -1$, is homogeneously distributed and is taken equal and opposite to the charge density +1 of the ion. The direct interaction of each electron with the other one is thereby canceled by the ion interaction,

$$\phi_{11}^{(0)} = \phi_{22}^{(0)} = -\phi_i. \tag{8}$$

The whole plasma system has one excess negative charge. This excess charge will become of minor importance for multistream models with many electrons. In a strictly neutral equilibrium, the equilibrium wavefunctions would be more complicated due to the presence of the binding potential of the ion background. Inserting (7) into the Poisson equation, one obtains the exchange potentials

$$\phi_{21}^{(0)} = \phi_{12}^{(0)*} = -\frac{1}{|\boldsymbol{p}_1 - \boldsymbol{p}_2|^2 + k_0^2}\,\psi_2^{(0)*}\psi_1^{(0)}. \tag{9}$$

Inserting (7) and (9) into the Hartree-Fock equations, one arrives at the energy-momentum relations

$$E_{1,2} = \frac{p_{1,2}^2}{2} - \frac{1}{|\boldsymbol{p}_1 - \boldsymbol{p}_2|^2 + k_0^2} . \tag{10}$$

The kinetic energy is reduced by the exchange energy. For $k_0 = 0$, the latter one becomes singular for $\boldsymbol{p}_1 \to \boldsymbol{p}_2$. To avoid such singularities the screening constant k_0 is assumed to be finite. It is noted that these single-electron energies can be summed up in the continuum limit and then lead to the well-known Hartree-Fock result for the energy of the free-electron gas.

2.3. Two-Stream Perturbations

Perturbations of the equilibrium state will be treated in linear-order. For this purpose the carrier-envelope waves (1) are written in the form

$$\psi_s = (1 + \chi_s(\boldsymbol{r}, t))\psi_s^{(0)}, \qquad \chi_s(\boldsymbol{r}, t) = h_s(\boldsymbol{r})e^{-i\omega t} . \tag{11}$$

The envelope perturbation $\chi_s(\boldsymbol{r}, t)$ may be viewed as the perturbation in the interaction picture, since it evolves relative to the unperturbed stationary state. The set of linearized equations can be Fourier-transformed with respect to the spatial coordinates. The Fourier-transforms of χ_s and χ_s^* within the volume V are defined by

$$\hat{\chi}_s = \frac{1}{V} \int d^3r \ \chi_s(\boldsymbol{r})e^{-i\boldsymbol{k}\cdot\boldsymbol{r}}, \qquad \check{\chi}_s = \frac{1}{V} \int d^3r \ \chi_s^*(\boldsymbol{r})e^{-i\boldsymbol{k}\cdot\boldsymbol{r}}. \tag{12}$$

Linearizing the Hartree-Fock equations (6) with these perturbations, one obtains the system of equations

$$(\omega - \Omega_1)\hat{\chi}_1 = -G(\boldsymbol{k})(\hat{\chi}_2 + \check{\chi}_2) + G(\boldsymbol{k} + 2\boldsymbol{p})(\hat{\chi}_1 + \check{\chi}_2) - G(2\boldsymbol{p})(\hat{\chi}_1 - \hat{\chi}_2), \tag{13a}$$

$$(\omega + \Omega_2)\check{\chi}_2 = -G(\boldsymbol{k})(\hat{\chi}_1 + \check{\chi}_1) + G(\boldsymbol{k} - 2\boldsymbol{p})(\check{\chi}_2 + \hat{\chi}_1) - G(2\boldsymbol{p})(\check{\chi}_2 - \check{\chi}_1), \tag{13b}$$

$$(\omega - \Omega_2)\check{\chi}_1 = G(\boldsymbol{k})(\check{\chi}_2 + \hat{\chi}_2) - G(\boldsymbol{k} - 2\boldsymbol{p})(\hat{\chi}_2 + \check{\chi}_1) + G(2\boldsymbol{p})(\check{\chi}_1 - \check{\chi}_2), \tag{13c}$$

$$(\omega + \Omega_1)\check{\chi}_2 = G(\boldsymbol{k})(\check{\chi}_1 + \hat{\chi}_1) - G(\boldsymbol{k} + 2\boldsymbol{p})(\hat{\chi}_1 + \check{\chi}_2) + G(2\boldsymbol{p})(\check{\chi}_2 - \check{\chi}_1), \tag{13d}$$

with

$$\Omega_{1,2} = k(p \pm \frac{k}{2}), \qquad G(\boldsymbol{x}) = -\frac{1}{x^2 + k_0^2}.$$

The frequencies $\Omega_{1,2}$ are known as the frequencies of single-particle excitations, since they are asymptotic solutions for $G \to 0$. In general, the single-particle excitations are coupled by the direct interaction $G(\boldsymbol{k})$ and by the exchange interactions $G(\boldsymbol{k} \pm 2\boldsymbol{p}))$ and $G(2\boldsymbol{p})$.

For comparison with the Hartree and Vlasov theories, we comment on the corresponding equations. In the Hartree approximation one has only a direct interaction of each electron with the other one. This interaction is given by the terms

$$U_{H1,2} = \mp G(\boldsymbol{k})(\hat{\chi}_{2,1} + \check{\chi}_{2,1}). \tag{14}$$

The corresponding set of equations is obtained from (13) by omitting there the exchange terms, setting $G(\boldsymbol{k} \pm 2\boldsymbol{p})) = 0$ and $G(2\boldsymbol{p}) = 0$.

In the Vlasov approximation, the same terms are omitted, however, the self-interaction is not excluded. The equilibrium charge density is exactly zero, since both electrons contribute to the effective electrostatic potential. Setting the corresponding ion charge to +2, the effective interaction for both electrons is the average of the two effective interactions in the Hartree model,

$$U_V = \frac{1}{2}\left[U_{H1} + U_{H2}\right]. \tag{15}$$

In the following, we will discuss results from the solution of the Hartree-Fock equations (13) and make comparison with the corresponding Hartree and Vlasov results derived with the reduced interactions (14) and (15), respectively.

3. Results

The basic equations (13) for small amplitude perturbations will now be analysed in turn for the Vlasov, the Hartree and the Hartree-Fock approximations.

3.1. Vlasov Model

The two-stream instability has been analysed in the past for the Vlasov approach [3] and we will briefly summarize these results. Using the perturbation equations (13) with the effective interactions given by (15), one can easily obtain a single equation

$$D_V \Sigma_V = 0, \tag{16}$$

with

$$\Sigma_V = \hat{\chi}_1 + \check{\chi}_1 + \hat{\chi}_2 + \check{\chi}_2, \tag{17a}$$

$$D_V = 1 - \frac{1}{k^2}\left(\frac{\Omega_1}{\omega^2 - \Omega_1^2} - \frac{\Omega_2}{\omega^2 - \Omega_2^2}\right). \tag{17b}$$

The variable Σ_V represents the Fourier-transformed density perturbation, which is also proportional to the Fourier-transformed electrostatic potential. The mode frequencies are obtained from the solubility condition $D_V = 0$, which can be written as a quadratic equation in ω^2,

$$\omega^4 - P\omega^2 + Q = 0. \tag{18}$$

Its zeros are given by

$$\omega_{1,2}^2 = \frac{P}{2} \pm \sqrt{\frac{P^2}{4} - Q}, \tag{19a}$$

$$P = \Omega_1^2 + \Omega_2^2 + 1 = 1 + 2k^2\left(p^2 + \frac{k^4}{4}\right), \tag{19b}$$

$$Q = \Omega_1\Omega_2(\Omega_1\Omega_2 - 1) = k^2\left(p^2 - \frac{k^2}{4}\right)\left[k^2\left(p^2 - \frac{k^2}{4}\right) - 1\right]. \tag{19c}$$

The parameter region of the two-stream instability can be readily determined by looking at (19) and convincing that $P > 0$ and $P^2 > 4Q$. The upper branch ω_1^2 is always positive, while the lower branch ω_2^2 becomes negativ for $Q < 0$. The instability criterion is therefore

$$Q < 0 \iff 0 < \Omega_1\Omega_2 < 1, \iff p^2 < \frac{1}{k^2} + \frac{k^2}{4} \quad \text{and} \quad p > \frac{k}{2}. \tag{20}$$

The quantum Vlasov result (19) reduces to the classical one by the substitution $\Omega_{1,2} = kp$, thereby omitting all the terms with the factor $1/4$ in the definitions of P and Q. In the classical limit the instability criterion becomes $p < \frac{1}{k}$. This classical instability boundary is the asymptote to the upper quantum-mechanical boundary for small wave numbers. The quantum version has another lower boundary $p = k/2$ that is related to the zero of the single-particle excitation energy $\Omega_2 = 0$, when the wave with momentum $\boldsymbol{k}$ can excite transitions $\boldsymbol{p} \longleftrightarrow -\boldsymbol{p}$ between counterpropagating beams of equal energy. These instability boundaries in the kp-plane are shown in Figure 1. One can recognize, that the quantum treatment leads to a second unstable k-interval above the minimum $k = \sqrt{2}$, $p = 1$ of its upper boundary.

For small wavenumbers, the two solution branches can be classified as a stable plasma mode and as the two-stream instability

$$\omega_1^2 \to 1 + 3k^2p^2, \qquad \omega_2^2 \to -k^2p^2, \qquad \text{for} \qquad k \to 0. \tag{21}$$

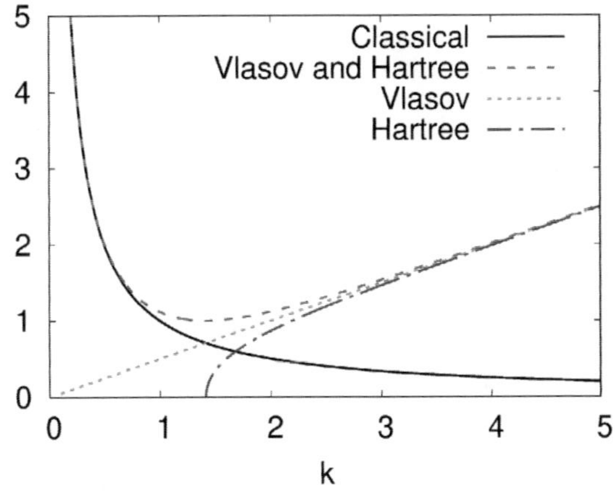

Figure 1. Stability boundaries of the two-stream instability with beam momenta $\pm p$ and wave momentum k. Classically the unstable region is the area below the boundary $p = 1/k$ (solid). Quantum-mechanically, there is another instability region near the line $p = k/2$ where the single-particle excitation frequency Ω_2 vanishes. The instability regions of the quantum Vlasov and quantum Hartree models have the same upper boundary (dashed) but different lower boundaries (dotted and dotted-dashed).

For large wavenumbers, outside the instability region, both branches approach stable single-particle excitations

$$\omega_{1,2}^2 \to \Omega_{1,2}^2, \qquad \text{for} \qquad k \to \infty. \tag{22}$$

For large wavenumbers, the instability region near $\Omega_2 = 0$ can be defined by $p - \frac{k}{2} = \frac{c}{k^3}$, with the parameter c in the interval $0 < c < 1$. The parameters P and Q then can be expanded as $P = k^4 + O(1)$, $Q = c(c-1) + O(k^{-4})$. The unstable branch for large wavenumbers then can be found to be

$$\omega_2^2 = \frac{Q}{P} = -\frac{c(1-c)}{k^4}, \qquad \text{for} \qquad k \to \infty, \quad 0 < c < 1. \tag{23}$$

3.2. Hartree Model

Next, we discuss results of the Hartree model. In this model each electron experiences only the average field of the other electron as given by (14). The analysis of this model is instructive, since it represents the limit of the Hartree-Fock model for strong screening of the exchange interaction.

Eliminating the variables of one electron from (13) , one obtains for the other electron the condition

$$D_H \Sigma_H = 0, \qquad \Sigma_H = \hat{\chi}_1 + \check{\chi}_1, \tag{24a}$$

$$D_H = 1 - \frac{1}{(\omega^2 - \Omega_1^2)(\omega^2 - \Omega_2^2)}. \tag{24b}$$

The solubility condition $D_H = 0$ leads again to a quadratic equation of the form of (18). Its coefficients are given by

$$P = \Omega_1^2 + \Omega_2^2 = 2k^2p^2 + \frac{1}{2}k^4, \tag{25}$$

$$Q = \Omega_1^2\Omega_2^2 - 1 = \left[k^2\left(p^2 - \frac{k^2}{4}\right)^2\right]^2 - 1. \tag{26}$$

The instability criterion is again given by $Q < 0$ with the corresponding boundaries

$$p^2 = \pm\frac{1}{k^2} + \frac{k^2}{4}. \tag{27}$$

The upper boundary is the same as in the Vlasov case. The lower boundary is now shifted below the line $k = 2p$ and a finite unstable k-interval is obtained for $p \to 0$ (Figure 1). The two frequency branches can be obtained as

$$\omega_{1,2}^2 = k^2p^2 + \frac{k^4}{4} \pm \sqrt{1 + p^2k^6}. \tag{28}$$

The asymptotes for small and large wavenumbers become

$$\omega_{1,2}^2 \to \pm 1 + k^2p^2, \quad \text{for} \quad k \to 0, \tag{29a}$$

$$\omega_{1,2}^2 \to \Omega_{1,2}^2 \quad \text{for} \quad k \to \infty. \tag{29b}$$

The instability has now a finite growth rate equal to the plasma frequency for $k \to 0$.

3.3. Hartree-Fock Model

The analysis of the full Hartree-Fock system (13) is more cumbersome. We have calculated the determinant of the coefficient matrix and obtain again a polynomial of second order in ω^2 as given by (18). Its coefficients are defined by

$$P = \Omega_{HF,1}^2 + \Omega_{HF,2}^2 + 2b^2 - d^2 - c^2, \tag{30a}$$

$$Q = (\Omega_{HF,1}^2 - c^2)(\Omega_{HF,2}^2 - d^2) + 2b^2(\Omega_{HF,1}\Omega_{HF,2} - cd) + b^4, \tag{30b}$$

with the Hartree-Fock frequencies of the single-particle excitations

$$\Omega_{HF,1} = \Omega_1 + c - b, \qquad \Omega_{HF,2} = \Omega_2 + b - d$$

and some effective-potential parameters

$$b = G(2\boldsymbol{p}) - G(\boldsymbol{k}), \qquad c = G(2\boldsymbol{p} + \boldsymbol{k}) - G(\boldsymbol{k}), \qquad d = G(2\boldsymbol{p} - \boldsymbol{k}) - G(\boldsymbol{k}).$$

We have investigated the instability region in the kp-plane by evaluating the coefficients P and Q numerically. The resulting instability diagram is shown in Figure 2 for different values of the screening constant k_0. The criterion for stability is

$$P > 0 \quad \text{and} \quad 0 \le Q \le \left(\frac{P}{2}\right)^2. \tag{31}$$

These stable regions are colored in blue. Unstable regions are found as before for $P > 0$ and $Q < 0$. These areas are colored in red. Furthermore there is an instability region for $P > 0$ and $Q > (P/2)^2$, which is colored in yellow. Within the red region, the unstable frequency is purely imaginary and therefore the perturbation amplitude shows monotonic exponential growth. Within the yellow region, the unstable frequencies have both nonvanishing real and imaginary parts. The perturbation amplitudes show oscillatory growth. For sufficiently large screening, the Hartree-Fock result converges to the Hartree result, as can be seen in a) for $k_0 = 2.4$. For smaller screening constants ($k_0 = 1.5, 1.0, 0$) , the instability region for large k is completely shifted below the line $k = 2p$. In addition, one can observe that the two unstable areas around the lines $k = 0$ and $k = 2p$ become disconnected. In between these disconnected areas there appears an island of solutions with oscillatory growth. This island is a unique feature of the exchange interaction in the Hartree-Fock model.

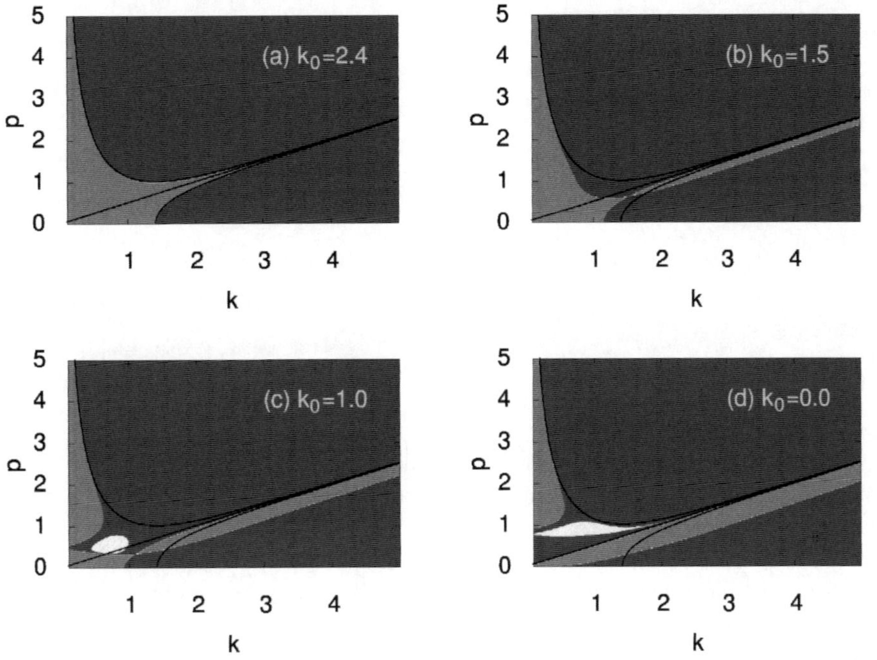

Figure 2. Instability regions in the Hartree-Fock model as a function of the screening constant k_0. Black lines indicate the Vlasov and Hartree boundaries as shown in Figure 1. The colors indicate stable regions (blue), non-oscillatory unstable regions (red) and oscillatory unstable regions (yellow). a) For $k_0 = 2.4$ exchange effects are marginal. b) With decreasing screening, the exchange interaction leads to two disconnected read areas. c) The red areas become separated by a yellow island. d) Below this island, the instability is stabilized for small wavenumbers by the exchange interaction.

An evaluation of the two branches $\omega_{1,2}^2(k)$ is shown in Figure 3 for $p = 0.5$ and different screening constants $k_0 = 2.4, 1.1, 0.6, 0.2$. For $k_0 = 2.4$, one can observe the stable plasmon branch and the unstable two-stream branch of the Hartree model. For $k_0 = 1.1$, one can see that the two branches have merged. There is now an interval, where both branches have the same real part and different non-zero imaginary parts. For $k_0 = 0.6$, the two branches to the left of the island are both positive and therefore stable. Note that they have interchanged their physical significance. The plasmon branch with the asymptote $\omega^2 \to 1$ for $k \to 0$ is now the lower branch. Finally, for $k_0 = 0.2$, this interchange is nearly complete. The plasmon branch now starts as the lower branch on the left side and continues as the upper branch on the right side of the island. The other branch approaches a first-order pole at $k = 2p$. Thereby the whole region $k < 2p_0$ has become stable and the instability has become shifted to a region above the pole $k = 2p$.

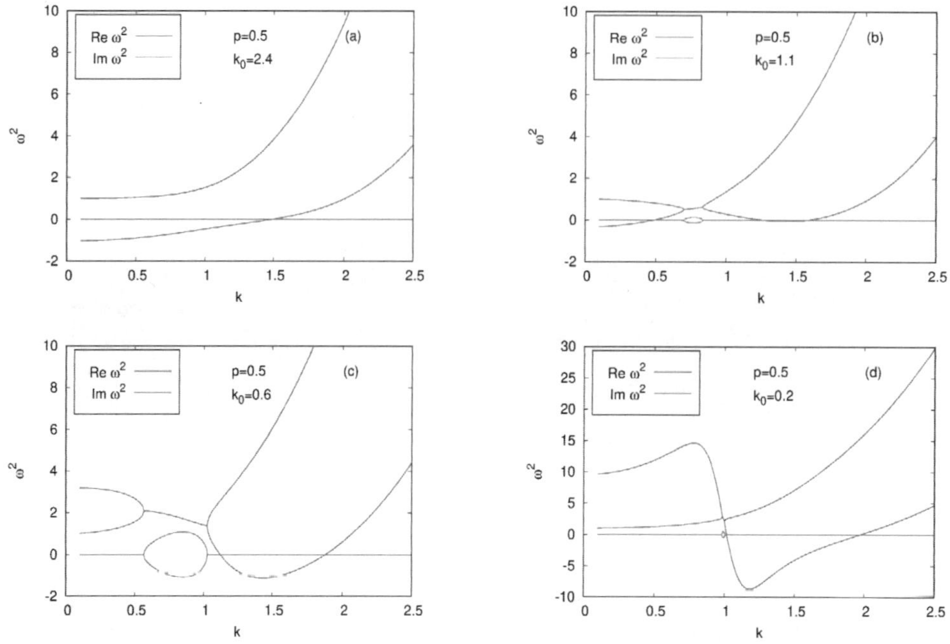

Figure 3. Real and imaginary parts of the branches $\omega^2(k)$. a) For sufficiently strong screening of the exchange interaction one has two disconnected branches for the plasmon excitation (upper branch) and the two-stream instability (lower branch). b) With decreasing screening, the two branches merge and thereby non-zero imaginary parts appear. c) For still weaker screening, the branches on the left side have become interchanged, the plasmon branch being the lower one. d) Finally, the upper branch approaches a pole singularity at $k = 2p = 1$, that changes strongly the instability behavior.

4. Conclusions

In this work, we have investigated the role of the exchange interaction on the quantum two-stream instability. As the exchange interaction is mostly neglected in self-consistent field treatments of plasmas, some validation for such a procedure appeared appropriate. Looking at the two-stream instability of the unscreened Coulomb interaction, it became evident that the exchange contribution is quite substantial and leads to completely different modes and instability regions in comparison to the more familiar Vlasov results. Although the two-stream model is highly idealized, this conclusion will also hold for multistream models and their thermodynamic limit. The simple reason is that the effective exchange potentials of the long-range Coulomb interaction are quite singular and become dominant in various parameter regions. However, with increasing screening, these differences become marginal and then the results of the simpler Hartree model have been approached. The present differences between the Hartree and the Vlasov models will also become negligible for many-electron systems. The screening constant of metals under normal conditions is such that k_0 varies between about 2 and 2.5. Under these conditions the two-stream model indicates, that the exchange interaction is actually a small correction.

References

[1] Kuzelev M and Rukhadze A 1999 *Physics-Uspekhi* **42** 603–605

[2] Kuzelev M V 2010 *JETP* **110**(4) 710–721

[3] Haas F, Manfredi G and Feix M 2000 *Phys. Rev. E* **62** 2763–2772

[4] Vladimirov S and Tyshetskiy Y O 2011 *Physics-Uspekhi* **54** 1243–1256

[5] Schmidt-Bleker A, Gassen W and Kull H J 2011 *EPL* **95** 55003

[6] Dawson J M 1960 *Phys. Rev.* **118**(2) 381–389

[7] Steffen W and Kull H J 2016 *Phys. Rev. E* **93**(3) 033207

[8] Anderson D, Hall B, Lisak M and Marklund M 2002 *Phys. Rev. E* **65**(4) 046417

[9] Ali S and Shukla P 2007 *Eur. Phys. J. D* **41** 319–324

[10] Haas F, Bret A and Shukla P K 2009 *Phys. Rev. E* **80**(6) 066407

[11] Zeba I, Yahia M, Shukla P and Moslem W 2012 *Physics Letters A* **376** 2309 – 2313

[12] Ashcroft N W and Mermin N D 1976 *Solid State Physics* (Philadelphia: Saunders College)

[13] Lozovik Y and Klyuchnik A 1989 *The dielectric function of condensed systems* (Amsterdam: Elsevier Science Publishers B.V., North-Holland Physics)

[14] Koonin S E, Davies K T R, Maruhn-Rezwani V, Feldmeier H, Krieger S J and Negele J W 1977 *Phys. Rev. C* **15** 1359–1374

[15] Kull H J and Pfirsch D 2000 *Phys. Rev. E* **61** 5940–5947

Plasmon effects in photoemission

J. Z. Kamiński, F. Cajiao Vélez and K. Krajewska

Institute of Theoretical Physics, Faculty of Physics, University of Warsaw, Pasteura 5, 02-093 Warszawa, Poland

E-mail: `Jerzy.Kaminski@fuw.edu.pl`

Abstract. Recently, we observe an increasing interest in the laser-assisted processes that take place at the solids surfaces or at the interfaces of semiconductor heterostructures. These processes are affected by the plasmon effects induced by the interaction of quasi-free electrons in solids with laser fields. The plasmon effects lead to a significant enhancement of the oscillating in time electric field in a vicinity of the surface. The latter strongly modifies processes such as the high-order harmonic generation from atoms passing close to the surface, the electron photoemission from metal tips or the tunneling processes in semiconductor heterostructures. In our paper, we focus on the photoemission phenomenon accounting for the space-dependence of an oscillating in time, effective electric field created at the surface.

1. Introduction

The aim of this paper is to investigate the photoemission from a semiconductor heterostructure occurring in an arbitrary space-dependent scalar potential and a time- and space-dependent vector potential. The vector potential is periodic in time and describes a laser field. Its space-dependence results from the interaction of the laser field with electrons in solids. Such conditions are met, for example, in semiconductor nano-structures [1, 2, 3, 4, 5] (like quantum wires or wells), photoemission from a metal tip [6, 7], carbon nano-tubes or graphene [8, 9, 10] or in surface physics [11, 12, 13, 14, 15, 16, 17, 18, 19, 20]. To make our presentation as clear as possible we shall restrict ourselves to the one-space-dimensional case, although extension of the presented method to systems of higher dimensionality is possible (see, e.g. [21, 22]).

This paper is organized as follows. In Sec. II the most general solution of the Schrödinger equation is introduced. The transfer-matrix method and matching conditions are analyzed in Sec. III, whereas the reflection and transition probabilities are introduced in Sec. IV. These probabilities must sum up to 1, which puts a very strong check for the accuracy of our numerical calculations. The most important part of this paper, i.e., the concept of the scattering-matrix method, is discussed in Sec. V. It is shown there why the scattering-matrix algorithm has to be introduced, instead of a much simpler transfer-matrix algorithm. Numerical illustrations of the applicability of this algorithm are presented in Sec. VI and are followed by short conclusions.

Unless it is stated otherwise, atomic units are used in our numerical illustrations.

2. Solution of the Schrödinger equation

Let us start with the one-dimensional Schrödinger equation of the form [23],

$$\mathrm{i}\partial_t\psi(x,t) = \Big[\frac{1}{2}\Big(\frac{1}{\mathrm{i}}\partial_x - eA(x,t)\Big)\frac{1}{m(x)}\Big(\frac{1}{\mathrm{i}}\partial_x - eA(x,t)\Big) + V(x)\Big]\psi(x,t). \tag{1}$$

We assume that the space-dependent mass $m(x)$, scalar potential $V(x)$, and vector potential $A(x,t)$ are spatially constant in finite intervals. Their values in an interval (x_{i-1}, x_i) are denoted as m_i, V_i, and $A_i(t)$. We require also that the function $A(x,t)$ is periodic in time,

$$A(x, t+T) = A(x,t), \tag{2}$$

where $T = 2\pi/\omega$ and ω is the frequency of the oscillating in time electric field. Defining the probability density $\rho(x,t)$,

$$\rho(x,t) = |\psi(x,t)|^2, \tag{3}$$

and the probability current $j(x,t)$,

$$j(x,t) = \frac{1}{2}\psi^*(x,t)\frac{1}{m(x)}\Big(\frac{1}{\mathrm{i}}\partial_x - eA(x,t)\Big)\psi(x,t) + \frac{1}{2}\psi(x,t)\frac{1}{m(x)}\Big[\Big(\frac{1}{\mathrm{i}}\partial_x - eA(x,t)\Big)\psi(x,t)\Big]^*, \tag{4}$$

we show, using Eq. (1), that the probability is conserved. Indeed, assuming the above definitions, we obtain the continuity equation,

$$\partial_t\rho(x,t) + \partial_x j(x,t) = 0. \tag{5}$$

The space dependence of mass in Eq. (1) forces one to impose non-standard continuity conditions on any solution of this equation. It is now the wave-function $\psi(x,t)$ and the quantity

$$\frac{1}{m(x)}\Big(\frac{1}{\mathrm{i}}\partial_x - eA(x,t)\Big)\psi(x,t) \tag{6}$$

that have to be continuous at points of discontinuity of mass $m(x)$ and both potentials, $V(x)$ and $A(x,t)$ [23, 24, 25, 26]. We denote a general solution $\psi(x,t)$ of Eq. (1) in a given interval (x_{i-1}, x_i) as $\psi_i(x,t)$. Note that, due to time periodicity of the Hamiltonian, $\psi_i(x,t)$ can be chosen such that the Floquet condition,

$$\psi_i(x, t+T) = \mathrm{e}^{-\mathrm{i}ET}\psi_i(x,t), \tag{7}$$

is satisfied, where E is the so-called quasi-energy. In this case, $\psi_i(x,t)$ takes the form [28, 29, 30, 31],

$$\psi_i(x,t) = \sum_{M=-\infty}^{\infty} \exp\big(-\mathrm{i}(E + M\omega)t\big) \sum_{\sigma=\pm}\sum_{N=-\infty}^{\infty} C_{iN}^{\sigma}\mathcal{B}_{M-N}(\sigma p_{iN})\exp(\mathrm{i}\sigma p_{iN}x), \tag{8}$$

where C_{iN}^{σ} are arbitrary complex numbers to be determined and

$$p_{iN} = \sqrt{2m_i(E + N\omega - V_i - U_i)}, \tag{9}$$

with $U_i = e^2\langle A_i^2(t)\rangle/2m_i$ being the ponderomotive energy. Here, $\langle A_i^2(t)\rangle$ means the time-average of $A_i^2(t)$ over the laser-field oscillations.

The components for which p_{iN} are purely imaginary are called the closed channels. These channels are not observed for a particle in initial or final states, but they have to be taken into account in order to satisfy the unitary condition of the time evolution. In a general case, $\mathcal{B}_{M-N}(\sigma p_{iN})$ is a component of the Fourier expansion,

$$\exp\big(\mathrm{i}\Phi_{iN}^{\sigma}(t)\big) = \sum_{M=-\infty}^{\infty} \exp(-\mathrm{i}M\omega t)\mathcal{B}_{M-N}(\sigma p_{iN}), \tag{10}$$

provided that the vector potential $A(x,t)$ is periodic in time. Functions $\Phi_{iN}^{\sigma}(t)$ are defined as follows:

$$\Phi_{iN}^{\sigma}(t) = \int_0^t \Big[\frac{\sigma e}{m_i}A_i(t)p_{iN} - \frac{e^2}{2m_i}\big(A_i^2(t) - \langle A_i^2(t)\rangle\big)\Big]\mathrm{d}t. \tag{11}$$

One can understand from these equations that the $\mathcal{B}_{M-N}(\sigma p_{iN})$ functions depend on the form of the vector potential $A(x,t)$ describing the laser field.

3. Matching conditions and transfer matrix

Continuity conditions discussed above and applied to a general solution (8) of the Schrödinger equation (1) lead to an infinite chain of equations connecting constants C^{σ}_{iN} in the neighboring domains. These matching conditions can be written in the matrix form,

$$B(i-1, x_{i-1})C_{i-1} = B(i, x_{i-1})C_i, \tag{12}$$

where $C^{\pm}_{iN} = [C^{\pm}_i]_N$ are the components of the columns $C^{\pm}_i$. The matrices $B(i,x)$ and C_i are defined as follows,

$$B(i,x) = \begin{pmatrix} B^+(i,x) & B^-(i,x) \\ B'^+(i,x) & B'^-(i,x) \end{pmatrix}, \; C_i = \begin{pmatrix} C^+_i \\ C^-_i \end{pmatrix}. \tag{13}$$

The elements of $B(i,x)$ can be computed in the following way.

For an arbitrary function $A(x,t)$ which satisfies Eq. (2), we have that

$$A(x,t) = \sum_{n=-\infty}^{\infty} b_n(x) \exp(-in\omega t). \tag{14}$$

The coefficients $b_n(x)$ assume constant values in the interval (x_{i-1}, x_i), which we shall denote as $b_{i,n}$. Using the continuity condition for the wave-function $\psi_i(x,t)$ at the point x_{i-1}, we compute the elements of the matrices B^+ and B^-,

$$B^{\pm}(i,x)_{M,N} = \mathcal{B}_{M-N}(\pm p_{i,N})\mathrm{e}^{\pm \mathrm{i} p_{i,N} x}. \tag{15}$$

On the other hand, elements of the B' matrix can be evaluated by substituting a general solution (8) to the expression (6) and applying the continuity condition at x_{i-1}. After some algebraic manipulations we finally obtain the expression for the B'-matrices,

$$B'^{\pm}(i,x)_{M,N} = \pm\frac{1}{m_i}\mathcal{B}_{M-N}(\pm p_{i,N})p_{i,N}\mathrm{e}^{\pm \mathrm{i} p_{i,N} x} - \frac{1}{m_i}\sum_{n=-\infty}^{\infty} e b_{i,n}\mathcal{B}_{M-N-n}(\pm p_{i,N})\mathrm{e}^{\pm \mathrm{i} p_{i,N} x}, \tag{16}$$

and a set of equations for vectors C_i,

$$C_i = B_i C_{i-1}, \tag{17}$$

where

$$B_i = [B(i, x_{i-1})]^{-1} B(i-1, x_{i-1}). \tag{18}$$

These relations allow to connect a solution in a given domain $x_{i-1} < x < x_i$ with an analogous solution in any other domain $x_{j-1} < x < x_j$,

$$C_j = B_j B_{j-1}, \ldots, B_{i+1} C_i = \mathcal{T}_{ji} C_i, \tag{19}$$

where $\mathcal{T}_{ji}$ is the so-called transfer matrix [27, 28, 25].

4. Reflection and transition probabilities

It is clear now that, on the basis of Eq. (19), we can connect solutions in the boundary domains $(-\infty, x_0)$ and (x_{L-1}, ∞). Values of mass $m(x)$, scalar potential $V(x)$ and vector potential $A(x,t)$ in these domains will be denoted as m_0, V_0, $A_0(t)$ and m_L, V_L, $A_L(t)$, respectively. We can then write down solutions of (1) for each of these domains. These solutions represent incident ($\psi_{\rm inc}$), reflected ($\psi_{\rm ref}$), and transmitted ($\psi_{\rm tr}$) waves,

$$\psi_{\rm inc}(x,t) = \sum_{M=-\infty}^{\infty} {\rm e}^{-{\rm i}Et}{\rm e}^{-{\rm i}M\omega t}\mathcal{B}_M(p_0){\rm e}^{{\rm i}p_0 x}, \tag{20}$$

$$\psi_{\rm ref}(x,t) = \sum_{N,M=-\infty}^{\infty} C^{-}_{0,N}{\rm e}^{-{\rm i}Et}{\rm e}^{-{\rm i}M\omega t}\mathcal{B}_{M-N}(-p_N){\rm e}^{-{\rm i}p_N x}, \tag{21}$$

$$\psi_{\rm tr}(x,t) = \sum_{N,M=-\infty}^{\infty} C^{+}_{L,N}{\rm e}^{-{\rm i}Et}{\rm e}^{-{\rm i}M\omega t}\mathcal{B}_{M-N}(q_N){\rm e}^{{\rm i}q_N x}, \tag{22}$$

where

$$p_N = \sqrt{2m_0(E+N\omega-V_0-U_0)}, \quad q_N = \sqrt{2m_L(E+N\omega-V_L-U_L)}. \tag{23}$$

Constants $C^{-}_{0,N}$ and $C^{+}_{L,N}$ will be denoted from now on as R_N and T_N, respectively. Using continuity conditions for the functions defined above, we get the probability conservation equation for reflection and transition amplitudes, R_N and T_N,

$$\sum_{N\geqslant N_{\rm ref}} \frac{p_N}{p_0}|\mathrm{R}_N|^2 + \sum_{N\geqslant N_{\rm tr}} \frac{m_0 q_N}{m_L p_0}|\mathrm{T}_N|^2 = 1, \tag{24}$$

where summations are over such N for which p_N and q_N are real, i.e., over the open channels. This equation permits us to interpret

$$P_{\rm R}(N) = \frac{p_N}{p_0}|\mathrm{R}_N|^2 \tag{25}$$

and

$$P_{\rm T}(N) = \frac{m_0 q_N}{m_L p_0}|\mathrm{T}_N|^2 \tag{26}$$

as reflection and transition probabilities for a tunneling process in which absorption ($N>0$) or emission ($N<0$) of energy $N\omega$ by electrons occurred [28, 26]. In the case of a monochromatic laser field, this process can be interpreted as absorption or emission of N photons from the laser field.

The unitary condition (24) can be also interpreted as the conservation of the electric charge. If we define the quantities proportional to the density of electric currents,

$$J_{\rm inc} = \frac{p_0}{m_0}, \quad J_{\rm ref} = \sum_{N\geqslant N_{\rm ref}} \frac{p_N}{m_0}|\mathrm{R}_N|^2, \quad J_{\rm tr} = \sum_{N\geqslant N_{\rm tr}} \frac{q_N}{m_L}|\mathrm{T}_N|^2, \tag{27}$$

then Eq. (24) has the form of the first Kirchhoff law,

$$J_{\rm inc} = J_{\rm ref} + J_{\rm tr}. \tag{28}$$

Using Eq. (19) we can calculate constants $C^{-}_{0,N} = \mathrm{R}_N$ and $C^{+}_{L,N} = \mathrm{T}_N$ appearing in equations (20) - (22). Indeed, since

$$C_L = \mathcal{T}C_0, \tag{29}$$

where the transfer matrix $\mathcal{T} = \mathcal{T}_{L0}$, and because $\mathcal{T}$, C_0, and C_L have the following block forms,

$$\mathcal{T} = \begin{pmatrix} \mathcal{T}^{++} & \mathcal{T}^{+-} \\ \mathcal{T}^{-+} & \mathcal{T}^{--} \end{pmatrix}, \quad C_0 = \begin{pmatrix} C_0^+ \\ \mathrm{R} \end{pmatrix}, \quad C_L = \begin{pmatrix} \mathrm{T} \\ 0 \end{pmatrix}, \tag{30}$$

we arrive at

$$\mathrm{T} = \mathcal{T}^{++}C_0^+ + \mathcal{T}^{+-}\mathrm{R}, \quad 0 = \mathcal{T}^{-+}C_0^+ + \mathcal{T}^{--}\mathrm{R}. \tag{31}$$

Here, R and T denote the columns of R_N and T_N, and $[C_0^+]_N = \delta_{0,N}$. Thus, after some algebraic manipulations, we obtain

$$\mathrm{R} = -(\mathcal{T}^{--})^{-1}\mathcal{T}^{-+}C_0^+, \quad \mathrm{T} = (\mathcal{T}^{++} - \mathcal{T}^{+-}(\mathcal{T}^{--})^{-1}\mathcal{T}^{-+})C_0^+, \tag{32}$$

which allows us to determine the quantities R_N and T_N for a given transfer matrix $\mathcal{T}$. For open channels, these quantities are the amplitudes of reflection (R_N) and transition (T_N) probabilities, from which one can compute reflection and transition probabilities using equations (25) and (26).

5. The scattering matrix

It follows from equations (15) and (16) that each of the B_i matrices that constitute the transfer matrix $\mathcal{T}_{ji}$ contain elements $\exp(\pm \mathrm{i}p_{i,N}x_i)$. The latter depend on the x_i coordinates at which the discontinuities appear. For closed channels, that is when the $p_{i,N}$ momenta are purely imaginary, these numbers are real and may assume arbitrary values, depending on the x_i coordinates. The number of B_i matrices is equal to the number of discontinuity points, i.e., it depends on how we divide the space into short intervals in order to make our potential tractable by our algorithm. It may, therefore, turn out that in order to compute the transfer matrix $\mathcal{T}_{ji}$, we have to multiply a large number of the B_i matrices, each containing both very small and very large numbers. It is clear that such a procedure is numerically unstable. We have to find a way to modify our method of calculations in order to compute the elements of each B_i matrix at the same point $x = 0$ independently of where the 'real' x_i is. This would eliminate "dangerous" $\exp(\pm \mathrm{i}p_{i,N}x_i)$ elements (turning them to 1), however at the cost of appearing somewhere else. We shall see later that these 'left-overs' appear only as differences $x_{i+1} - x_i$ and, therefore, do not cause any harmful side-effects.

It follows from Eq. (19) that in the neighboring domains, (x_{i-2}, x_{i-1}) and (x_{i-1}, x_i), we have,

$$C_i = \mathcal{T}_{i,i-1}C_{i-1}. \tag{33}$$

Although the elements of the transfer matrix $\mathcal{T}_{i,i-1}$ have been computed from the continuity conditions at point x_{i-1}, one can compute them at any other point, for example at $x = 0$. To this end, let us note what follows from the solution (8). Translation of the system by a certain distance δ along the x-axis causes only multiplication of each member of the sum over N in (8) by a constant $\exp(\mathrm{i}\sigma p_{iN}\delta)$. These constants can be included into the coefficients C_{iN}^{σ}. In this way, we obtain a new set of constants which we shall denote as $\tilde{C}_{iN}^{\sigma}$,

$$\tilde{C}_{iN}^{\sigma} = \exp(\mathrm{i}\sigma p_{iN}\delta)C_{iN}^{\sigma}. \tag{34}$$

We shall interpret these constants as coefficients in the solution (8), given by the continuity conditions at a point $x_{i-1} - \delta$. Eq. (34), written in the matrix form, becomes

$$\tilde{C}_i = \mathcal{P}_i(\delta)C_i, \tag{35}$$

where

$$\mathcal{P}_i(\delta) = \begin{pmatrix} P_i^+(\delta) & 0 \\ 0 & P_i^-(\delta) \end{pmatrix} \tag{36}$$

and

$$C_i = \begin{pmatrix} C_i^+ \\ C_i^- \end{pmatrix}, \tilde{C}_i = \begin{pmatrix} \tilde{C}_i^+ \\ \tilde{C}_i^- \end{pmatrix}. \tag{37}$$

In the equation above, $P_i^\sigma(\delta)$ is a diagonal matrix,

$$[P_i^\sigma(\delta)]_{NN'} = \delta_{NN'} \exp(\mathrm{i}\sigma p_{iN}\delta), \tag{38}$$

whereas $C_i^\pm$ and $\tilde{C}_i^\pm$ are the columns consisting of the constants $C_{iN}^\pm$ and $\tilde{C}_{iN}^\pm$, respectively, that is $[C_i^\pm]_N = C_{iN}^\pm$ and $[\tilde{C}_i^\pm]_N = C_{iN}^\pm$. It follows from the form of the matrix $\mathcal{P}_i(\delta)$ that the following relations are satisfied,

$$\mathcal{P}_i^{-1}(\delta) = \mathcal{P}_i(-\delta), \quad \mathcal{P}_i(\delta_1)\mathcal{P}_i(\delta_2) = \mathcal{P}_i(\delta_1 + \delta_2). \tag{39}$$

Let us also note that the translation of the system defined above modifies the transfer matrix $\mathcal{T}_{i,i-1}$. We have

$$\mathcal{P}_i^{-1}\tilde{C}_i = C_i = \mathcal{T}_{i,i-1}C_{i-1} = \mathcal{T}_{i,i-1}\mathcal{P}_{i-1}^{-1}(\delta)\mathcal{P}_{i-1}(\delta)C_{i-1}. \tag{40}$$

Thus,

$$\tilde{C}_i = \mathcal{P}_i(\delta)\mathcal{T}_{i,i-1}\mathcal{P}_{i-1}^{-1}(\delta)\tilde{C}_{i-1} \tag{41}$$

and we can write it down as

$$\tilde{C}_i = \tilde{\mathcal{T}}_{i,i-1}\tilde{C}_{i-1}, \tag{42}$$

where

$$\tilde{\mathcal{T}}_{i,i-1} = \mathcal{P}_i(\delta)\mathcal{T}_{i,i-1}\mathcal{P}_{i-1}^{-1}(\delta). \tag{43}$$

Matrix elements denoted with the tilde symbol refer to the translated system. Using the method defined above and the relation (19), we can connect now the solution in the domain $(-\infty, x_0)$ with the solution in any other domain (x_{i-1}, x_i). In this way the elements of the transfer matrix, which until now have been computed at the points of discontinuity $x_0 \ldots x_{i-1}$, are computed each time at the same point $x = 0$. Let us illustrate this method for a special case of $i = 3$

$$\begin{aligned} C_3 &= \mathcal{T}_{3,2}\mathcal{T}_{2,1}\mathcal{T}_{1,0}C_0 = \mathcal{P}_3^{-1}(x_2)\mathcal{T}_{3,2}^0\mathcal{P}_2(x_2)\mathcal{P}_2^{-1}(x_1)\mathcal{T}_{2,1}^0\mathcal{P}_1(x_1)\mathcal{P}_1^{-1}(x_0)\mathcal{T}_{1,0}^0\mathcal{P}_0(x_0)C_0 \\ &= \mathcal{P}_3^{-1}(x_2)\mathcal{T}_{3,2}^0\mathcal{P}_2(x_2 - x_1)\mathcal{T}_{2,1}^0\mathcal{P}_1(x_1 - x_0)\mathcal{T}_{1,0}^0\mathcal{P}_0(x_0)C_0. \end{aligned} \tag{44}$$

Eq. (44) connects constants C_0 and C_3 using the matrices $\mathcal{T}_{j,j-1}^0$ calculated at $x = 0$ independently of j, and the diagonal matrices $\mathcal{P}_j(\delta_j)$, given by the relations (36) and (38), where $\delta_j = x_j - x_{j-1}$. The edge matrices $\mathcal{P}_0(x_0)$ and $\mathcal{P}_3^{-1}(x_2)$ in Eq. (44) can be omitted while computing the transmission and reflection probability amplitudes, as their only role is to multiply the amplitudes by phase quotients which disappear while computing the probabilities. Although these matrices lead to significant modifications of the closed channels in the domains of $x < x_0$ and $x > x_3$, these channels do not influence the reflection and transition amplitudes. Transmission and reflection probabilities can thus be computed using a modified transfer matrix,

$$\mathcal{T}_{3,0}^0 = \mathcal{T}_{3,2}^0\mathcal{P}_2(x_2 - x_1)\mathcal{T}_{2,1}^0\mathcal{P}_1(x_1 - x_0)\mathcal{T}_{1,0}^0. \tag{45}$$

The matrices $\mathcal{T}_{i,i-1}^0$ are equal to the matrices B_i in Eq. (18) calculated, however, for $x_{i-1} = 0$. This speeds up numerical calculations since now the matrix $B(i, x = 0)$ in Eq. (18) have to be inverted only once. Further, we shall omit the superscript 0 in $\mathcal{T}$ and the tilde over C in order to simplify our notation.

The method presented above is still numerically unstable. The reason for this instability is the existence of large numerical values of elements of $\mathcal{P}_i^-(\delta)$ for imaginary momenta p_{iN}. In other words, for

$$C_i = \begin{pmatrix} C_i^+ \\ C_i^- \end{pmatrix} = \mathcal{T}_{i,i-1}C_{i-1} = \begin{pmatrix} \mathcal{T}_{i,i-1}^{++} & \mathcal{T}_{i,i-1}^{+-} \\ \mathcal{T}_{i,i-1}^{-+} & \mathcal{T}_{i,i-1}^{--} \end{pmatrix} \begin{pmatrix} C_{i-1}^+ \\ C_{i-1}^- \end{pmatrix}, \tag{46}$$

the source of numerical instabilities are matrix elements $\mathcal{T}_{i,i-1}^{--}$ that contain large numbers. There is, however, a chance for improving the stability, if only its inverse will be used, $\left(\mathcal{T}_{i,i-1}^{--}\right)^{-1}$. This is possible provided that in our numerical algorithm only the so-called scattering matrix will be applied. For this reason we will show below how to compute the scattering matrix, $\mathcal{S}_{j,i}$, using only elements of the transfer matrix, $\mathcal{T}_{j,i}$. For the transfer matrix $\mathcal{T}_{j,i}$ we have,

$$\begin{pmatrix} C_j^+ \\ C_j^- \end{pmatrix} = \begin{pmatrix} \mathcal{T}_{j,i}^{++} & \mathcal{T}_{j,i}^{+-} \\ \mathcal{T}_{j,i}^{-+} & \mathcal{T}_{j,i}^{--} \end{pmatrix} \begin{pmatrix} C_i^+ \\ C_i^- \end{pmatrix}. \tag{47}$$

Thus,

$$C_j^+ = \mathcal{T}_{j,i}^{++} C_i^+ + \mathcal{T}_{j,i}^{+-} C_i^-, \quad C_j^- = \mathcal{T}_{j,i}^{-+} C_i^+ + \mathcal{T}_{j,i}^{--} C_i^-. \tag{48}$$

On the basis of (48) we now want to compute the elements of the $\mathcal{S}_{j,i}$ matrix. This matrix is supposed to connect the coefficients $C_i^{\pm}$ and $C_j^{\pm}$ in the following way,

$$\begin{pmatrix} C_i^- \\ C_j^+ \end{pmatrix} = \begin{pmatrix} \mathcal{S}_{j,i}^{++} & \mathcal{S}_{j,i}^{+-} \\ \mathcal{S}_{j,i}^{-+} & \mathcal{S}_{j,i}^{--} \end{pmatrix} \begin{pmatrix} C_i^+ \\ C_j^- \end{pmatrix}. \tag{49}$$

Using the set of linear equations (48), we easily compute the coefficients C_i^- and C_j^+ on the left-hand side of equation (49) as functions of the coefficients C_j^- and C_i^+. We obtain then the following relations,

$$\begin{aligned} C_i^- &= (\mathcal{T}_{j,i}^{--})^{-1}(C_j^- - \mathcal{T}_{j,i}^{-+} C_i^+), \\ C_j^+ &= (\mathcal{T}_{j,i}^{++} - \mathcal{T}_{j,i}^{+-}(\mathcal{T}_{j,i}^{--})^{-1}\mathcal{T}_{j,i}^{-+})C_i^+ + \mathcal{T}_{j,i}^{+-}(\mathcal{T}_{j,i}^{--})^{-1}C_j^-. \end{aligned} \tag{50}$$

Finally, we compute the elements of the matrix $\mathcal{S}_{j,i}$,

$$\begin{aligned} \mathcal{S}_{j,i}^{++} &= -(\mathcal{T}_{j,i}^{--})^{-1}\mathcal{T}_{j,i}^{-+}, & \mathcal{S}_{j,i}^{+-} &= (\mathcal{T}_{j,i}^{--})^{-1}, \\ \mathcal{S}_{j,i}^{-+} &= \mathcal{T}_{j,i}^{++} - \mathcal{T}_{j,i}^{+-}(\mathcal{T}_{j,i}^{--})^{-1}\mathcal{T}_{j,i}^{-+}, & \mathcal{S}_{j,i}^{--} &= \mathcal{T}_{j,i}^{+-}(\mathcal{T}_{j,i}^{--})^{-1}. \end{aligned} \tag{51}$$

As expected, the matrix $\mathcal{S}_{j,i}$ contains only numerically stable elements $(\mathcal{T}_{j,i}^{--})^{-1}$.

It follows from Eq. (19) that the transfer matrix $\mathcal{T}_{j,i}$ can be written as the product of two transfer matrices, $\mathcal{T}_{j,k}$ and $\mathcal{T}_{k,i}$ $(i < k < j)$,

$$\mathcal{T}_{j,i} = \mathcal{T}_{j,k}\mathcal{T}_{k,i}, \tag{52}$$

where matrices $\mathcal{T}_{j,k}$ and $\mathcal{T}_{k,i}$ are defined as follows,

$$C_k = \mathcal{T}_{k,i}C_i, \quad C_j = \mathcal{T}_{j,k}C_k. \tag{53}$$

Applying the method presented above, for each of the transfer matrices $\mathcal{T}_{j,k}$ and $\mathcal{T}_{k,i}$ we can now construct a scattering matrix, $\mathcal{S}_{j,k}$ and $\mathcal{S}_{k,i}$, respectively. The elements of the scattering matrix $\mathcal{S}_{j,i}$ can be computed using only elements of $\mathcal{S}_{j,k}$ and $\mathcal{S}_{k,i}$. Using the notation above, we obtain the following expressions for the elements of the $\mathcal{S}_{j,i}$ matrix,

$$\begin{aligned} \mathcal{S}_{j,i}^{++} &= \mathcal{S}_{k,i}^{++} + \mathcal{S}_{k,i}^{+-}(1 - \mathcal{S}_{j,k}^{++}\mathcal{S}_{k,i}^{--})^{-1}\mathcal{S}_{j,k}^{++}\mathcal{S}_{k,i}^{-+}, & \mathcal{S}_{j,i}^{+-} &= \mathcal{S}_{k,i}^{+-}(1 - \mathcal{S}_{j,k}^{++}\mathcal{S}_{k,i}^{--})^{-1}\mathcal{S}_{j,k}^{+-}, \\ \mathcal{S}_{j,i}^{--} &= \mathcal{S}_{j,k}^{--} + \mathcal{S}_{j,k}^{-+}\mathcal{S}_{k,i}^{--}(1 - \mathcal{S}_{j,k}^{++}\mathcal{S}_{k,i}^{--})^{-1}\mathcal{S}_{j,k}^{+-}, & \mathcal{S}_{j,i}^{-+} &= \mathcal{S}_{j,k}^{-+}(1 - \mathcal{S}_{j,k}^{++}\mathcal{S}_{k,i}^{--})^{-1}\mathcal{S}_{k,i}^{-+}. \end{aligned} \tag{54}$$

It is clear that the $\mathcal{S}_{j,i}$ matrix is not merely a product of two matrices $\mathcal{S}_{j,k}$ and $\mathcal{S}_{k,i}$, but rather a complicated nonlinear composition of them. It is important, however, to note that, despite its complexity, such a construction of the scattering matrix is numerically stable. This is in contrast to the transfer matrix method which fails if a system with a large number of discontinuity points x_i is considered. Stability of such an algorithm has been proven in our numerical investigations by checking that the condition (24) is satisfied with an error smaller than 10^{-14}. Such an accuracy can never be achieved for systems with a large number of discontinuity points if the transfer matrix is applied.

6. Photoemission

In our studies, we focus on essential features of the solid-vacuum interface, as demonstrated by the Sommerfeld model in which the band structure is neglected. This simplification allows us to consider a quite general form of the laser field. To be more specific, the solid surface is described by a continuous step potential,

$$V(x) = V_0 g(x/w_0), \quad g(x) = 1/(1+\mathrm{e}^{-x}). \tag{55}$$

The parameter w_0 determines the skin depth of a surface. For $w_0 = 0$, the surface potential represents the step function, commonly used in the Sommerfeld model. In our illustrations, we take $w_0 = 5$. We apply our theory to the gold surface and assume that the electron effective mass is close to the free electron mass. The work function and the Fermi energy for the gold metal are equal to 5.1 eV and 5.53 eV, respectively. This means that the constant V_0 above (as the sum of the work function and the Fermi energy) equals 10.63 eV.

The surface potential described above can be generalized further to meet conditions suitable for other solids. In particular, one can take into account the space-dependent effective mass of electrons in semiconductor heterostructures or metals with effective masses different from the free electron mass.

On the other hand, the form of the laser field is assumed to depend on both space and time coordinates. Since, for laser pulses of duration $\sim$ 30 fs and 800 nm wavelength, the monochromatic approximation works well, we use the following form of the laser electric field:

$$\mathcal{E}(x,t) = \mathcal{E}_0(x)\sin(\omega t) = \mathcal{E}_0 f_L(x)\big(1+\epsilon f_P(x)\big)\sin(\omega t), \tag{56}$$

where

$$f_L(x) = g(x/\zeta_L - a_L)g(b_L - x/\mu_L), \quad f_P(x) = g(x/\zeta_P - a_P)g(b_P - x/\mu_P). \tag{57}$$

The parameter ϵ defines the plasmon-enhanced part of the laser field. We choose the Ti:sapphire laser beam of frequency $\omega = 1.5498$ eV ($\lambda = 800$ nm). This means that, inside the solid, the laser field intensity averaged over the time period decays exponentially, $I(x) \sim \mathrm{e}^{2x/\zeta_L}$. On the other hand, in vacuum, it stays constant close to the surface, and then again decays exponentially. In this way, we can mimic a real physical situation in which the radiation-filled space is finite. In our illustrations, we take $\zeta_L = 40$, which means that the penetration depth of the laser field equals $\zeta_L/2 = 20$. The parameter $a_L\zeta_L$ describes the distance in a solid at which the intensity is not reduced substantially. On the other hand, $b_L\mu_L$ corresponds to the laser focus diameter in vacuum, whereas μ_L alone determines the intensity reduction rate outside the focus. Similar parameters with the subscript P refer to the plasmon-enhanced part of the laser field. The remaining parameters have been chosen as follows: $a_L = 3$, $b_L = 20$, $\mu_L = 100$, $a_P = 1$, $\zeta_P = 8$, $b_P = 4$, $\mu_P = 20$, and $\epsilon = 0,\dots,5$. All dimensional parameters are in atomic units.

In our discussion below, the laser field intensity is characterized by the dimensionless parameter $\xi = U_p/\omega$, where $U_p = \mathcal{E}_0^2/(4\omega^2)$ is the ponderomotive energy of electrons in the monochromatic plane wave of frequency ω; hence $\mathcal{E}_0 = 2\omega\sqrt{\omega\xi}$. In Fig. 1 we draw the space-dependence of the continuous step potential $V(x)$ and the electric field amplitude $\mathcal{E}_0(x)$ for $\epsilon = 5$ and $\xi = 0.1$.

The total photoemission probability is equal to

$$P_{\mathrm{T}} = \sum_{N\geqslant N_{\mathrm{tr}}} \frac{m_0 q_N}{m_L p_0}|\mathrm{T}_N|^2. \tag{58}$$

We plot it in Fig. 1 as a function of the electron kinetic energy for $\xi = 0.1$ and for six values of ϵ. We clearly see the multi-photon structure in this distribution, i.e., the total probability jumps

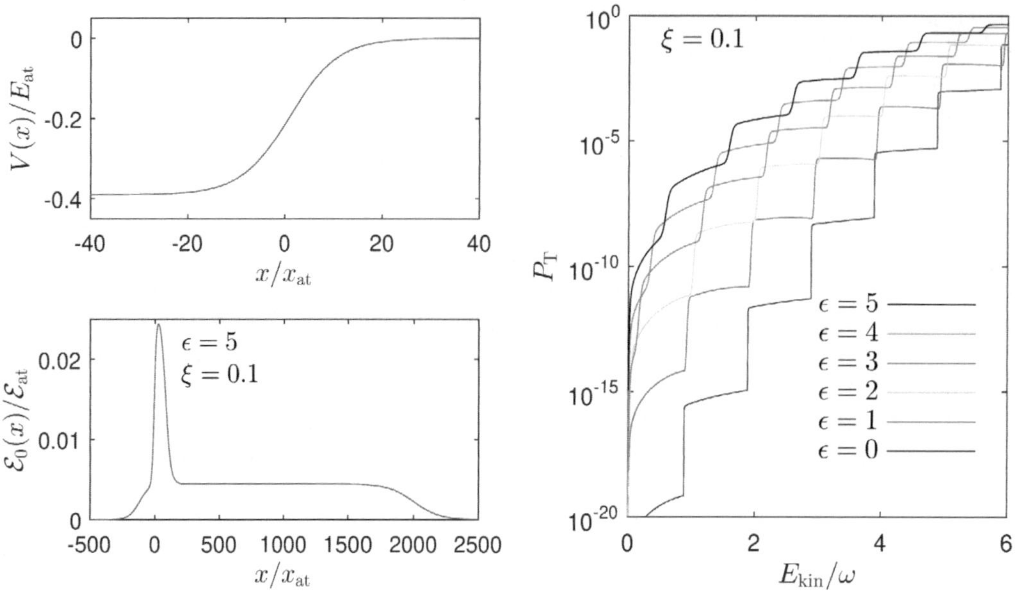

Figure 1. The continuous step potential (upper panel on the left) and the space-dependent electric field amplitude of the laser field (lower panel on the left). The atomic units of length, energy, and electric field strength are $x_{\mathrm{at}} \approx 0.053\,\mathrm{nm}$, $E_{\mathrm{at}} \approx 27.21\mathrm{eV}$, and $\mathcal{E}_{\mathrm{at}} \approx 5.14 \times 10^{11}\,\mathrm{V/m}$, respectively. In the right panel, we show total photoemission probabilities as functions of the kinetic energy of electrons for $\xi = 0.1$ and for six values of ϵ.

by few orders of magnitude if a smaller number of laser photons is sufficient for photoemission. As expected, the plasmon effect usually increases the photoemission probability. Moreover, the energy of the multi-photon channel opening increases with increasing ϵ, which is due to the increase of the space-dependent ponderomotive energy of the laser field. The significance of this effect for the tunneling phenomena is going to be discussed in due course.

7. Conclusions

As mentioned above, our numerical algorithm is convergent provided that a sufficient number of discrete points is introduced. For systems considered here, this number should not be smaller than 100. If the laser field is very weak, this does not create significant numerical problems, except that calculations become longer. However, when the laser field is sufficiently intense, the algorithm based on the transfer matrix is unstable. This instability is due to the existence of closed channels, which introduce into numerical calculations very small and very large numbers at the same time. The augmenting precision significantly slows down the calculation and does not diminish the problem. We have found that it is possible to make this algorithm numerically stable by applying non-linear matrix transformations, without introducing higher precisions.

Illustrations presented in this paper show that photoemission of electrons can be changed significantly by applying nonperturbative, oscillating in time and space-dependent electric fields. The efficiency of the numerical algorithm opens up the possibility of investigating surface phenomena in the presence of more realistic laser pulses that gradually decrease within solids and extend on a mesoscopic scale in vacuum.

Acknowledgements

We would like to thank Dr S. Varró for stimulating discussions during the LPHYS'16 conference.

References

[1] Hamilton K E, Kovalev A A, De A and Pryadko L P 2015 *J. Appl. Phys.* **117** 213103
[2] Faisal F H M and Genieser R 1989 *Phys. Lett. A* **141** 297
[3] Kamiński J Z 1993 *Acta Phys. Pol. A* **83** 495
[4] Klinovaja J, Stano P and Loss D 2016 *Phys. Rev. Lett.* **116** 176401
[5] Puviani M and Manghi F 2016 *Phys. Rev. B* **94** 161111(R)
[6] Kim S, Jin J, Kim Y-J, Park I-Y, Kim Y and Kim S-W 2008 *Nature* **453** 757
[7] Krüger M, Schenk M and Hommelhoff P 2011 *Nature* **475** 78
[8] Zhou Y and Wu M W 2011 *Phys. Rev. B* **83** 245436
[9] Khosravi H, Daneshfar N and Bahari A 2009 *Optics Lett.* **34** 1723
[10] Hsu H and Reichl L E 2006 *Phys. Rev. B* **74** 115406
[11] Mahmood F, Chan C-K, Alpichshev Z, Gardner D, Lee Y, Lee P A and Gedik N 2016 *Nat. Phys.* **12** 306
[12] Wang Y H, Steinberg H, Jarillo-Herrero P and Gedik N 2013 *Science* **342** 453
[13] Wang Z-B, Jiang H, Liu H and Xie X C 2015 *Solid State Commun.* **215-216** 18
[14] Faraggi M, Aldazabal I, Gravielle M S, Arnau A and Silkin V M 2009 *J. Opt. Soc. Am. B* **26** 2331
[15] Saathoff G, Miaja-Avila L, Aeschlimann M, Murnane M M and Kapteyn H C 2008 *Phys. Rev. A* **77** 022903
[16] Faraggi M N, Gravielle M S and Mitnik D M 2007 *Phys. Rev. A* **76** 012903
[17] Dombi P, Krausz F and Farkas G 2006 *J. Mod. Opt.* **53** 163
[18] Varró S and Kroó N 2011 *Appl. Phys. B* **105** 509
[19] Kroó N, Rácz and Varró S 2015 *EPL* **110** 67008
[20] Kroó N, Varró S, Rácz and Dombi P 2016 *Phys. Scr.* **91** 053010
[21] Faisal F H M, Kamiński J Z and Saczuk E 2005 *Phys. Rev. A* **72** 023412
[22] Faisal F H M, Kamiński J Z and Saczuk E 2006 *Laser Phys.* **16** 272
[23] Lévy-Leblond J-M 1992 *Eur. J. Phys.* **13** 215
[24] Kamiński J Z and Ehlotzky F 1999 *J. Phys. B* **32** 3193
[25] Moiseyev N and Lefebvre R 2001 *Phys. Rev. A* **64** 052711
[26] Saczuk E and Kamiński J Z 2003 *Phys. Stat. Sol. (b)* **240** 603
[27] Tsu R and Esaki L 1973 *Appl. Phys. Lett.* **22** 562
[28] Kamiński J Z 1990 *Z. Phys. D* **16** 153
[29] Faisal F H M and Kamiński J Z 1997 *Phys. Rev. A* **56** 748
[30] Faisal F H M and Kamiński J Z 1996 *Phys. Rev. A* **54**, R1769
[31] Faisal F H M and Kamiński J Z 1998 *Phys. Rev. A* **58**, R19

IOP Conf. Series: Journal of Physics: Conf. Series **826** (2017) 012014 doi:10.1088/1742-6596/826/1/012014

Quantum-Mechanical Description of Ionization-Induced Generation of Tunable Mid-Infrared Pulses

A A Silaev[1,2], **A A Romanov**[1,2], **V A Kostin**[1,2], **N V Vvedenskii**[1,2]

[1]Institute of Applied Physics, Nizhny Novgorod 603950, Russia

[2]Lobachevsky State University of Nizhny Novgorod, Nizhny Novgorod 603950, Russia

E-mail: silaev@appl.sci-nnov.ru, romanov.alexander.al@gmail.com, vk1@appl.sci-nnov.ru, vved@appl.sci-nnov.ru

Abstract. The work is devoted to the analytical study of the excitation of low-frequency current density (with frequencies much smaller than the frequency of the optical wave) which radiates in mid-infrared band in the gas ionization by two-color laser pulses. We calculate the low-frequency current density using the analytical solution of time-dependent Schrödinger equation on the basis of strong-field approximation. Closed-form analytical formulas for the low-frequency current density are obtained for laser-pulse parameters corresponding the tunneling and multiphoton ionization regimes. The features of excitation of low-frequency current density are analyzed in both regimes.

1. Introduction

The process of gas ionization by two-color laser pulses, containing the components with the frequency ratio close to two, is accompanied by excitation of the low-frequency current density (with frequencies much smaller than the frequency of the optical wave) [1–5]. The frequency detuning from the exact "synchronism" determines the frequency of the low-frequency current, which can be located in mid-infrared band [3–5]. This can be used to generate short pulses of mid-infrared radiation [4]. The low-frequency current density excited during gas ionization by a two-color laser pulse has been previously calculated analytically and numerically with the help of semiclassical and quantum-mechanical approaches [1–9]. The semiclassical approach is based on the solution of the hydrodynamic equation for the electron current density and the equation for the density of free electrons with a quasi-static probability of tunneling ionization per unit time [10]. The quantum-mechanical approach is based on the solution of the three-dimensional time-dependent Schrödinger equation for the electron wave function [9, 11]. The range of applicability of the semiclassical approach is limited by the laser pulse parameters corresponding to the tunneling ionization regime, at which the Keldysh parameter $\gamma = I_p/2U_p$ [12] is much less than unity (here I_p is the ionization potential of an atom, and U_p is the ponderomotive energy of an electron in the laser field) [5, 11, 13]. For $\gamma \gg 1$, the electron release occurs at a time of the order of the field period and greater, and to adequately calculate the low-frequency current density it is necessary to apply the quantum-mechanical approach.

This work is devoted to the derivation of closed-form analytical expressions for the low-frequency current density on the basis quantum-mechanical approach using strong-field approximation [12, 14]. Previously, such analytic expression was obtained for the laser-pulse parameters corresponding to tunneling regime of ionization [8], where the absolute coincidence with the results of semiclassical approach is achieved [5]. In this paper, we also find an analytic expression for the low-frequency current density for multiphoton ionization regime. The main features of the formation of low-frequency current density in tunneling and multiphoton regimes are discussed.

2. Analytical model

Let the electric field $\mathbf{E}(t)$ of an ionizing two-color laser pulse be linearly polarized along the z axis and be consisted of an intense component at the main frequency ω_0 and a weak additional field at a frequency $\omega_1 = 2\omega_0 + \Delta\omega$. The presence of the detuning can be considered as a change in time in the phase shift φ between the fields with frequencies ω_0 and $2\omega_0$:

$$\begin{cases} \mathbf{E}(t) = -\frac{1}{c}\frac{d\mathbf{A}}{dt}, \quad \mathbf{A}(t) = -\hat{\mathbf{z}}\frac{cE_0}{\omega_0}a(t), & (1) \\ a(t) = f(t)\left(\sin[\omega_0 t] + \frac{\alpha}{2}\sin[2\omega_0 t + \varphi(t)]\right), & (2) \end{cases}$$

where

$$\varphi(t) = \varphi_0 + \Delta\omega t. \tag{3}$$

Here, $\hat{\mathbf{z}}$ is the unit vector along the z axis, E_0 is the peak amplitude of the main field, $\alpha \ll 1$ is the small ratio of the amplitudes of the additional and main fields, ω_0 is the main frequency, φ_0 is the constant part of phase shift, $f(t)$ is the pulse envelope, and c is the speed of light in vacuum. We neglect the interaction of atoms with each other assuming that the gas density is sufficiently low. In addition, we do not take into account the polarization response of plasma assuming that the maximum density of plasma is much less than the critical density and plasma frequency is $\omega_p \ll \tau_p^{-1}$. For simplicity, we use atomic units in which $|e| = \hbar = m_e = 1$, where $\hbar$ is the reduced Planck constant, e and m_e are the electron charge and mass, respectively.

The quantum-mechanical approach for calculation of the low-frequency current density is based on the solution of time-dependent Schrödinger equation for the electron wave function ψ:

$$i\frac{\partial\psi}{\partial t} = \left(-\frac{1}{2}\nabla^2 + U(\mathbf{r}) + zE_z\right)\psi. \tag{4}$$

Here, E_z is the projection of the electric field strength on the z axis and $U(\mathbf{r})$ is the potential of interaction with the parent ion. For simplicity, we assume that the gas consists of hydrogen atoms. In this case, the potential of the ion is the Coulomb potential, $U(\mathbf{r}) = -1/r$. Due to cylindrical symmetry of the system, the excited electron current density is directed along the z axis.

Let us assume that the pulse duration is sufficiently great, so that the concentration of free electrons increases during many periods of the electric field. This makes it possible to present the low-frequency current density as the integral $\overline{j}(t) = \int_{-\infty}^{t}(\partial\overline{j}/\partial t)dt$, where time derivative of the low-frequency current density is

$$\frac{\partial\overline{j}}{\partial t} = -(N_g - \overline{N})\int p_z\overline{W}(\mathbf{p}, t)d^3p. \tag{5}$$

Here, $\overline{W}(\mathbf{p}, t)$ is the time-averaged momentum distribution of the ionization probability per unit time, N_g is the initial concentration of gas atoms, and $\overline{N}$ is the time-averaged concentration of electrons, which satisfies the approximate equation

$$\frac{\partial\overline{N}}{\partial t} = (N_g - \overline{N})\int \overline{W}(\mathbf{p}, t)d^3p. \tag{6}$$

Since the characteristic time of variation of $\overline{W}(\mathbf{p},t)$ is much larger than the field period, in order to calculate $\overline{W}(\mathbf{p},t)$ we can assume that the electric field has constant envelope and phase shift [8] and that the depletion of the ground state of an atom for the period of the laser pulse field is negligible. The latter assumption is valid even at a very high peak intensity, because in this case, the depletion of the ground state of the atom takes place mainly at the leading edge of the laser pulse at an intensity that is close to the breakdown threshold. Within the framework of the strong-field approximation [12], in which we neglect the interaction of photoelectrons with the parent ion, $\overline{W}(\mathbf{p},t)$ is represented as the sum of the probabilities of n-photon processes:

$$\overline{W}(\mathbf{p},t) = \frac{\omega_0^2}{2\pi} |L(\mathbf{p})|^2 \sum_{n=m}^{\infty} \delta\left(\Delta E - n\omega_0\right). \tag{7}$$

Here, $m = \langle \tilde{I}_p/\omega_0 + 1 \rangle$ is the minimum number of photons required for ionization (the expression $\langle x \rangle$ denotes the integer part of the number x), $\tilde{I}_p = I_p + U_p$, $U_p = U_{p0}(1+\alpha^2/4)$ is the ponderomotive energy in the two-color field, $U_{p0} = E_{f0}^2/4\omega_0^2$ is the ponderomotive energy in the field of main field, $E_{f0}(t) = E_0 f(t)$ is the main field amplitude, $\Delta E = p^2/2 + \tilde{I}_p$ is the energy for ionization and acceleration of the electron, and $\delta(x)$ is Dirac delta function. Thus, the momentum distribution of photoelectrons is concentrated near spheres with radii corresponding to the final momentum of the photoelectrons which satisfies the energy conservation law, $\Delta E = n\omega_0$, where n is natural number.

The function $L(\mathbf{p})$ in (8) describes the envelope of the momentum distribution of the ionization probability per unit time. Assuming that the photon energy is much less than the ionization energy (i.e. $n_0 = I_p/\omega_0 \gg 1$), the function $L(\mathbf{p})$ is

$$L(\mathbf{p}) = \sum_{t_s} e^{iS(\mathbf{p},t_s)+iS_C(\mathbf{p},t_s)} \left(\pi \left.\frac{\partial^2 S}{\partial t^2}\right|_{t_s}\right)^{-1/2}. \tag{8}$$

Here,

$$S(\mathbf{p},t) = \int_0^t \left(\frac{(\mathbf{p}+\mathbf{A}/c)^2}{2} + I_p\right) dt' \tag{9}$$

is the part of the action of a free electron which does not depend on the coordinates, t_s are the stationary points of $S(\mathbf{p},t)$ and s is an index numbering the stationary points. The t_s values satisfy the equation

$$\partial S(\mathbf{p},t)/\partial t|_{t_s} = 0 \tag{10}$$

and have a positive imaginary part and a real part which lies in the range $[0, 2\pi/\omega_0)$. S_C is correction of action associated with the interaction of photoelectrons with the parent ion [15]. Further we take into account the zero-order correction of action which is calculated neglecting the influence of additional-field component. Accounting of this correction in the exponent in (8) leads to the well-known Coulomb factor [15, 16], which is valid for arbitrary values of Keldysh parameter $\gamma = \sqrt{I_p/2U_{p0}}$,

$$Q_C = \left|e^{iS_C}\right|^2 = (2/E_{f0})^2(1+2\gamma/e)^{-2}. \tag{11}$$

In order to calculate the sum in Eq. (8) we will take into account the action of the second harmonic field by considering terms linear with α. Under the conditions $\alpha \ll \gamma, \gamma^{-1}$ the stationary points of action are found in the absence of additional field:

$$t_1^0 = \omega_0^{-1} \arcsin\left(\gamma p_z + i\gamma\sqrt{1+p_\perp^2}\right), \tag{12}$$

$$t_2^0 = \pi/\omega_0 - (t_1^0)^*. \tag{13}$$

Here p_z is the projection of momentum on the z axis, $p_\perp$ is the module of the transverse momentum. Following the work [12] we assume that the main contribution in $\overline{W}(\mathbf{p},t)$ is given by the small values of final electron momentum, $p^2 \ll 1$. It makes possible to neglect the momentum dependence in the pre-exponential factor and use a Taylor series expansion of p up to the quadratic terms in the exponential factor:

$$\overline{W}(\mathbf{p}) \approx \frac{\omega_0\gamma Q_C}{\pi^2\sqrt{1+\gamma^2}} \exp(2n_0 b(\mathbf{p}))\left[\cosh\left(2n_0 a(\mathbf{p})\right)+\varepsilon(\mathbf{p},\gamma)\right] \sum_{n=-\infty}^{\infty} \delta\left(\Delta E - n\omega_0\right). \qquad (14)$$

Here,

$$a(\mathbf{p}) = \alpha\gamma\cos\varphi\left(\frac{2}{3}+p_\perp^2\right), \qquad (15)$$

$$b(\mathbf{p}) = \frac{\gamma p_z}{\sqrt{\gamma^2+1}}\left(p_z-\alpha\gamma\sin\varphi\right)+\frac{\sqrt{\gamma^2+1}}{2\gamma}-\left(p^2+1+\frac{1}{2\gamma^2}\right)\operatorname{arcsh}\gamma. \qquad (16)$$

The term $\varepsilon(\mathbf{p},\gamma)$ in Eq. (14) is associated with the intercycle interference of two electron trajectories originating from electron ionization from neighboring half-cycles [17]. Therefore, when calculating the integral characteristics such as average ionization probability and current density, one can neglect the term ε. The rest of the function $\overline{W}(\mathbf{p})$ is a product of the smooth envelope and the sum of delta functions corresponding the spheres defined by the energy conservation law. The maximum of the smooth envelope corresponds to some optimal momentum

$$p_{opt} = \frac{\alpha\gamma\sin\varphi}{2\left(1-\sqrt{1+\gamma^{-2}}\operatorname{arcsh}\gamma\right)}. \qquad (17)$$

It can be seen that for $\alpha = 0$ the optimal momentum is $p_{opt} = 0$ and the function $\overline{W}(\mathbf{p})$ is symmetric in the longitudinal momentum. Therefore, in the absence of the second harmonic the average photoelectron momentum is zero. The addition of a small field at the doubled frequency breaks the symmetry of the momentum distribution of the ionization probability. It leads to the excitation of non-zero low-frequency current density.

Substituting Eq. (14) in Eq. (5) we obtain that the low-frequency current and free-electron density can be represented as a sum over n-photon processes:

$$\frac{\partial\bar{j}}{\partial t} = -C_1\sum_{n=m}^{\infty}\int_{-p_n}^{p_n} F_n(p_n,p_z)p_z dp_z, \qquad (18)$$

$$\frac{\partial\overline{N}}{\partial t} = C_1\sum_{n=m}^{\infty}\int_{-p_n}^{p_n} F_n(p_n,p_z)dp_z, \qquad (19)$$

$$C_1 = (N_g-\overline{N})\frac{2\omega_0\gamma Q_C}{\pi\sqrt{1+\gamma^2}}\exp\left[-\frac{2\tilde{I}_p}{\omega_0}\left(\operatorname{arcsh}\gamma-\gamma\frac{\sqrt{\gamma^2+1}}{2\gamma^2+1}\right)\right]. \qquad (20)$$

Here, $p_n = [n/n_0-(1+1/2\gamma^2)]^{1/2}$ is the momentum corresponding to n-photon absorption, and

$$F_n = \exp(-2n_0p_n^2\operatorname{arcsh}\gamma)\cosh\left[2n_0\alpha\gamma\cos\varphi(2/3+p_n^2-p_z^2)\right]\exp\left[\frac{2n_0\gamma}{\sqrt{\gamma^2+1}}\left(p_z^2-\alpha\gamma\sin\varphi p_z\right)\right]. \qquad (21)$$

The obtained expression is rather complicated. However, it can be significantly simplified in two limiting cases: in the tunneling regime of ionization ($\gamma \ll 1$) and multiphoton regime of ionization ($\gamma \gg 1$).

IOP Conf. Series: Journal of Physics: Conf. Series **826** (2017) 012014 doi:10.1088/1742-6596/826/1/012014

2.1. Tunneling limit

In the case $\gamma \ll 1$, the characteristic scale of the envelope distribution of photoelectrons $\overline{W}(\mathbf{p})$ is much wider than the distance between the spheres of location of function $\overline{W}(\mathbf{p})$. This allows one to replace the summation by integration in Eq. (18). As a result, time derivative of low-frequency current density is represented as a product of the averaged ionization rate and the most probable momentum

$$\frac{\partial \overline{j}}{\partial t} = -p_{opt}\frac{\partial \overline{N}}{\partial t}, \tag{22}$$

where

$$p_{opt} = -\frac{3}{2}\frac{\alpha E_{f0}}{\omega_0}\sin\varphi. \tag{23}$$

The equation for the averaged electron density becomes

$$\frac{\partial \overline{N}}{\partial t} = (N_g - \overline{N})\overline{w}_0(E_{f0})\cosh\left(\frac{2\alpha}{3E_{f0}}\cos\varphi\right), \tag{24}$$

where

$$\overline{w}_0(E) = 4\sqrt{\frac{3}{\pi E}}\exp\left(-\frac{2}{3E}\right) \tag{25}$$

is the averaged tunneling ionization probability in the main field of the laser pulse.

2.2. Multiphoton limit

In the multiphoton regime of ionization the distance between the spheres (at which the momentum distribution is concentrated) is comparable with the characteristic scale of the distribution. As a result, the main contribution to the low-frequency current density is provided by the electrons which absorb the minimum possible number of photons $n = m$:

$$\frac{\partial \overline{j}}{\partial t} = \left[C_2\sqrt{2/n_0}\sinh\left(\alpha p_m E_{f0}^{-1}\sin\varphi\right) - \frac{1}{2}\alpha\gamma\sin\varphi\frac{\partial \overline{N}}{\partial t}\right], \tag{26}$$

$$\frac{\partial \overline{N}}{\partial t} = C_2\sum_{s=\pm 1}\exp\left[s\alpha p_m E_{f0}^{-1}\sin\varphi\right]\mathcal{F}\left[\sqrt{2n_0}(p_m + \alpha\gamma s\sin\varphi/2)\right], \tag{27}$$

$$C_2 = (N_g - \overline{N})2^{1-2m}\pi^{-1}e^{m+2}\omega_0^{-2m-1/2}E_{f0}^{2m}\cosh\left(\frac{2\alpha}{3E_{f0}}\cos\varphi\right). \tag{28}$$

Here $\mathcal{F}(x) = \int_0^x \exp(y^2 - x^2)dy$ is the Dawson integral.

3. Discussions and conclusions

The obtained expressions reveal that in the tunneling regime of ionization the time derivative of the low-frequency current density is a sine function of the phase shift φ between the carriers of the laser-pulse harmonics. Therefore, when the frequency of additional field is detuned from $2\omega_0$, the maximum of low-frequency current spectrum corresponds to the detuning frequency $\Delta\omega$ (in accordance with the results of the semiclassical model [3–5]). As can be seen from the formula (26), in the multiphoton regime $\partial\overline{j}/\partial t$ is a complex function of φ. The form of this function depends on the relationship between the parameters α, γ, and n_0. In particular, in the multiphoton regime the factor $\cosh(2\alpha\cos\varphi/3E_{f0})$ becomes significant. It is due to the fact that the probability of atom ionization dramatically increases when peaks of the laser-pulse components are superimposed (it is implemented for the phase shift $\varphi = 0$). As a result, the form of the spectrum of low-frequency current can differ from the corresponding spectrum in tunneling regime. In particular, the maximum of spectrum can be shifted from the detuning frequency. However, the strong-field approximation used in this paper may give inaccurate results in finding the dependence of low-frequency current density on the phase. It is due to the fact that the Coulomb correction to the action depends essentially on the final momentum of the photoelectron in multiphoton regime. The account of this dependence can be made on the basis of imaginary time method used in [7, 9] for the calculation of residual current density excited by two-color laser pulse.

Unlike the phase dependences, the analytical result for maximum of time derivative of low-frequency current density (corresponding to the optimum phase) is, apparently, accurate throughout the full range of the Keldysh parameter. This is associated with the fact that accounting for the zero-order Coulomb correction to the electron action leads to a quantitatively correct values of the probability of ionization per unit time [16]. As can be seen from the formula (26), the dependence of the maximum $\partial\bar{j}/\partial t$ on the intensity of the main field is non-monotonic function for $\gamma \gg 1$. It has minima close to certain intensity values corresponding to the thresholds of closing of ionization channels. When the intensity increases, the distance between the minima is reduced, and they become less pronounced. The investigation of threshold effects in the formation of low-frequency current density can become interesting topic of research, since it is important for understanding the physical bases of the ionization-induced transformation of two-color laser pulses into low-frequency radiation.

Acknowledgments

The development of the general analytical expressions for ionization probability and low-frequency current density was supported by the Russian Foundation for Basic Research (RFBR) under Grant No. 16-32-60200. The comparison with the semiclassical results was supported by the RFBR under Grant No. 16-32-60166. The analysis of excitation of low-frequency current in the regimes of tunneling and multiphoton ionization was supported by the Russian Science Foundation (Grant No. 15-12-10033).

References

[1] Kim K Y, Taylor A J, Glownia J H and Rodriguez G 2008 *Nat. Photonics* **2** 605–609
[2] Clerici M, Peccianti M, Schmidt B E, Caspani L, Shalaby M, Giguére M, Lotti A, Couairon A, Légaré F, Ozaki T, Faccio D and Morandotti R 2013 *Phys. Rev. Lett.* **110** 253901
[3] Vvedenskii N V, Korytin A I, Kostin V A, Murzanev A A, Silaev A A and Stepanov A N 2014 *Phys. Rev. Lett.* **112** 055004
[4] Balčiūnas T, Lorenc D, Ivanov M, Smirnova O, Zheltikov A, Dietze D, Unterrainer K, Rathje T, Paulus G, Baltuška A *et al.* 2015 *Opt. Exp.* **23** 15278
[5] Kostin V, Laryushin I, Silaev A and Vvedenskii N 2016 *Phys. Rev. Lett.* **117** 035003
[6] Chen W, Huang Y, Meng C, Liu J, Zhou Z, Zhang D, Yuan J and Zhao Z 2015 *Phys. Rev. A* **92** 033410
[7] Popruzhenko S and Tulsky V 2015 *Phys. Rev. A* **92** 033414
[8] Romanov A, Silaev A and Vvedenskii N 2015 *Problems of Atomic Science and Technology* **98** 286
[9] Vvedenskii N V, Romanov A and Silaev A A 2016 *Quant. Electr.* **46** 426
[10] Gildenburg V B and Vvedenskii N V 2007 *Phys. Rev. Lett.* **98** 245002
[11] Silaev A A and Vvedenskii N V 2009 *Phys. Rev. Lett.* **102** 115005
[12] Keldysh L V 1965 *Sov. Phys. JETP* **20** 1307
[13] Silaev A A and Vvedenskii N V 2015 *Phys. Plasmas* **22** 053103
[14] Popov V S 2004 *Phys. Usp.* **47** 855
[15] Perelomov A and Popov V 1967 *Sov. Phys. JETP* **25** 207
[16] Popruzhenko S, Mur V, Popov V and Bauer D 2008 *Phys. Rev. Lett.* **101** 193003
[17] Karnakov B M, Mur V D, Popruzhenko S V and Popov V S 2015 *Phys. Usp.* **58** 3

Comparative evaluation of the effects of high-intensity and low- intensity laser radiation on microcirculation among patients with knee arthritis

D B Kulchitskaya[1], T V Konchugova[1, 2], N E Fedorova[1]

[1] Federal State Institution "Russian Scientific Center of Medical Rehabilitation and Balneology" of Ministry of Health of the Russian Federation, Moscow, Russia
[2] I M Sechenov First Moscow State Medical University, Moscow, Russia

n.fedorova@mail.ru

Abstract. Sixty patients with knee arthritis aged from 40 to 75 years old were examined. The patients were randomly divided into two groups: 1st group (30 patients) received high intensity laser radiation; 2nd group (20 patients) received low intensity laser radiation. As a result of the conducted research it was found that high intensity laser radiation is more efficient and leads to more vivid positive changes in the microcirculation of patients with knee arthritis. The changes in microcirculation were based on the normalization of the myogenic and neurogenic tonus of the arterioles, strengthening oscillation of the endothelial range. As a result of local mechanisms activation of tissue blood flow there occurs adequate modulation of the microcirculatory bloodstream, which is aimed at the elimination of congestive phenomena in the capillary and venular level of the microcirculation bloodstream. We should note that in the long-term more significant were the positive changes in the state of the venular level of the microcirculation bloodstream. in constructing both.

1. **Introduction** Arthritis of the knee joint is one of the most common causes of knee pain. There are different types of arthritis that can affect the knee joint. Osteoarthritis is the most common type of knee arthritis. Also called degenerative joint disease, osteoarthritis is characterized by progressive wearing away of the cartilage in the joint. The knee becomes swollen, and activities become increasingly painful. The most common symptoms of knee arthritis include: 1-pain with activities, 2-limited range of motion, 3- stiffness of the joint, 4-swelling of the joint, 5-deformity of the joint. Osteoarthritis is often the result of increased strain on the joint due to weight, or repeated injury or damage. Knee arthritis is definitely more common among middle aged or elderly persons. It is treated to provide pain relief and to increase mobility and strength in the knee. Treatment options include exercises, medications, physical therapy, surgery [1,2,3].Usually all other methods of knee arthritis treatments are exhausted before surgery is considered. Physical therapy is one of them.
Clinical and basic research studies have demonstrated the physiological effects and medical applicability of low-level laser therapy (LLLT) [4, 5]. Ohshiro also demonstrated a positive effect on microcirculation and verified changes by thermography in parallel with the reduction of pain [1, 6]. In addition to cartilage damage and bone metabolism, pathological alterations are also known to exhibit reduced circulation in the vessels of the joint parallel to the degenerative changes. Numerous authors have reported increased micro-vascularization as a histological effect of the laser beam [7, 8].

LLLT reduces pain in knee arthritis and improves microcirculation in the irradiated area. In summary, low-level laser represents an effective treatment tool for short-term improvement in patients suffering from painful knee arthritis [9]. In recent years, new physiotherapy methods, particularly the usage of high intensity laser radiation (HIL-therapy), have been applied in medical practice [10, 11]. However, to date there is little research on the scientific justification of the use of this factor.
The goal of this research is in a comparative study of the effect of low intensity and high intensity laser radiation on the microcirculation among patients with knee arthritis.

2. Methods of studying

The study involved 60 patients with knee arthritis aged between 40 and 75 years. Patients were divided randomly into two groups:
- Group 1 (30 patients) received high-intensity laser pulses delivered by a BTL-6000 HIL device, wavelength of 1064 nm, pulse mode, frequency of 25 Hz, a dose of 10 J/cm^2, onto the projection area of joint spaces and popliteal region, 4 minutes per field, alternate days, a course of 10 treatment sessions;
- Group 2 (30 patients) received low intensity laser impact using a "Mustang 2000" device, wavelength of 890 nm, pulse mode, frequency of 80 Hz, pulse power of 40W, onto the projection area of joint spaces and popliteal region, 4 minutes per field, alternate days, a course of 10 treatment sessions.

In this study, the state of the capillary blood flow was evaluated by laser Doppler flowmetry (LDF), which was carried out using a "LAKK-M" laser capillary blood flow analyzer. The skin of the anterior surface of the knee joint in supine position was studied, at room temperature not less than 20°C.
In the course of the study, the following indicators of LDF-signal were recorded and calculated: mean value of microcirculation index (MI) and its standard deviation (σ).
Analysis of the amplitude and frequency of the rhythmic components of flux motions (fluctuations in the flow of red blood cells measured by LDF) was conducted based on the use of the mathematical apparatus technique of the wavelet transform.
Using wavelet analysis, we calculated and analyzed the amplitude and frequency of rhythmic components, such as :
I-rhythms caused by the secretory activity of the endothelium (E);
II-neurogenic fluctuations arising from the sympathetic adrenergic effects on the smooth muscle in the arterioles and meta-arterioles (N);
III-myogenic rhythms caused by the own internal activity of myocytes by pacemaker mechanism (M);
IV-respiratory rhythms (D);
V-cardiac rhythms (C).
Evaluation of the amplitude (A) indices of each rhythm depended on the strength of LDF-signal (M): A rhythm/M ×100%, and against the value of its maximum spread (σ): A rhythm/3σ ×100%.

3. Results and Discussion

Before the treatment, most patients had distinct clinical manifestations of knee arthritis, the most important of which were pain during movement and pain after getting up or a period of inactivity. Increase in the affected joint circumference compared with the symmetric healthy one, sensitivity to palpation, rough crepitus during movement, restrain of movements of various degrees, changes in gait were objectively found.
According to the LDF, disorders of microcirculation (MC) were revealed, which allowed dividing all patients into two groups according to the classification proposed by V I Makolkin:
Group A (91%) - Patients with congestive-stagnant type;
Group B (9%) - Patients with normocirculatory type of microcirculation.
Group A patients had an increase in the amplitude of myogenic and neurogenic oscillations by 18% ($p < 0.05$) and 16% ($p < 0.05$), respectively. These data were confirmed by the low myogenic and neurogenic tone of arterioles. A decrease in the amplitude of endothelial oscillations by 12% ($p < 0.05$) and presence of congestive phenomena in the capillary and venular links of the microvasculature were found. The mentioned changes were confirmed by an increase in the index characterizing the

Table 1. Dynamics of LDF-factors under the influence of prolonged treatment effects of high intensity and low intensity laser radiation; statistical significance of P before and after treatment: * — < 0.05, ** — < 0.01, *** — < 0.001.

Amax / 3σ x 100%	**E**	**N**	**M**	**D**	**C**
Normal range	14.1±0.9%	17.1±0.8%	15.0±0.9%	7.9±0.8%	5.7±0,7%
Group 1 before treatment	12.39±0.21%	19.74±0.7%	17.74±0.5%	12.3±0.8%	8.1±0.3%
Group 1 after treatment	13.87±0.1% ***	17.5±0.4% **	15.2±0.3% ***	9.1±0.4% **	7.0±0.2% **
Group 2 before treatment	12.37±0.22%	19.76±0.7%	17.91±0,6%	12.7±0.8%	8.0±0.3%
Group 2 after treatment	13.0±0.11%**	18.1±0.4% *	16.03±0.4% *	9.9±0.5% **	7.3±0.4% **

contribution of pulse and respiratory oscillations (Ac/3σ × 100%, and Ad/ 3σ × 100%, respectively) (Table 1).

The patients in Group B had LDF indicators within the normal range.

Following the laser therapy, the examined patients had positive dynamics in LDF indicators. However, more significant changes were observed in patients treated with high-intensity laser radiation. Thus, this category of patients showed improvement in the endothelial function - Ac/3σ 100% indicator changed by 10% ($p<0.001$) in comparison to the initial one, whereas the second group of patients had an increase in this indicator that was 2 times less – by 5% ($p <0.01$) to the initial one. A significant decrease in Am/3σ × 100% indicator by 18% ($p <0.001$) and by 12% ($p <0.05$) against the initial values was discovered in groups, respectively, which implies normalization of the myogenic tone of arterioles.

As a result of applying sessions of high intensity laser radiation and low intensity laser radiation among patients with knee arthritis, positive dynamics of An/3σ × 100% indicator, the state of the neurogenic tone of arterioles, was observed. It decreased by 14% ($p <0.01$) and by 9% ($p <0.05$) against the initial values in the groups, respectively. The above changes were accompanied by improvement in blood filling in the capillary and venular parts of the microcirculation bloodstream. The data obtained by using LDF correlated with the regression of clinical symptoms and signs. Thus, the majority of patients experienced a significant reduction in the severity of pain during activity. It was manifested in a significant decrease in VAS among patients of Group 1 during the second and third treatment sessions by 39.4% and at the end of treatment - by 46.7%. In Group 2, the value decreased by 17.6% in the middle of the treatment, and by 25.1% at the end of the treatment. The intensity of pain at rest changed similarly. The patients of Group 1 experienced a significant decrease in pain intensity by 46.45% against the initial data after the second and third sessions, and by 78.78% by the end of treatment. In Group 2, pain intensity decreased by 19.97% in the middle of treatment and by 34.68% at the end of treatment.

Table 2. Dynamics of LDF-factors 12 months after prolonged treatment with high intensity laser radiation; statistical significance of P— in comparison with values before treatment: * — < 0.05, ** — < 0.01, *** — < 0.001.

Amax / 3σ x 100%	**E**	**N**	**M**	**D**	**C**
Normal range	14.1±0.9%	17.1±0.8	15.0±0.9	7.9±0.8%	5.7±0.7
Group 1 before treatment	12.39±0.21	19.74±0.7	17.74±0.5	12.3±0.8	8.1±0.3
Group 1 after treatment	13.87±0.1 %	17.5±0.4%	15.2±0.3 %	9.1±0.4%	7.0±0.2 %
Group 1 12 months after treatment	12.97±0.2 %	18.7±0.3%	16.1±0.6 %	8.2±0.8%	6.1±0.3 %

4. Conclusion

Summarizing the results of the studies relating to the effect of high intensity laser radiation and low intensity laser radiation on the state of microcirculation among patients with knee arthritis, it may be noted that positive changes in the microcirculatory hemodynamics were more significant when using high-intensity laser therapy. The changes in microcirculation were due to normalization of the myogenic and neurogenic tonus of arterioles, and increased oscillations of the endothelial range. Activation of the local mechanisms of tissue blood flow resulted in an adequate modulation of blood flow aimed at the elimination of congestion in the capillary and venular levels of the microcirculation bloodstream.

Analysis of afterhistory, according to LDF, confirmed that 12 months after the treatment with high intensity laser radiation the myogenic and neurogenic tonus decreased, but remained above the initial values (Table 2). Conspicuous is the fact that in the long-term period positive changes in the state of the venular level of the microcirculation bloodstream were more significant. Ad/3σ × 100% indicator improved by 26% by treatment sessions of high intensity laser radiation, whereas 12 months after the treatment it improved by 34%.

5. References

[1] Konchugova T V, Orekhova E M, Kulchitskaya D B 2013 *Issues of Balneology, Physiotherapy and Exercise Therapy* **90** 26-31

[2] Dashina T A, Grigorieva V D, Sidorov V D 2011 *Issues of Balneology, Physiotherapy and Exercise Therapy* № 1 16-20

[3].Titskaya E V, Levitsky E F, Zaripova T N, Reshetova G G, Kozlov S V, Barabash L V, Alaytseva S V, Dostovalova O V, Shakhova S S 2014 *Physiotherapy, Balneology and Rehabilitation* № 6 31-38

[4].Alfredo P P, Bjordal J M, Dreyer S H et al 2012 *Clininal Rehabilitation* **26** 523-533

[5].Alghadir A, Omar M T, Al-Askar A B, Al-Muteri N K 2014 *Lasers in Medical Science* **29** 749-755

[6] Ohshiro T 1988 *Low-Level Laser Therapy: A Practical Introduction* (New York: John Willey and Son) pp. 56–62

[7] Tumilty S, Munn J, McDonough S, Hurley D A; Basford J R; Baxter G D 2010 *Photomedicine and Laser Surgery* **28** 3-16

[8] Kulchitskaya D B, Konchugova T V, Kolbaya L I 2012 *Issues of Balneology, Physiotherapy and Exercise Therapy* №1 14-16

[9]. Hegedűs B, Viharos L, Gervain M, and Gálfi M 2009 *Laser Surgery* **27** 577–584

[10]. Thabet A A M M, Mohamed M E S, Ali M M I, Helal O F 2013 *Energy for Health* **10** 16-21

[11]. Viliani T, Carabba C, Mangone G, Pasquetti P 2012 *Energy for Health* **9** 18-22

New method in medical tomography based on vibrating wire: bench-test experiment on laser beam

M.A. Aginian[1], J Alonso[3], S.G.Arutunian[1*], M. Chung[2*], A.V. Margaryan[1], E.G. Lazareva[1], L.M. Lazarev[1], L.A. Shahinyan[1]

1 Yerevan Physics Institute, Alikhanian Br. St. 2, Yerevan, 0036, Armenia
2 Ulsan National Institute of Science and Technology, Ulsan, 689-798, Korea
3 Massachusetts Institute of Technology, Cambridge MA, USA

*Corresponding authors: S. G. Arutunian (femto@yerphi.am), M. Chung (mchung@unist.ac.kr)

Abstract. A new method for fast transverse beam profiling, where a vibrating wire is served as a resonant target, has been developed. The speed of scan up to a few hundred mm/s provides opportunity to make a set of beam profiles at different directions of the scan within a reasonable measurement time. This profile set allows us to reconstruct 2D beam profile by filtered back-projection algorithm. The new method may be applied for proton, X-ray, gamma, and neutron beams, and can also be of interest in tomography including medical applications. The method has been tested experimentally by means of laser beams.

1. Resonant target method

Wire scanners widely used for profile measurements of various types of beams, are based on the detection of secondary particles/radiation generated when the beam particles interact with the wire. To pick out this beam signal from the high level background, we propose using vibrating wire as a target whose oscillation frequency serves as a reference to separate signal from noise in the measurements [1-3]. The principle of operation of the proposed method is described in Fig. 1.

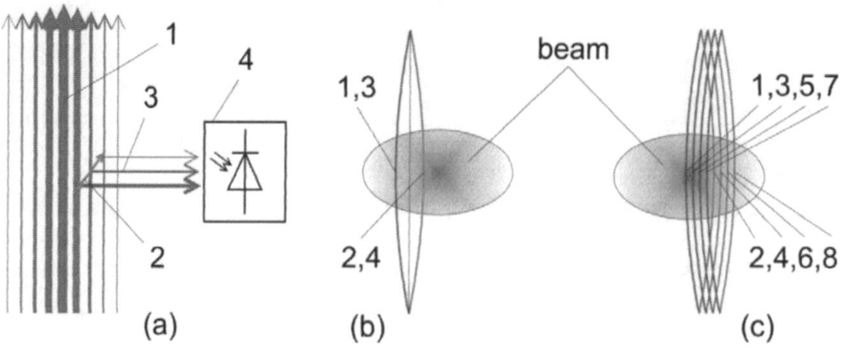

Figure 1.1. The main principle of resonant target method is based on the measurements of scattered/reflected particles/photons in opposite positions of the wire during oscillation process. In

figure: (a) the nonuniform beam flux is represented by vertical lines with different thicknesses (1), vibrating wire is stretch perpendicular to the figure plane and wire center is moving along segment (2), (3) - scattered/reflected particles/photons are detected by detector (4) with synchronism with wire oscillation frequency: (b) plane of unmovable vibrating wire, (1), (3) - positions of the wire at "left" measurements, (2), (4) - at "right" measurements, beam is presented by ellipse with nonuniform distribution; (c) plane of movable vibrating wire, (1), (3), (5), (7) - positions of the wire at "left" measurements, (2), (4), (6), (8) - at "right" measurements. Each step of wire feed at one period is less than oscillation amplitude. All schematics are not to scale.

Serial subtraction of measurements [1] of the photons reflected from the opposite positions of the vibrating wire oscillations (mentioned in Fig. 1 as "left" and "right") eliminates a high level of background noise falling on the photodiode during the photon measurements and also minimizes the noise in the measurement circuits.

In Fig. 1.2 we present the main results of laser beam 1D profiling [3] made by this method for the case where besides the main measurement object (i.e., laser beam), a photodiode also measures a 50 Hz lighting and considerable reflection of the laser beam from the holder of vibrating wire. The proposed measurement method allows reconstructing the laser beam profile by acquiring profile gradients. Speed of scan was about 12 mm/s, the scan of about 3 mm wide laser beam took about 250 ms.

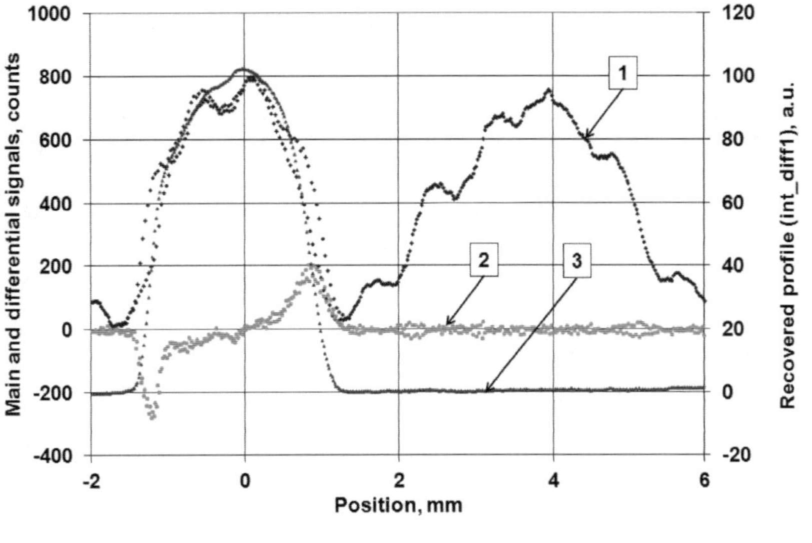

Figure 1.2. Laser beam 1D profiling made by RT method separate the photon reflections on the vibrating wire from the reflections on holder of the monitor and 50 Hz background: 1 - ADC measurements by fast photodiodes in synchronism with wire oscillations, the left peak is originated by the reflections from the wire while the right is reflections from the monitor holder; 2 - differential signal of wire half-period measurements, 3 - reconstructed profile provide only laser beam. All scan from -2 mm to 6 mm lasted about 600 ms.

2. Tomography mathematics

The aim of tomography is the reconstruction of distribution of a 2D object from a set of its 1D projections (line integrals along a finite number of lines of known directions). From many of mathematical methods of such a reconstruction we chose the filtered back projection method described in details in [4-6] and proposed in [2] for recovering of complex proton beam profile by mean of vibrating wire monitors. In the case the vibrating wire accumulate the information of particles penetrating the wire either by measuring of natural frequency shift caused by temperature increase or measuring of scattered/reflected particles in known wire position. Such measurements give the integral of particles distribution along the wire (one point in 1D profile at given scan with fixed direction of the wire). Cause by thermal type of interaction of the beam with vibrating wire (wire heating at beam impact) the each 1D scan requires tens of seconds. Usage of vibrating wire as a resonant target allows substantially speed up the fulfilment of the 1D scan and solve the tomography problem at reasonable time.

We assume that the beam intersects the investigated object in so-called reconstruction region and work in following geometry: the wire direction remains the same (assumed horizontal) while the investigated object is rotating. In discrete space we define as reconstruction region square (N_{scr}, N_{scr}) (numbering of indexes according to Visual Basic mathematics that we would like apply started from 0). By first index of arrays in this space we present the x -coordinate and by second index – y -coordinate. By Beam (N_{scr}, N_{scr}) we denote the array of initial beam profile at rotation angle $\varphi = 0$. Array of beam projections along the x -axis we denote as Pro_y(N_{φ}, N_{scr}) (first index – angle φ of projection, second index – y-coordinate). The array of reconstructed beam profile we denoted as RezBeam(N_{scr}, N_{scr}).

2.1. Reconstruction algorithm

When there is no a priori knowledge of the object, we always assume that the intensity of the object is uniform along the ray path. In other words, we distribute the projection intensity evenly among all pixels along the ray path. This process leads to the concept of backprojection.
The projection process and following trivial back projection scheme is presented in Fig. 2.1.

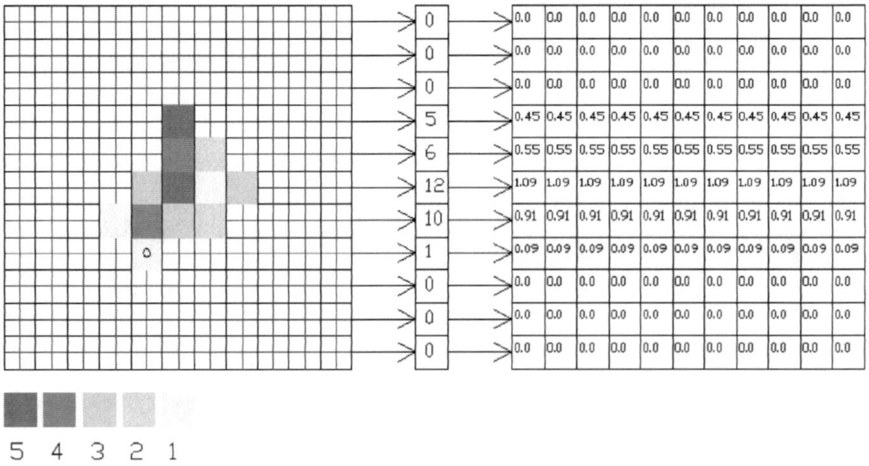

Figure 2.1. "Trivial" reconstruction of one projection deposition (backprojection). Given projection corresponds to $\varphi = 0$, $N_{scr} = 10$. Left - the model beam in discrete reconstruction region, middle column - the projection for angle $\varphi = 0$, right - the uniformly spread of projection value along the corresponding horizontal rows.

Cause to chosen geometry the aim is to reconstruct the beam profile for $\varphi = 0$. So the procedure of reconstruction is following: for given angle $\varphi > 0$ we make back-projection then rotate it back on the angle $-\varphi$ and accumulate into reconstruction array. For this purposes we introduce two auxiliary arrays Rec_φ(N_{scr}, N_{scr}) and Rec_0(N_{scr}, N_{scr}) - first for reconstructed back-projection for angle $\varphi > 0$ and the second for back-projection array rotated back on the angle $-\varphi$.
The whole algorithm for "trivial" back-projection can be presented as following

$$\text{Rec\_}\varphi(\text{i\_x}, \text{i\_y}) = \text{Pro\_y}(\text{i\_}\varphi, \text{i\_y})/N_{scr}, \text{ for i\_x} = 0 \ldots N_{scr}, \text{ i\_y} = 0 \ldots N_{scr}, \quad (1)$$
$$\text{Rec\_0}(N_{scr}, N_{scr}) = \text{Rotate}(\text{Rec\_}\varphi(N_{scr}, N_{scr}), -\varphi), \quad (2)$$
$$\text{RezBeam}(N_{scr}, N_{scr}) => \mathit{RezBeam}(N_{scr}, N_{scr}) + \text{Rec\_0}(N_{scr}, N_{scr}), \quad (3)$$

where $\text{Rotate}(\text{Array}(N_{scr}, N_{scr}), \varphi)$ procedure returns rotated array $\text{RotArray}(N_{scr}, N_{scr})$ of $\text{Array}(N_{scr}, N_{scr})$.

Algorithm is repeated for each projection angle $\varphi = \Delta\varphi * \text{i\_}\varphi$, where $\Delta\varphi = 2\pi/N_{\varphi}$, $\text{i\_}\varphi = 0 \ldots N_{\varphi}$. This algorithm however is not mathematically correct and leads to blurring of recovered profile. The double Fourier transformation provides a straightforward solution for tomographic reconstruction, but it presents some problems in actual implementation [4] (e.g. an error produced on a single sample in Fourier space affects the appearance of the entire image). Alternative implementation of the Fourier method is the so-called filtered backprojection (FBP) algorithm. In FBP reconstruction process, each projection is first convoluted with a specific and suitable filtering function [7]. Corresponding convolution function called a convolution kernel, or a filter, or a transfer function [8]. We follow to J.Alonso's approach [9] based on the transfer function algorithm [10-13], where the specific cell in projection column is spread in reconstruction region not only to the corresponding row but also with some weighs in neighbor rows as presented in Fig. 2.2.

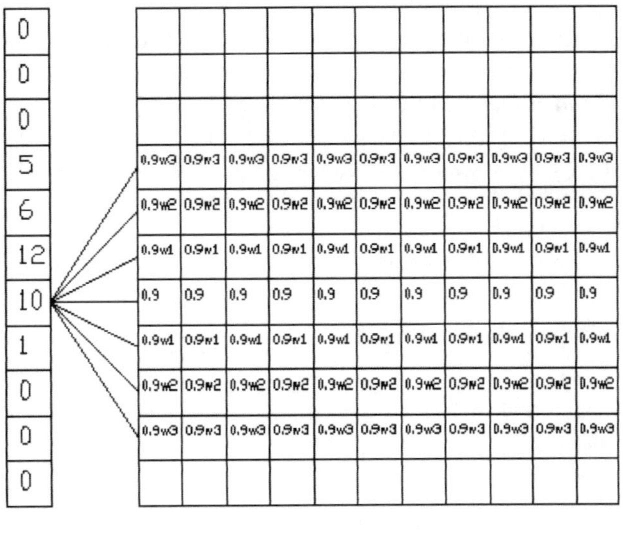

Figure 2.2. Back-projection with filtering on three rows from specific cell of projection column with weighs w_1, w_2, w_3. According to Cho [13] we choose the optimal set should be $w_1 = -0.5232$, $w_2 = 0.1016$, $w_3 = -0.0531$. Procedure will be repeated for all cells of projection. Presented projection corresponds to $\varphi = 0$ of a test beam of Fig. 2.1.

The whole filtered back-projection method can be rewritten in form:

$$\text{Rec\_}\varphi(\text{i\_x}, \text{i\_y}) = \text{Pro\_y}(\text{i\_}\varphi, \text{i\_y})/N_{scr}, \text{ for i\_x} = 0 \ldots N_{scr}, \text{ i\_y} = 0 \ldots N_{scr}, \quad (4)$$
$$\text{Rec\_0}(N_{scr}, N_{scr}) = \text{Rotate}(\text{Rec\_}\varphi(N_{scr}, N_{scr}), -\varphi), \quad (5)$$
$$\text{RezBeam}(\text{i\_x}, \text{i\_y}) => \mathit{RezBeam}(i\_x, i\_y) + \mathit{Rec\_0}(i\_x, i\_y), \quad (6)$$
$$\text{RezBeam}(\text{i\_x}, \text{i\_y}) => \mathit{RezBeam}(i\_x, i\_y) + w_1\text{Rec\_0}(\text{i\_x}, \text{i\_y} \pm 1), \quad (7)$$
$$\text{RezBeam}(\text{i\_x}, \text{i\_y}) => \mathit{RezBeam}(i\_x, i\_y) + w_2\text{Rec\_0}(\text{i\_x}, \text{i\_y} \pm 2), \quad (8)$$

$$\mathrm{RezBeam}(\mathrm{i\_x},\mathrm{i\_y}) \Rightarrow RezBeam(i\_x,i\_y) + \mathrm{w}_3\mathrm{Rec\_0}(\mathrm{i\_x},\mathrm{i\_y} \pm 3), \quad (9)$$

for $\mathrm{i\_x} = 0\ldots\mathrm{N_{scr}}$, $\mathrm{i\_x} = 0\ldots\mathrm{N_{scr}}$ (indexes outside of the space are ignored). Same as for (1-3) the procedure should be repeated for each projection angle $\varphi = \Delta\varphi * \mathrm{i\_}\varphi$, where $\mathrm{i\_}\varphi = 0\ldots\mathrm{N}_\varphi$. One can see that the method is very similar to "trivial" backprojection except the equation (3) that transforms to set of equations (6-9).

2.2. Rotation algorithms

In filtered back-projection method, huge role plays rotation procedure (function Rotate in (5)). Generally rotation described by the usual rotation matrix

$$\begin{pmatrix} x^* \\ y^* \end{pmatrix} = \begin{pmatrix} cos\varphi & sin\varphi \\ -sin\varphi & cos\varphi \end{pmatrix}\begin{pmatrix} x \\ y \end{pmatrix}, \quad (10)$$

where x,y are the coordinates of pixel before rotation on angle φ and $\mathrm{x}^*,\mathrm{y}^*$ are coordinates after rotation. However, in discrete rotation there is present so called aliasing problem – rotated pattern of pixel does not match to the discrete space. In some pixels of space do not set not a single rotated pixel, and in the some two. In [2] proposed in [9] algorithm for discrete object rotation was used. After rotation each pixel of rotating pattern was rotated according the pixel center backward on global rotation angle to match pixel direction to the main coordinate system of reconstruction region. Usually the center of pixel is not lie on the digital space grid. To generate the discrete pixel system each intersection of re-rotated pixel of rotated pattern with grid is spread to the corresponding pixel of reconstruction region. However this algorithm leads to smearing of the reconstructed view.

In this paper we use another algorithm that preserves each rotating pixel value – so called rotation by three shears. Mathematically it means that instead of rotation matrix (10) we use three shearing matrices (y –share on $-\mathrm{tg}(\varphi/2)$, x –share on $\sin\varphi$ and – y -share on $-\mathrm{tg}(\varphi/2)$) [7, 8]:

$$\begin{pmatrix} \mathrm{x}^* \\ \mathrm{y}^* \end{pmatrix} = \begin{pmatrix} 1 & 0 \\ -\mathrm{tg}(\varphi/2) & 1 \end{pmatrix}\begin{pmatrix} 1 & \sin\varphi \\ 0 & 1 \end{pmatrix}\begin{pmatrix} 1 & 0 \\ -\mathrm{tg}(\varphi/2) & 1 \end{pmatrix}\begin{pmatrix} \mathrm{x} \\ \mathrm{y} \end{pmatrix}. \quad (11)$$

The process of such rotation on 27 deg is illustrated in Fig. 2.3.

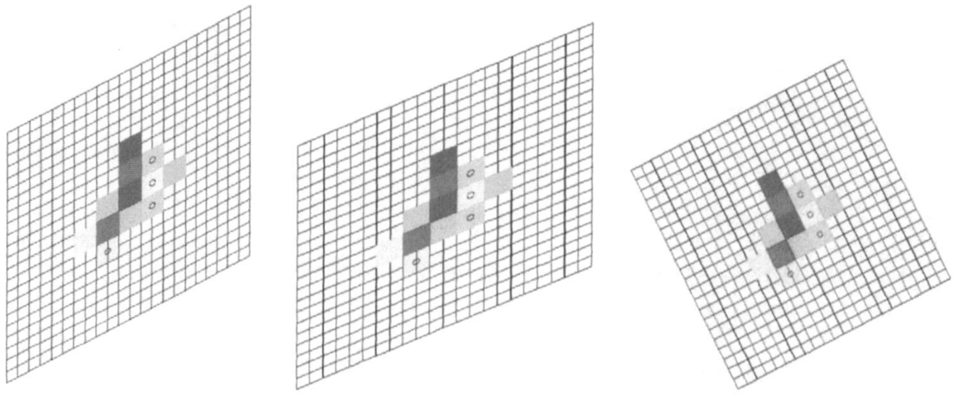

Figure 2.3. Rotation by three shears (to show algorithm we use Paint Image/Stretch/Skew instruments).

To allow vertical shears the discrete space in vertical direction should be more than $\mathrm{N_{scr}}$. If the rotation angles are limited by $\pi/2$ the size of vertical space (temporary value, only for rotation procedure) can be chosen as $2N_{scr}$. This is a reason when operate with angles $> \pi/2$ transform them to rotation of mirrored images on angles $< \pi/2$.

One can see that the shears along the y –axis (first and the third matrices) contains $tg(\theta/2)$ divergated at $\theta = \pi$. To avoid big values of $tg(\varphi/2)$ should be better rotation angles limited in range $|\varphi| < \pi/2$. In formal way the projection procedure gives the same result for mirrored along the axis perpendicular to projection axis. This allows us when rotate to angles more than angles $>\pi/2$ used mirrored objects. Thus the rotation of object on the angle $\pi/2 < \varphi < \pi$ by mean of projection gives the same result as rotation of the same object mirrored along x-axis and rotated on angle $0 < \pi - \varphi < \pi/2$.

2.3. Numerical simulation

Based on the (4-9) a special numerical program was developed that operates either with model test beams or with experimental data. To allow use data of rotation set up to 2π the corresponding algorithm with mirror and rotation by shears is introduced. Special instruments to prepare test beam profiles are developed. An example of test beam 2D profile is presented in Fig. 2.4.

Figure 2.4. Test beam profile made by program graphical instruments.

The corresponding for this model set of projections is presented in Fig 2.5a (here horizontal axis present the rotation angle and vertical axis correspond to y-axis of reconstruction region). The whole pattern of reconstructed profile is presented in Fig. 2.5b.

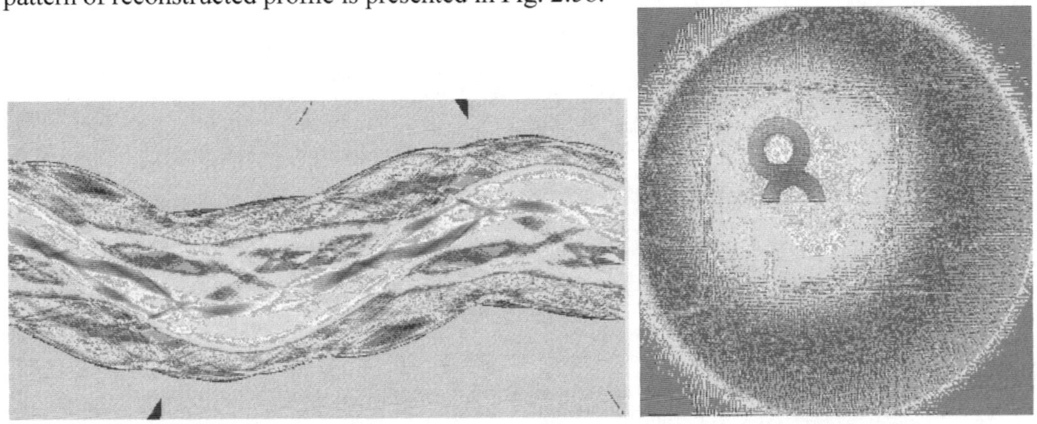

Figure 2.5. Projections (left) and reconstruction (right) of beam of a model presented in Fig. 2.4.

3. Tomography experiment

The aim of the experiment is to reconstruct complicated 2D profile of laser beam. As a scanning mechanism we use VWM mounted on the shaft rotating with uniform angular velocity. The vibrating wire is directed perpendicular to the rotating plane. Radius of the wire trajectory is 91.5 mm, the rotation frequency is about 0.166 Hz, so the linear scanning velocity is about 95.8 mm/s (it is much more than speed about 12 mm/s used in the previous experiment described in Sec. 1. The only small part of the vibrating wire trajectory penetrates the laser beam, so the measurement process starts by signal from opto-interrupter which reacts on the hole of the disk conjugated with the rotating shaft.

Parameters of the laser used in the experiment are following: maximum output power < 200 mW, wavelength 532 nm. To make a nonuniform distribution we used a nozzle with wire along the diameter.

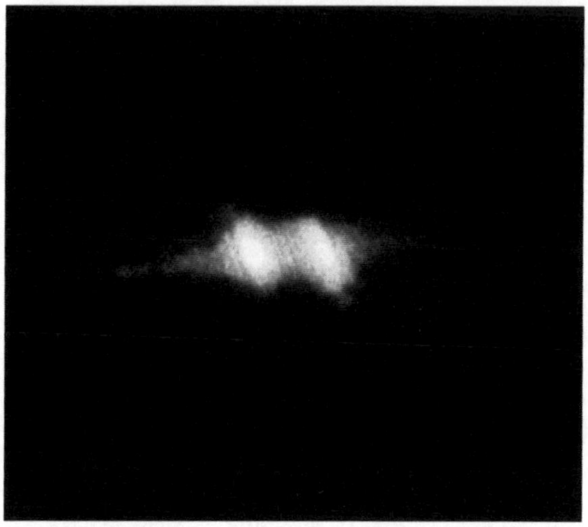

Figure 3.1. Photograph of laser beam.

The main view of laser beam cross section is presented in Fig. 3.1. The photograph of beam was made by digital CCD camera in manual mode operation after reflection of the beam from one surface of optical orange filter which absorbs all transient parts of laser radiation (by this procedure the flux of laser beam has been reduced so that we can adjust CCD camera range without saturation).

The layout of the experiment is presented in Fig. 3.2.

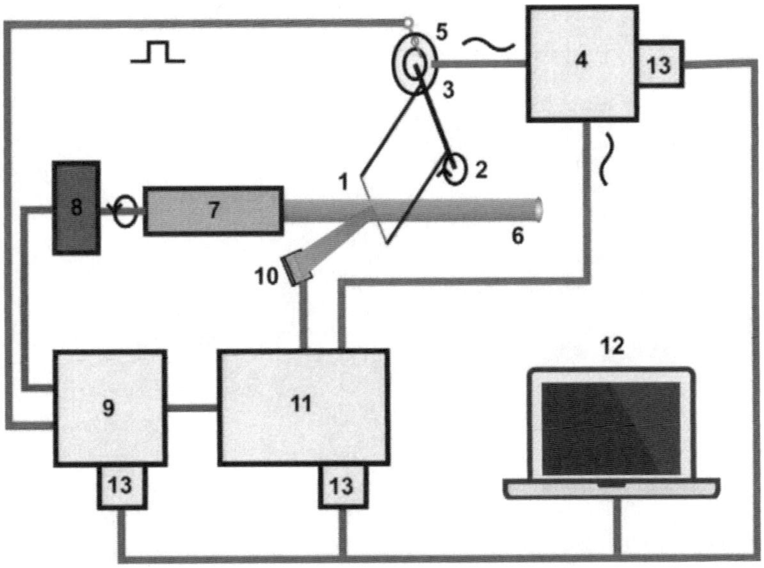

Figure 3.2. Vibrating wire 1 mounted on the rotating shaft with uniform angular speed axis 2, disk 3 with two contact pileups provide the electrical connection with autogeneration electronics 4, opto interrupter reacts on the hole 5 of the contact disk and generates the measurement cycle at the moment before the wire is submerged into the laser beam 6; laser 7 is mounted on the axis of stepper motor 8, which makes a definite number of steps after end of measurement cycle (electronic unit 9); fast photodiode 10 collects the laser photons reflected from the wire and measurements are done with ADC based electronic unit 11; measurements are done in short time in synchronism with wire oscillations (two measurements in oscillation period); the reflected photon measurement results, vibrating wire frequency and laser rotating stepper motor steps are transferred, visualized and stored in computer 12; the RS232 interface 13.

Photograph of the experimental layout is presented in Fig. 3.3.

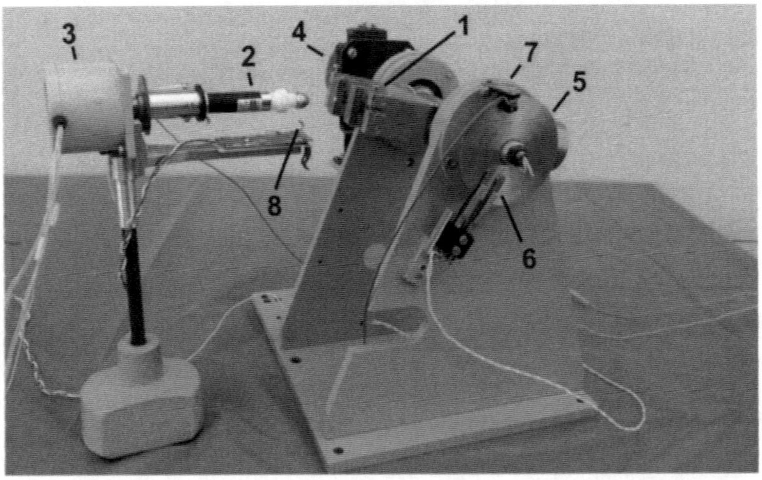

Figure 3.3. Layout of experiment: 1- vibrating wire monitor, 2 - laser mounted on the stepper motor 3 rotating shaft, 4 - DC motor with fixed angular speed rotates the VWM, 5 - contact disk with two pileups 6, 7 - opto interrupter, 8 - fast photodiode with front-end electronics.

During the one tomography experiment we made 200 projections in 0 ~ π range. Each scan needed 120 ms. Some of the projection data are plotted in Fig. 3.4.

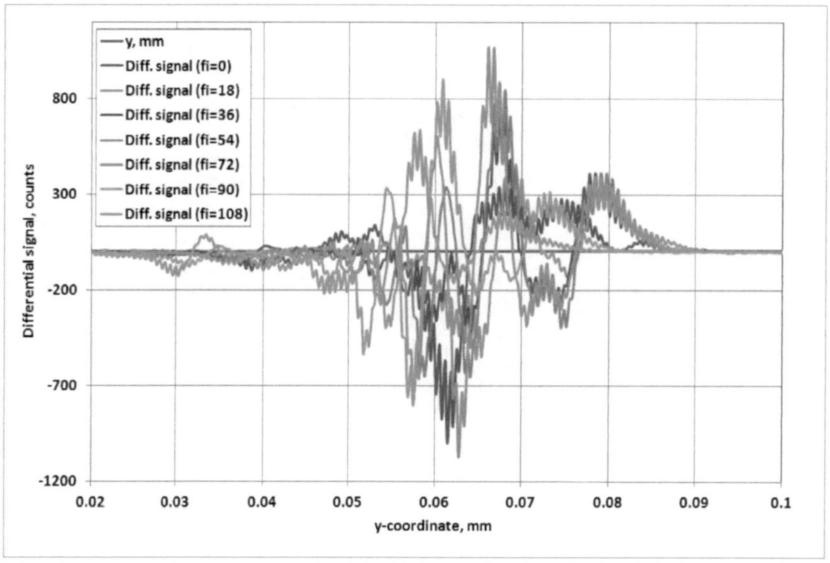

Figure 3.4. Plots of some projection data. From the original photodiode signals (ranged in counts 0-4096), we calculated the differential signals of each pair of the measurement points).

The complete set of projections is presented in Fig. 3.5.

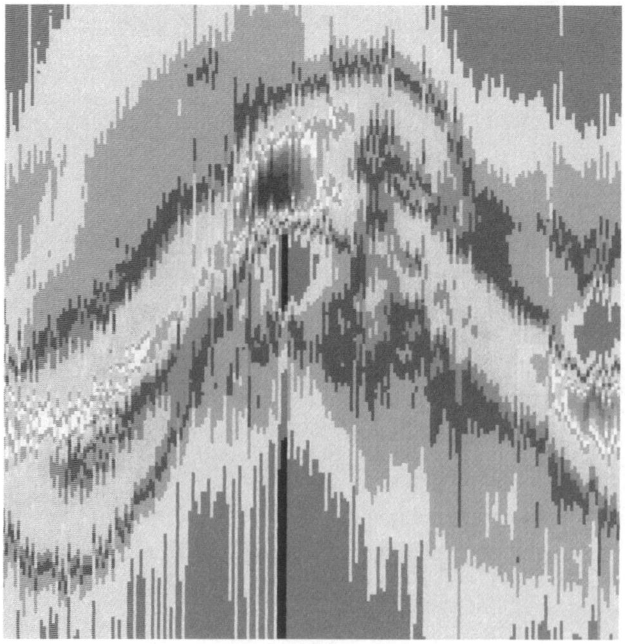

Figure 3.5. The set of projections obtained from the experiment in the same format as in Fig. 2.5a.

The 1D profiles for each rotation angle are obtained by the integration of the corresponding differential signal over the y-coordinate. The final reconstruction of 2D profile of laser beam was made by equations (4-9) and is presented in Fig. 3.6.

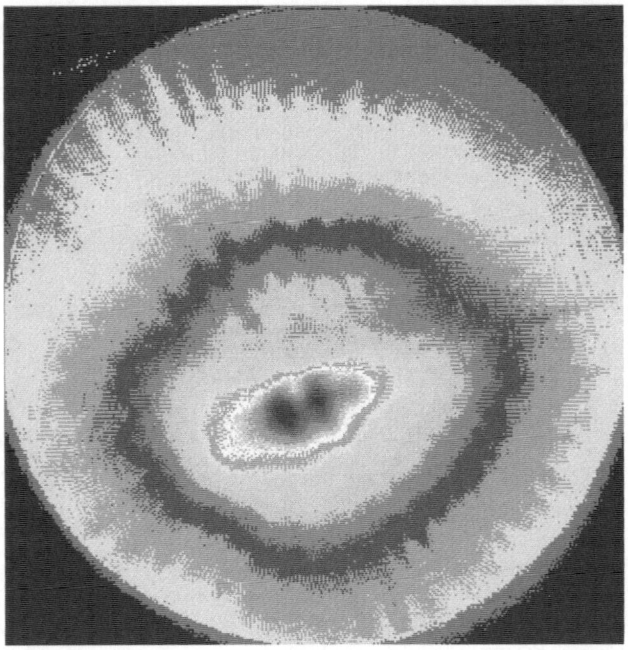

Figure 3.6. Reconstruction of the laser beam by experimental data presented in Fig. 3.5. The achieved pattern approximately coincides with the direct photograph of the laser beam.

4. Conclusion

The main advantage of the proposed method is that it is applicable for beams of different origins - charged particles (electrons, protons, and ions), neutrons and photons in wide range of energies. Compared to the direct method, which is based on the measurements of wire's temperature increase (or corresponding frequency shifts), we achieved much faster operation speed with the resonant target method, leading to a rapid decrease in scan time. This gives an opportunity to apply the method in different types of tomography including the medical tomography.

Acknowledgments
Authors are grateful to J. Bergoz for offer to extend the range of usage of the VWM, and R. Reetz for his many years support in VWM development. We would also like to express special thanks to D. Choe for help during the work and G.M. Davtyan for offering the resources for the experiment.

References

[1] Arutunian S.G., Margaryan A.V., Oscillating wire as a "Resonant Target" for beam, Proceedings of International Particle Accelerator Conference IPAC2014 (Dresden, Germany, 2014), ISBN 978-3-95450-132-8, pp. 3412-3414.
[2] J. Alonso, S.G. Arutunian, COMPLEX BEAM PROFILE RECONSTRUCTION, A NOVEL ROTATING ARRAY OF VIBRATING WIRES, International Part. Acc. Conf., 2014.
[3] S.G. Arutunian, M. Chung, G.S. Harutyunyan, A.V. Margaryan, E.G. Lazareva, L.M. Lazarev, and L.A. Shahinyan, Fast resonant target vibrating wire scanner for photon beam, Review of Scientific Instruments 87, 023108 (2016).

[4] J. Hsieh, Computed tomography: principles, design, artifacts and recent advances, 2nd edition, Wiley interscience, SPIE Press, 2009.
[5] A.C. Kak, M.Slaney, Principles of computerized tomographic imaging, Classics in Applied Mathematics, 33, SIAM, 2001.
[6] G.T. Herman, Fundamentals of comuterized tomography, Image reconstruction from projections, Second ed., Springer, 2009.
[7] F. Kharfi, Mathematics and Physics of Computed Tomography (CT): Demonstrations and Practical Examples - Math_43595.pdf.
[8] Y. Nievergelt, Elementary inversion of the exponential X-ray transfirm, IEEE Transactions on Nuclear Science, 1991, 38, 2, 873-876.
[9] J. Alonso, Private communication.
[10] Z.H. Cho, I.S. Ahn, Computer algorithm for the tomographic image reconstruction with X-ray transmission scans, Computers and Biomedical Research (1975), 8, 8-25.
[11] Z.H. Cho, I.S. Ahn, C.M. Tsai, Computer algorithm and detector electronicas for the transmission X-ray tomography, (1974) NS-21, 218-227.
[12] Z.H. Cho, K. Chan, A comparative study of 3-D image reconstruction algorithms with reference to number of projections and noise filtering, (1975) NS-22, 344-358.
[13] Z.H. Cho, I. Ahn, C. Bohm, G. Huth, Computerized image reconstuction methods with multiple photon/X-ray transmission scanning, Phys. Med. Biol (1974), 19, 4, 511-522.

THz Pulse Duration Influence on High Energy Level Excitation Due to Cascade Mechanism

Vyacheslav A. Trofimov, Dmitry Yu. Zagursky, Irina G. Zakharova
Lomonosov Moscow State University, Moscow, 119992, Russia
E-mail: vatro.cs.msu.ru, zagurski@physics.msu.ru, zaharova@physics.msu.ru

Abstract. We study influence of an incident broadband THz pulse duration on the spectral features of a signal transmitted through/reflected from a substance covered by a disordered structure by means of computer simulation. It is well-known that under real conditions, the results of a standard THz TDS undergo various factors. For example, a substance under investigation can be put into a bulk medium with ordinary properties. This often results in the distortion of the reflected/transmitted pulse spectra and hence, one may reveal additional absorption frequencies which can be thought as belonging to a dangerous substance. An issue from this situation may be a substance emission spectrum using. As we showed the emission frequencies appear due to the cascade mechanism of higher energy level excitation. In this paper we study the incident THz pulse duration influence on the emission frequencies manifestation.

1. Introduction

At present, the development of computer and optical technology offers novel opportunities for the creation of new sensors and systems providing a sufficient accuracy for practice in infrared (IR) and terahertz (THz) ranges of frequencies. Their usage is promising way since it provides a visualization and spectral analysis of an object under investigation without contact with it (so called, nondestructive control), and therefore, devices based on using the IR and THz ranges of frequencies can be applied for quality control in various industries and for security applications [1]. It is well-known that an IR spectroscopy has been till now for the quality control of meat products [2, 3], milk [4], other food [5] and for a control of human health. It is also employed in other areas, for example to analyze the composition of the air [6], analysis of composite materials [7]. Because the characteristic frequencies of various explosives and dangerous substances lie in the THz and IR regions of the spectrum [8–14], then this opens many opportunities for new security systems developing.

There are examples of practical realization of Raman spectroscopy devices measuring IR spectra of substances concealed by covers [15–17]. However, an improvement of devices for the nondestructive detection of substances is of great importance. In connection with that the THz radiation high penetration ability through clothes and ordinary materials is worth attention [18, 19] because the THz range of frequencies allows us to perform measurements through comparably dense or thick covers. Nevertheless, it is desirable to extend distances between an object under investigation and measurement device.

To overcome the crucial difficulties of a remote substance identification it is urgent to solve some problems. One of them arises from a fact that the THz radiation wavelength lies in the range of 0.3-3 mm. However, the macroscopic peculiarities of the sample structure, such as density fluctuations or granule sizes, belongs also to this range. This leads to a strong modulation of the THz pulse interaction with the sample. Moreover, an object surface roughness can result in non-uniform scattering depending on the radiation wavelength [20,21]. One more difficulty is the presence of the object dielectric permittivity. For example, there are layers (or domains) with different densities and, consequently, different permittivities in the object under investigation. This can result, obviously, in frequency-dependent reflection from the whole object. Let us note that a paper, which often covering the sample under investigation, can act as disordered structure [22]. Therefore, the covering distorts THz spectra, when a remote measurement is made to provide the identification and detection of substance [23].

Early, we have demonstrated the existence of the cascade mechanism of high energy level excitation at a THz pulse action on substance and this leads to the higher frequency appearance in the spectrum of a pulse transmitted through or reflected from a substance [24, 25]. These emission frequencies gives us a more reliable information about the chemical structure of the irradiated object. Therefore, a detailed analysis of the various factors, which influences on the emission frequency brightness, is of the great interest. In this paper we numerically investigate an influence of one key factor - the pulse duration - on the cascade mechanism of high energy level excitation for a medium with multiple transition frequencies.

2. Problem statement

We carry out computer simulation of the THz pulse propagation in a finite medium layer possessing non-linear non-instantaneous response. Initially, the pulse is located in a vacuum, then it falls on the left boundary of the medium, transmits through it, and goes out of the medium in a vacuum. This process is accompanied by a partial reflection of the pulse from the medium boundaries, and a part of the THz pulse appears to be reflected in the opposite direction. Below we discuss the spectra of the pulses transmitted through the medium as well as reflected from it. To describe an electromagnetic pulse propagation in a multilevel medium we use the multilevel Maxwell-Bloch system of equations in dimensionless form:

$$\frac{\partial H}{\partial t} = -\frac{\partial E}{\partial z}, \tag{1}$$

$$\frac{\partial D}{\partial t} = -\frac{\partial H}{\partial z}, \tag{2}$$

$$D = E + 4\pi P, \tag{3}$$

$$\frac{\partial \rho_{mn}}{\partial t} = -(\gamma_{mn} + i\omega_{mn}) + i\alpha E \sum_q (d_{mq}\rho_{qn} - \rho_{mq}d_{qn}), \tag{4}$$

$$\frac{\partial \rho_{mm}}{\partial t} = -\sum_q (W_{mq}\rho_{mm} - W_{qm}\rho_{qq}) + i\alpha E \sum_q (d_{mq}\rho_{qm} - \rho_{mq}d_{qm}), \tag{5}$$

$$P = \sum_{mn} (d_{mn}\rho_{nm}). \tag{6}$$

Here, E and H are the electric and magnetic field strength correspondingly. D is the electric field induction, ρ_{mn} is a density matrix element. Diagonal elements of the density matrix characterize the number of medium molecules belonging to the corresponding energy level. Non-diagonal elements describe correlations between different energy levels and define the polarization of the medium (6). ω_{mn} is the matrix of frequencies corresponding to the energy level transition between m and n levels. d_{mn} is the dipole moment of the corresponding transition. γ_{mn} is the relaxation rate of a non-diagonal matrix element, which defines the medium polarization relaxation. This parameter is varied below with the aim to study its influence on the cascade mechanism of molecule high energy level excitation. W_{mn} is the population relaxation rate. Typically W_{mn} is several orders of magnitude smaller than γ_{mn} and thus, below we consider it to be equal to 0. α is a dimensionless constant, which characterizes the electromagnetic field-matter interaction (that is a medium polarizability) and it depends on the dimensional physical parameters of both the pulse and the medium:

$$\alpha = \frac{4\pi M d_0^2}{\hbar\omega}. \quad (7)$$

In (7) M is the concentration of molecules in the medium, d_0 and ω_0 are the characteristic values of the dipole moment and the carrier frequency of a wave packet measured in Debye and THz units correspondingly. The choice of the parameter values will be discussed below.

The pulse propagation through the vacuum is described by the classic Maxwell equations (1-3 with $P = 0$).

The incident pulse shape is taken as a Gaussian one with cosine filling:

$$E(z, t = 0) = H(z, t = 0) = E_0 \exp\left(-\frac{(z - z_0)^2}{\tau_p^2}\right) \cos(\omega_p(z - z_0)), \quad (8)$$

which falls on the medium. Here E_0 is the initial field amplitude, z_0 is the pulse centre position, which is chosen so that the whole pulse is located in vacuum initially, τ_p is the pulse duration (equal to its length in dimensionless units), ω_p is the carrier frequency of the wave packet. Such initial conditions describe the pulse propagation in the positive direction of z-axis. The length of the pulse is chosen so that it contains about 5 or 6 oscillation periods in it.

Molecules of the medium are initially in the ground state:

$$\rho_{11} = 1; \quad \rho_{mn} = 0, (mn) \neq (11). \quad (9)$$

These conditions correspond to the absolute zero of temperature. However, for investigation of the cascade mechanism appearance we consider this case corresponding to "pure conditions" of its observation. At the room temperature the effect will become more pronounced, but the interpretation of our results would be much more complicated because of a number of possible energy level transitions.

3. Computer simulation results and discussion

We consider the medium with six energy levels, three of which have similar transition frequencies from the ground level of molecules. The incident pulse duration varies in such a way that the pulse spectra corresponding to various pulse durations differ noticeably.

The computer simulation dimensionless parameters are given in the table 1. These parameters do not associate with a particular substance but they are similar to characteristic values commonly used in the THz range of frequencies. Namely, they correspond to a pulse with the carrier cyclic frequency of 0.87 THz, duration of about 6 ps and peak intensity about of $10^6 W/cm^2$ propagating through a thin (2 mm) medium layer with transition frequencies up to 2 THz.

The incident pulses and their spectra are depicted in the figures 1 and 2. The frequencies of energy level transitions are marked in the same pictures. It can be seen that the spectra corresponding to different pulse durations have various amplitudes at the transition frequencies, most notably, at the frequencies marked 1-2, 1-3, 1-4.

The chosen parameters correspond to negligible reflection from the medium. Therefore, the effective dielectric permittivity is close to that of the vacuum. Thus, the reflected signal consists only of the medium re-emission signal. Obviously, this situation corresponds to our

Table 1. The parameters of the numerical simulations.

Parameter	Value
ω_{mn}	$\begin{pmatrix} 0 & -4 & -4.5 & -5.5 & -11 & -15 \\ 4 & 0 & -0.5 & -1.5 & -7 & -11 \\ 4.5 & 0.5 & 0 & -1 & -6.5 & -10.5 \\ 5.5 & 1.5 & 1 & 0 & -5.5 & -9.5 \\ 11 & 7 & 6.5 & 5.5 & 0 & -4 \\ 15 & 11 & 10.5 & 9.5 & 4 & 0 \end{pmatrix}$
d_{mn}	$0.0075 \cdot \begin{pmatrix} 0 & 1 & 1 & 1 & 1 & 1 \\ 1 & 0 & 1 & 1 & 5 & 5 \\ 1 & 1 & 0 & 1 & 5 & 5 \\ 1 & 1 & 1 & 0 & 5 & 5 \\ 1 & 5 & 5 & 5 & 0 & 25 \\ 1 & 5 & 5 & 5 & 25 & 0 \end{pmatrix}$
γ_{mn}	$\frac{\omega_{mn}}{50}$
L_s	6
α	67.54
E_0	0.05
ω_p	5
τ_p	1.5 1.8 2.7 4.5

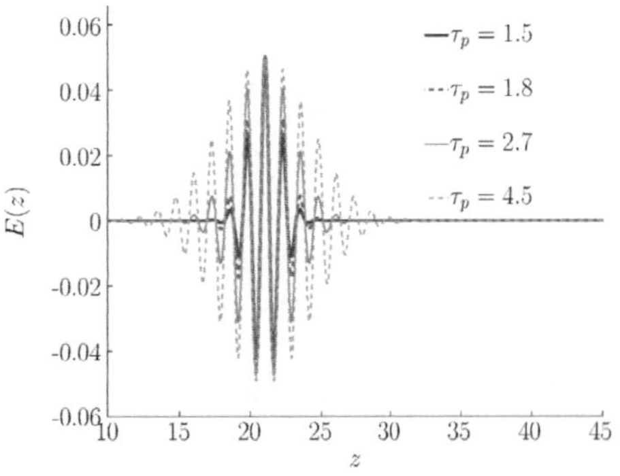

Figure 1. Incident pulses with different durations.

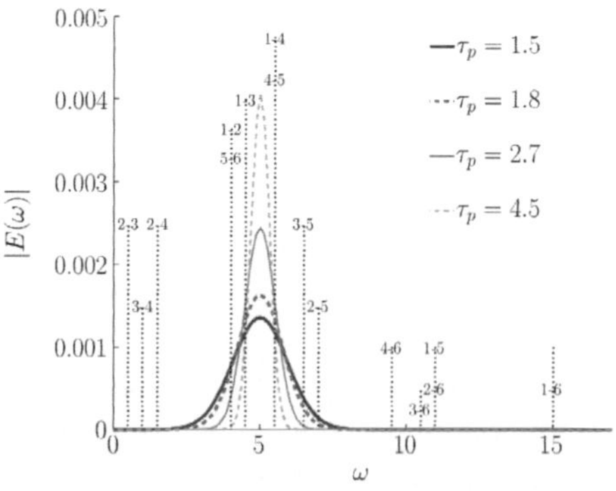

Figure 2. The incident pulse spectra. Vertical lines mark the transition frequencies between energy levels of the medium.

aim. Nevertheless, it would be possible to add a non-resonant permittivity to our model. But in this case the computation result interpretation will become complicate. Hence, we pay our attention to the signal which transmit through the medium layer and then it propagates in positive direction of z - axis.

In the figure 3 the spectra of the reflected signals are shown. We see that the reflected pulse spectrum possesses maxima with highly varied amplitudes in different ranges of frequencies. Therefore, the figure 4 depicts the spectra in the central area of the spectrum $\omega = [3, 7]$. In this area of the spectrum the incident pulse spectra are localized. As a result of the pulse interaction

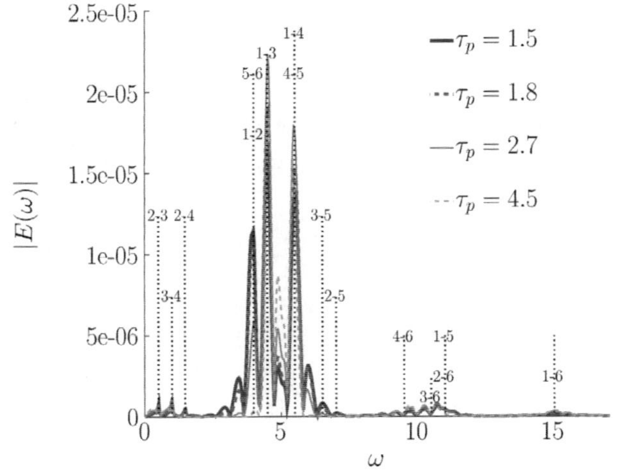

Figure 3. Spectra of the reflected signals obtained in the simulations. Vertical lines mark the transition frequencies in the medium.

with the medium, the spectra are transformed in comparison with their initial shapes, namely minima (absorption lines) appear. In spectroscopy the appearance of these lines can be an argument for determining the chemical structure and classifying the substance.

The pulses of all durations exhibit these minima at the frequencies of 4.5 and 5.5, but the maximal one ($\tau_p = 4.5$) has the spectrum which is so narrow, that it only bends and no minima are observed in it. Since the spectra of the shorter pulses are broad enough, the absorption line at the frequency 4 is present in each of them.

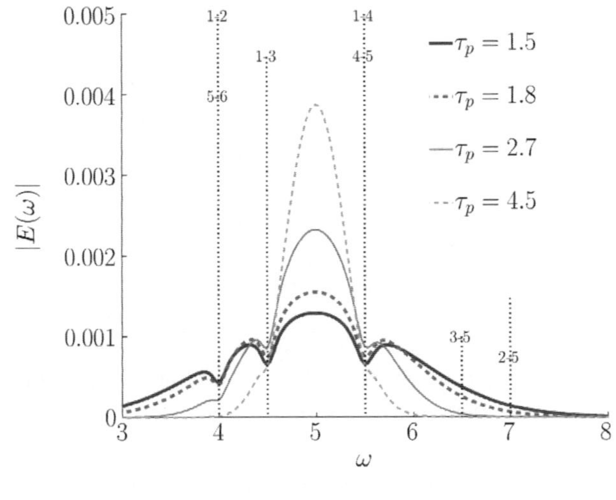

Figure 4. The spectra of transmitted pulses. Central frequency range. Vertical dotted lines mark transition frequencies in the medium.

As a result of the electromagnetic field energy absorption, the medium is excited, its molecules transit to higher energy levels from the ground one due to the cascade mechanism of their excitation. From higher energy levels molecules transit to lower ones. This process is accompanied by the emission of radiation at a corresponding frequency. In a multilevel medium there is a lot of possible transition series. Thus, the medium can re-emit the absorbed energy at frequencies different from those they were absorbed.

The figure 5 depicts the pulse spectra in the range $[0, 3]$ where such spectral intensity maxima are located. It is worth noticing that the dependence of maximal spectral intensities on the duration of the incident pulse has a non-monotonous character. Namely, for the energy level transitions 2 − 3 and 2 − 4 we see the highest spectral intensity if the pulse with the shortest duration falls on a medium. However, the spectral intensity corresponding to transition 3 − 4 achieves a maximum intensity if the pulse duration is equal to 2.7 dimensionless units. The fact is remarkable that the propagation of the pulse with maximal duration results in a spectrum with almost negligible peak 2 − 4.

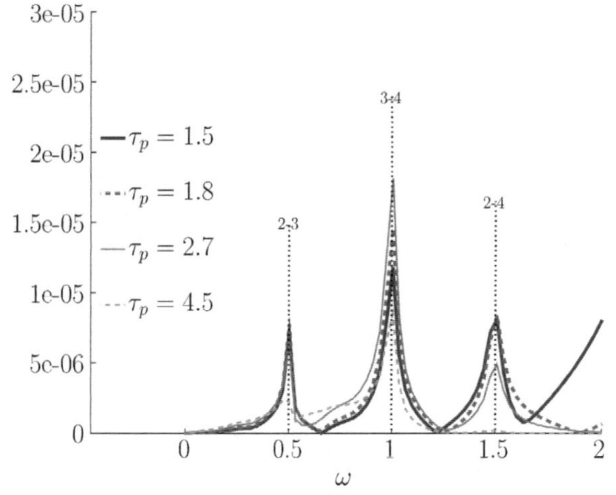

Figure 5. The spectra of transmitted pulses. Low frequency range. Vertical dotted lines mark transition frequencies in the medium.

Let us discuss the spectral line appearance in the range of frequencies which are higher than those belonging to the incident pulse spectrum. During the interaction process, molecules of the medium are excited to upper energy levels, and then they can relax down to neighboring energy levels and emits the radiation as discussed before. After that they can relax to the ground energy level and emit at the same frequencies at which the THz pulse energy absorption takes place (Fig. 3). On the other hand, the few-cycle pulse is characterized by its broad spectrum and, moreover, the energy levels in the THz frequency range are located quite densely, having transition frequencies rather close to each other. Thus, the absorption of THz photon sequence is not impossible in the medium under consideration.This is why we name this absorption mechanism as the cascade mechanism of high energy level excitation.

In the figure 6 the transmitted pulse spectra are shown in the range of frequencies [9 − 17]. The influence of the incident pulse duration is obvious. Between the frequencies 9 and 12 a pair of overlapping spectral maxima is noticeable. It can be seen that the maximal spectral intensity for these maxima changes with the incident pulse duration. It should be also emphasized that the rightmost of those peaks overlaps two different frequencies corresponding to two energy level transitions, namely, (10.5 and 11). Therefore, spectrum shape is dependent on the incident pulse duration. For the longer pulses the peak tends to be centred at the frequency 10.5. For shorter ones the peak center is closer to 11.

The last peak, corresponding to the transition between the energy levels 6 and 1, changes its height depending on the pulse duration as well, but due to its separate location, its shape is mostly unaffected.

4. Conclusion

We have studied the influence of the THz pulse duration on the pulse interaction with the multilevel medium layer. We demonstrated a significant influence of the pulse duration on the transmitted signal spectrum. Medium emission at higher and lower frequencies are the most

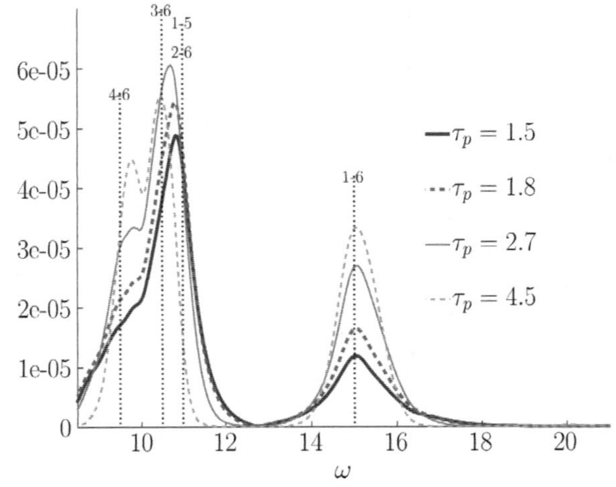

Figure 6. The spectra of transmitted pulses. Higher frequency range. Vertical dotted lines mark transition frequencies in the medium.

remarkable. Depending on the pulse duration the spectral lines can change its intensity or disappear altogether. Therefore, this dependence is not monotonic. Thus, the pulse must be taken into account in spectroscopic experiments in order to provide the correct characterization of the studied medium.

Acknowledgments
The research was carried out with the support of the Russian Science Foundation (Grant N 14-21-00081).

References

[1] Cialla D, März A, Böhme R, Theil F, Weber K, Schmitt M and Popp J 2012 *Analytical and Bioanalytical Chemistry* **403** 27–54 ISSN 1618-2650 URL http://dx.doi.org/10.1007/s00216-011-5631-x
[2] N M Anderson P N W 2003 *Transactions of the ASAE* **46**(1) 117–124
[3] Begley T H, Lanza E, Norris K H and Hruschka W R 1984 *Journal of Agricultural and Food Chemistry* **32** 984–987 (*Preprint* http://dx.doi.org/10.1021/jf00125a007) URL http://dx.doi.org/10.1021/jf00125a007
[4] Villar A, Gorritxategi E, Aranzabe E, Fernndez S, Otaduy D and Fernndez L A 2012 *Food Chemistry* **135** 2756 – 2760 ISSN 0308-8146 URL http://www.sciencedirect.com/science/article/pii/S0308814612011880
[5] Huang H, Yu H, Xu H and Ying Y 2008 *Journal of Food Engineering* **87** 303 – 313 ISSN 0260-8774 URL http://www.sciencedirect.com/science/article/pii/S0260877408000071
[6] Esler M B, Griffith D W T, Wilson S R and Steele L P 2000 *Analytical Chemistry* **72** 206–215 pMID: 10655655 (*Preprint* http://dx.doi.org/10.1021/ac9905625) URL http://dx.doi.org/10.1021/ac9905625
[7] Ospald F, Zouaghi W, Beigang R, Matheis C, Jonuscheit J, Recur B, Guillet J P, Mounaix P, Vleugels W, Bosom P V, Gonzlez L V, Lpez I, Edo R M, Sternberg Y and Vandewal M 2013 *Optical Engineering* **53** 031208 URL http://dx.doi.org/10.1117/1.OE.53.3.031208
[8] Lu M, Shen J, Li N, Zhang Y, Zhang C, Liang L and Xu X 2006 *Journal of Applied Physics* **100** 103104 URL http://scitation.aip.org/content/aip/journal/jap/100/10/10.1063/1.2388041
[9] Federici J F, Schulkin B, Huang F, Gary D, Barat R, Oliveira F and Zimdars D 2005 *Semiconductor Science and Technology* **20** S266 URL http://stacks.iop.org/0268-1242/20/i=7/a=018
[10] Chen J, Chen Y, Zhao H, Bastiaans G J and Zhang X C 2007 *Opt. Express* **15** 12060–12067 URL http://www.opticsexpress.org/abstract.cfm?URI=oe-15-19-12060
[11] Davies A G, Burnett A D, Fan W, Linfield E H and Cunningham J E 2008 *Materials Today* **11** 18 – 26 ISSN 1369-7021 URL http://www.sciencedirect.com/science/article/pii/S1369702108700166
[12] Puc U, Abina A, Rutar M, Zidanšek A, Jeglič A and Valušis G 2015 *Appl. Opt.* **54** 4495–4502 URL http://ao.osa.org/abstract.cfm?URI=ao-54-14-4495

[13] Lpez-Lpez M and Garca-Ruiz C 2014 *TrAC Trends in Analytical Chemistry* **54** 36 – 44 ISSN 0165-9936 URL http://www.sciencedirect.com/science/article/pii/S0165993613002550
[14] Hwang J, Choi N, Park A, Park J Q, Chung J H, Baek S, Cho S G, Baek S J and Choo J 2013 *Journal of Molecular Structure* **1039** 130 – 136 ISSN 0022-2860 URL http://www.sciencedirect.com/science/article/pii/S002228601300121X
[15] Stokes R J, Bailey M, Bonthron S, Stone T, Maskall G, Presly O, Roy E, Tombling C and Loeffen P W 2016 New capability for hazardous materials id within sealed containers using a portable spatially offset raman spectroscopy (sors) device URL http://dx.doi.org/10.1117/12.2241540
[16] Ali E M A, Edwards H G M and Cox R 2015 *Journal of Raman Spectroscopy* **46** 322–326 ISSN 1097-4555 jRS-14-0321.R2 URL http://dx.doi.org/10.1002/jrs.4632
[17] Izake E L, Sundarajoo S, Olds W, Cletus B, Jaatinen E and Fredericks P M 2013 *Talanta* **103** 20–27 ISSN 0039-9140 URL http://www.sciencedirect.com/science/article/pii/S0039914012008119
[18] Amenabar I, Lopez F and Mendikute A 2013 *Journal of Infrared, Millimeter, and Terahertz Waves* **34** 152–169 ISSN 1866-6906 URL http://dx.doi.org/10.1007/s10762-012-9949-z
[19] Jepsen P, Cooke D and Koch M 2011 *Laser & Photonics Reviews* **5** 124–166 ISSN 1863-8899 URL http://dx.doi.org/10.1002/lpor.201000011
[20] Ortolani M, Lee J S, Schade U and Hbers H W 2008 *Applied Physics Letters* **93** 081906 URL http://scitation.aip.org/content/aip/journal/apl/93/8/10.1063/1.2973403
[21] Palka N 2013 *Optical Engineering* **53** 031202–031202 URL http://dx.doi.org/10.1117/1.OE.53.3.031202
[22] Trofimov V A, Denisov A D and Tikhomirov V V 2013 Experimental realization of sda-method for the detection of substance at long distance URL http://dx.doi.org/10.1117/12.2029007
[23] Trofimov V A and Varentsova S A 2016 *Applied optics* **55** 9605–9618 ISSN 0003-6935; 1539-4522; 1559-128X
[24] Trofimov V, Zagursky D and Zakharova I 2015 *Frontiers in Optics 2015* (Optical Society of America) p JTu4A.9 URL http://www.osapublishing.org/abstract.cfm?URI=FiO-2015-JTu4A.9
[25] Trofimov V A, Zagursky D Y and Zakharova I G 2016 *Proceedings of SPIE - The International Society for Optical Engineering* **9917** Paper number 991730 (8 pages) ISSN 1996-756X; 0277-786X

Analytical consideration and computer simulation of DFWM

Vyacheslav A. Trofimov, Igor E. Kuchik

Lomonosov Moscow State University, Moscow 119992, Russia

E-mail vatro@cs.msu.ru

Abstract. Degenerate four-wave mixing (DFWM) for co-propagating femtosecond laser pulses is considered in the axial-symmetric case for bulk medium with cubic nonlinear response, if pump-wave amplitudes are being equal. Computer simulation is based on the set of nonlinear Schrödinger equations describing this process. For the analytical consideration we use the frame-work of both plane wave approximation and long pulse duration approximation taking into account the phase matching. In opposite to widely used approach, based on the pump-waves non-depletion, the problem invariants are used for analytical solution developing. The solution demonstrates various DFWM modes existence and allows us to provide full analysis of the problem in dependence of its parameters. Analytical solution and derived pulse interaction modes can explain complicated regime of DFWM, which may appear at different intensities of interacting waves.

1. Introduction

As it is well-known, a problem of four-waves mixing (FWM) is permanently under investigation [1-19]. In particular, such kind of laser pulse interaction is widely used for both a substance spectroscopy and a medium diagnostic. Another very important application of FWM is caused by a possibility of FWM using for optical frequency conversion to any wavelength with a careful fiber design, and therefore it is a very attractive.

Obviously there are many physical factors which influence on FWM modes. To understand the FWM features one needs to make computer simulation of this problem or various analytical approaches developing. Obviously, the last is more preferable because it allows to see a visual dependence of a frequency conversion efficiency. Usually the analytical consideration is based on the pump-waves non-depletion approximation which is valid at the initial stage of the pulses interaction. This approach possesses the essential restrictions. For example, the interacting wave energy does not preserve. Its more significant limitation arises out of neglecting of self-modulation and cross-modulation of the pumping waves. As a result of this, the phase difference between pumping waves and other two waves becomes wrong and therefore an evolution of the interacting waves is wrong also. To revise this situation we develop new approach for analytical solution of the problem: we use conservation laws of the FWM and both plane wave approximation and long pulse duration approximation. In this case we accurate take into account the interacting waves phase changing. Hence, the analytical solution of the problem becomes more adequate to a physical situation if pulse dispersion and beam diffraction is weak at a medium length under consideration. In this paper we compare the problem analytical solution made in [20, 21] with computer simulation results obtained using the solution of four nonlinear Schrödinger equations describing FWM of axially-symmetry beams with taking into consideration a second order dispersion of interacting pulses, their self- and cross–modulation and beam diffraction. Let us notice that early we have used with success the similar approach for analysis of frequency conversion efficiency when the laser pulses interact in a medium with quadratic nonlinear response [22-26]).

2. Problem statement

We consider the FWM of co-propagating laser pulses with current frequencies denoted as $\omega_1, \omega_2, \omega_3, \omega_4$ in a medium with cubic nonlinear response. In the framework of slowly varying envelope the considered process is governed by the set of dimensionless nonlinear Schrödinger equations with respect to complex amplitudes A_1, A_2, A_3, A_4 of interacting waves:

$$L_j A_j = i\alpha_j (F_j + 2A_2 A_{4-j}^* A_4 \cdot e^{i\Delta k\, z}),\ j=1,3,\quad L_j A_j = i\alpha_j (F_j + 2A_1 A_3 A_{6-j}^* \cdot e^{-i\Delta k\, z}),\ j=2,4, \quad (1)$$

$$L_j = \frac{\partial}{\partial z} + i\tilde{D}_{rj}\Delta_r + iD_j \frac{\partial^2}{\partial t^2},\ F_j = A_j(2(|A_1|^2+|A_2|^2+|A_3|^2+|A_4|^2) - |A_j|^2),\ 0<z\le L_z,\ 0<r<L_r,\ 0<t<L_t$$

with the following initial and boundary conditions for complex amplitudes

$$A_j(z=0,r,t) = A_{j0}\bar{A}_{j0}(r,t),\ 0\le r\le L_r,\ 0\le t\le L_t,\ A_j\big|_{t=0,L_t} = r\frac{\partial A_j}{\partial r}\bigg|_{r\to 0} = A_j\big|_{r=L_r} = 0,\ j=1-4. \quad (2)$$

We assume that there is a group velocity matching. In Eqs. (1) the dimensionless parameters and functions can be expressed through the physical ones in the following manner:

$$z = z'/l_{dif,1},\ t = t'/\tau_1,\ r = r'/a_1,\ l_{dif,j} = 2k_j a_1^2,\ \tilde{D}_{rj} = l_{dif,1}/l_{dif,j},\ D_j = l_{dif,1}/l_{dis,j},\ \alpha_j = {l_{dif,1}}\big/{l_{self,j}},$$

$$l_{dis,j} = -\frac{2\tau_1^2}{d^2k(\omega_j)/d\omega^2},\quad l_{self,j} = \frac{n(\omega_j)\cdot\lambda_j}{3\pi^2\chi^{(3)}A_{10}^2},\quad \Delta k = (k_1+k_3-k_2-k_4)\cdot l_{dif,1},\quad j=1-4.$$

Above z' is a physical longitudinal coordinate, along which the optical radiation propagates, z is measured in units of the diffraction length $l_{dif,1}$ of the laser beam with the frequency ω_1 (the first wave). Parameters τ_1, a_1 denote the pulse duration and the beam radius of this wave at input section, correspondingly. Variable t' denotes the physical time. Time t is measured in unit of the first wave pulse duration. L_t is a dimensionless time interval, during which the interaction of waves is analyzed, L_z is a dimensionless length of nonlinear medium. Variable r is a transverse coordinate normalized by the radius a_1, L_r is the maximum value of this coordinate, $\Delta_r = \frac{1}{r}\frac{\partial}{\partial r}\left(r\frac{\partial}{\partial r}\right)$ is the transverse Laplacian written in the axially symmetric case. Functions $A_{j0}(r,t)$ describe the temporal and spatial distributions of a complex amplitude at input section of a medium. k_j is a wave number of j-wave. Parameter Δk is a phase mismatching on the first wave diffraction length, A_j are complex amplitudes of waves, normalized on the square root of the first wave intensity $I_{10}^{0.5}$, $\bar{A}_{j0}(r,t)$ is a spatial-temporal distribution of the complex amplitude at the input section of a medium, A_{j0} are the dimensionless amplitudes of incident waves. We consider below that a wave with the frequency ω_4 possesses zero-value amplitude $(A_{40}=0)$, and the signal wave amplitude (A_{20}) (the frequency ω_2) is not equal to zero, and amplitudes of the pump waves are equal each other ($A_{10}=A_{30}$) as well as the nonlinear coefficients $\alpha_1 = \alpha_3$ because of an equality of the pump wave frequencies ($\omega_1 = \omega_3$). Coefficient D_j characterizes the group velocity dispersion of waves, $\tilde{D}_{rj}$ is a diffraction coefficient. It should be stressed that a dependence of a cubic susceptibility ($\chi^{(3)}$) of a medium on the optical frequency is neglected. The similar assumption is valid for the refractive index of interacting waves. Parameters α_j characterize self-action of the laser beams due to a cubic nonlinear response.

Below we will take into account the following relation for nonlinear coefficients of the problem:

$$\alpha_1 + \alpha_3 = \alpha_2 + \alpha_4. \tag{3}$$

Taking into account the equality of nonlinear coefficients for the pump waves, the equality (3) reduces to the next equality:

$$2\alpha_1 = \alpha_2 + \alpha_4. \tag{4}$$

3. Conservation laws (invariants) for the case of interacting waves phase matching ($\Delta k = 0$)

In this paper we compare computer simulation results with an analytical solution of the problem (1) in the case of a phase matching for the interacting waves. Therefore, we write invariants (conservation laws) for this case.

First invariant is energy one and it is written in the form:

$$I_1 = \int_0^{L_T}\int_0^{L_r}\left(\frac{|A_1|^2}{\alpha_1} + \frac{|A_2|^2}{\alpha_2} + \frac{|A_3|^2}{\alpha_3} + \frac{|A_4|^2}{\alpha_4}\right) rdrdt. \tag{5}$$

However, as it is well-known, there are other invariants which are similar to (5) and they are written as:

$$I_{14} = I_{34} = \int_0^{L_T}\int_0^{L_r}\left((\frac{|A_1|^2}{\alpha_1} + \frac{|A_4|^2}{\alpha_4})\right) rdrdt = \int_0^{L_T}\int_0^{L_r}\left((\frac{|A_3|^2}{\alpha_3} + \frac{|A_4|^2}{\alpha_4})\right) rdrdt,$$

$$I_{24} = \int_0^{L_T}\int_0^{L_r}\left((\frac{|A_2|^2}{\alpha_2} - \frac{|A_4|^2}{\alpha_4})\right) rdrdt.$$

The next invariant is Hamiltonian of the equation set (1) and it is seen as

$$I_3 = \int_0^{L_T}\int_0^{L_r}[\sum_{j=1}^{4}(\frac{D_j}{\alpha_j}\left|\frac{\partial A_j}{\partial t}\right|^2 + \frac{\tilde{D}_{rj}}{\alpha_j}\left|\frac{\partial A_j}{\partial r}\right|^2) + (4\cdot \mathrm{Re}(A_1 A_2^* A_3 A_4^*) +$$
$$+(|A_1|^2 + |A_2|^2 + |A_3|^2 + |A_4|^2)^2 - \frac{1}{2}(|A_1|^4 + |A_2|^4 + |A_3|^4 + |A_4|^4))]rdrdt. \tag{6}$$

To obtain the analytical solution we essentially use these invariants. Moreover, they are used for a control of computer simulation results: we develop conservative finite-difference scheme for a numerical solution of the problem (1).

4. Analytical solution f the problem in the frame-work of long pulse duration and plane wave approximation

Many special features of the FWM process can be clarified by using the long pulse and plane wave approximation. Taking into account the well-known representation of complex amplitude:

$A_j = a_j \exp(i\varphi_j)$, $j = 1-4$,

where an amplitude a_j and phase φ_j of the waves are real functions. In this case, the waves interacting can be described by the following equations:

$$\frac{da_1}{dz} + 2\alpha_1 \cdot a_2 a_3 a_4 \sin\varphi = 0,\ \frac{da_2}{dz} - 2\alpha_2 \cdot a_1 a_3 a_4 \sin\varphi = 0,$$

$$\frac{da_3}{dz} + 2\alpha_3 \cdot a_1 a_2 a_4 \sin\varphi = 0,\ \frac{da_4}{dz} - 2\alpha_4 \cdot a_1 a_2 a_3 \sin\varphi = 0, \tag{7}$$

$$\frac{d\varphi}{dz} + 2\cos\varphi \cdot (\alpha_1 \cdot \frac{a_2 a_3 a_4}{a_1} - \alpha_2 \cdot \frac{a_1 a_3 a_4}{a_2} + \alpha_3 \cdot \frac{a_1 a_2 a_4}{a_3} - \alpha_4 \cdot \frac{a_1 a_2 a_3}{a_4}) -$$

$$-\alpha_1 \cdot a_1^2 + \alpha_2 \cdot a_2^2 - \alpha_3 \cdot a_3^2 + \alpha_4 \cdot a_4^2 = 0,\quad \varphi = \varphi_2 - \varphi_1 + \varphi_4 - \varphi_3$$

with initial conditions

$$a_1\big|_{z=0} = a_3\big|_{z=0} = a_{10} = a_{30},\ a_2\big|_{z=0} = a_{20},\ a_4\big|_{z=0} = a_{40} = 0,\ \varphi\big|_{z=0} = \varphi_0.$$

In new notations, the invariants can be rewritten in the next form:

$$I_1 = a_1^2 + a_2^2 + a_3^2 + a_4^2,\ a_1^2 = \alpha_1 I_{14} - \frac{\alpha_1}{\alpha_4} a_4^2,\ a_2^2 = \alpha_2 I_{24} + \frac{\alpha_2}{\alpha_4} a_4^2,\ a_3^2 = \alpha_3 I_{34} - \frac{\alpha_3}{\alpha_4} a_4^2,$$

$$\tilde{I}_3 = 4a_1 a_2 a_3 a_4 \cos\varphi + a_4^2 \xi + a_4^4 \eta = 0,$$

where the invariant $\tilde{I}_3$ follows from the invariant I_3 and introduced parameters are defined as

$$\xi = \frac{\alpha_1\alpha_2}{\alpha_4}(I_{14} - I_{24}) + \frac{\alpha_2\alpha_3}{\alpha_4}(I_{34} - I_{24}) - \frac{\alpha_1\alpha_3}{\alpha_4}(I_{14} + I_{34}) + \alpha_1 I_{14} + \alpha_2 I_{24} + \alpha_3 I_{34}, \tag{8}$$

$$\eta = \frac{\alpha_2}{\alpha_4} - \frac{\alpha_1}{\alpha_4} - \frac{\alpha_3}{\alpha_4} - \frac{\alpha_1\alpha_2}{\alpha_4^2} + \frac{\alpha_1\alpha_3}{\alpha_4^2} - \frac{\alpha_2\alpha_3}{\alpha_4^2}.$$

Using the invariant $\tilde{I}_3$, one can write the algebraic equation, containing amplitudes of the interacting waves and $\cos\varphi$ instead the differential equation with respect to the phase difference φ . This allows us to eliminate a term $\sin\varphi$ in equation, concerning the interacting wave amplitudes, and write the fourth wave intensity ($p_4(z) = a_4^2(z)$) evolution in the form:

$$\alpha_4\sqrt{|\lambda|}z = \int_0^{p_4} \frac{dx}{\sqrt{f(x)}},\ f = \mathrm{sgn}(\lambda)x\left(x^3 + c_1 x^2 + c_2 x + c_3\right). \tag{9}$$

Above we introduced new notations:

$$\lambda = 16\frac{\alpha_1\alpha_2\alpha_3}{\alpha_4^3} - \eta^2,\ c_1 = (16\frac{\alpha_1\alpha_2\alpha_3}{\alpha_4^2}(I_{24} - I_{14} - I_{34}) - 2\eta\xi)/\lambda,$$

$$c_2 = (16\frac{\alpha_1\alpha_2\alpha_3}{\alpha_4}(I_{14}I_{34} - I_{14}I_{24} - I_{24}I_{34}) - \xi^2)/\lambda,\ c_3 = 16\alpha_1\alpha_2\alpha_3 I_{14}I_{24}I_{34}/\lambda. \tag{10}$$

The intensity $p_4(z)$ evolution is defined by roots of the equation $f(x) = 0$. We see from expression (9) and (10) that the root properties depend strongly on the problem parameters. One of the roots is always equal to 0 and others can be found from a cubic equation, which may possess either three real roots or one real root and two complex roots. Moreover, the fourth wave changing also depends on the sign of the parameter λ because of its influence on the radicand sign. Obviously, for the solution existence it is necessary a validity of the inequality: $f(x) = \mathrm{sgn}(\lambda)x(x^3 + c_1 x^2 + c_2 x + c_3) \ge 0$.

In general case, the fourth wave intensity changing is described by an elliptical function. Thus, the FWM mode is governed by the relationship between the parameters $(a_{10}, a_{20}, \alpha_2)$. Below we write only two different solutions because in our previous papers we describe various solutions of the problem in detail.

If the equation $f(x)=0$ has four real roots, satisfying the following relation $x_1=0<x_2<x_3<x_4$ and a sign of λ is negative, for zero-value of incident fourth wave intensity, the intensity p_4 belongs to the interval $[x_1=0, x_2]$ at the fourth wave propagation (this generation mode is named as low efficiency mode). In this case, the fourth wave intensity evolution is described by:

$$p_4(z)=|a_4(z)|^2=\frac{x_1(x_4-x_2)+x_4(x_2-x_1)\,sn^2(0.5\,\alpha_4\sqrt{|\lambda|(x_3-x_1)(x_4-x_2)}\,z,m)}{x_4-x_2+(x_2-x_1)\,sn^2(0.5\,\alpha_4\sqrt{|\lambda|(x_3-x_1)(x_4-x_2)}\,z,m)},\quad m=\sqrt{\frac{(x_3-x_2)(x_4-x_1)}{(x_3-x_1)(x_4-x_2)}},\tag{11}$$

where sn denotes the elliptical sinus.

If four real roots satisfy the following relation $x_1<x_2=0<x_3<x_4$ and a sign of λ is positive, then achievable value of the fourth wave intensity p_4 belongs to the interval $[x_2=0, x_3]$ if the incident intensity of the fourth wave satisfies the condition: $p_4(z=0)=0$. The fourth wave intensity p_4 evolution along the medium is written in the form of elliptical sinus (sn):

$$p_4(z)=|a_4(z)|^2=\frac{x_2(x_3-x_1)-x_1(x_3-x_2)sn^2(0.5\alpha_4\sqrt{|\lambda|(x_3-x_1)(x_4-x_2)}z,m)}{x_3-x_1-(x_3-x_2)sn^2(0.5\alpha_4\sqrt{|\lambda|(x_3-x_1)(x_4-x_2)}z,m)},\quad m=\sqrt{\frac{(x_4-x_3)(x_2-x_1)}{(x_3-x_1)(x_4-x_2)}}.\tag{12}$$

To clarify conditions of the used approach for the analytical solution developing we construct the conservative finite-difference scheme for the problem (1).

5. Conservative finite-difference scheme

The conservative finite-difference scheme for problem (1-2) is constructed as follows. In the domain $\Omega=\Omega_z\times\Omega_r\times\Omega_t=\{0\le z\le L_z\}\times\{0\le r\le L_r\}\times\{0\le t\le L_t\}$ we introduce the uniform grid $\omega=\omega_z\times\omega_r\times\omega_t$:

$$\omega_z=\{z_n=nh_z, n=\overline{0,N_z}, h_z=L_z/N_z\},\ \omega_r=\{r_k=(k+0.5)h_r, k=\overline{0,N_r}, h_r=L_r/(N_r+0.5)\},$$

$$\omega_t=\{t_l=l\tau, l=\overline{0,N_t}, \tau=L_t/N_t\}.$$

The grid functions $A_j=A_{jh}=A_{jh}(z_n,r_k,t_l)$ are defined on the grid ω. For brevity, we introduce the index-free notations:

$$\hat{A}_j=A_{jh}(z_{n+1},r_k,t_l),\ \overset{0.5}{A}_j=0.5(A_j+\hat{A}_j),\ |\overset{0.5}{A}_j|^2=0.5(|A_j|^2+|\hat{A}_j|^2),\ j=1-4.\tag{13}$$

The Laplacian with respect to t and r coordinates is approximated as

$$w_{\bar{t}t}=\frac{w(z_n,r_k,t_{l+1})-2w(z_n,r_k,t_l)+w(z_n,r_k,t_{l-1})}{\tau^2}$$

$$\Lambda_r w=\frac{1}{r_k(r_{k+1}-r_{k-1})}\left[(r_{k+1}+r_k)\frac{w(z_n,r_{k+1},t_l)-w(z_n,r_k,t_l)}{r_{k+1}-r_k}-(r_k+r_{k-1})\frac{w(z_n,r_k,t_l)-w(z_n,r_{k-1},t_l)}{r_k-r_{k-1}}\right]\tag{14}$$

at the internal point (z_n,r_k,t_l) of the grid ω. There a function w means one of the grid functions A_{jh}.

Using the notations introduced above, the finite-difference scheme for the problem (1) in the case of the phase matching can be written in the form:

$$\bar{L}_j u=\frac{\hat{u}-u}{h_z}+i\tilde{D}_{rj}\Lambda_r\overset{0.5}{u}+iD_j\overset{0.5}{u}_{\bar{t}t},$$

$$\bar{F}_j=\overset{0.5}{A}_j\left(2\left(\left|\overset{0.5}{A}_1\right|^2+\left|\overset{0.5}{A}_2\right|^2+\left|\overset{0.5}{A}_3\right|^2+\left|\overset{0.5}{A}_4\right|^2\right)-\left|\overset{0.5}{A}_j\right|^2\right),\tag{15}$$

$$\bar{L}_jA_j=i\alpha_j(\bar{F}_j+\overset{0.5}{A}_2\overset{0.5}{A}_4\overset{0.5}{A}{}^*_{4-j}),\ j=1,3,\qquad \bar{L}_jA_j=i\alpha_j(\bar{F}_j+\overset{0.5}{A}_1\overset{0.5}{A}_3\overset{0.5}{A}{}^*_{6-j}),\ j=2,4$$

with corresponding approximation of the boundary conditions and initial definition of the mesh functions.

Since the equations (15) are nonlinear ones, then we apply an iteration process:

$$\bar{L}_j \overset{S+1}{A}_j = i\alpha_j (\overset{S}{\bar{F}}_j + \overset{0.5}{A_2}\,\overset{0.5}{A_4}\,\overset{0.5}{A^*_{4-j}}),\ j=1,3,\quad \bar{L}_j \overset{S+1}{A}_j = i\alpha_j (\overset{S}{\bar{F}}_j + \overset{0.5}{A_1}\,\overset{0.5}{A_3}\,\overset{0.5}{A^*_{6-j}}),\ j=2,4. \qquad (16)$$

At boundaries points, the mesh functions are taken from s+1 iteration. At zero iteration ($s=0$) they are taken as

$$\overset{s=0}{\hat{A}_j} = A_j,\ j=1-4$$

and the iteration process is terminated if the following inequality:

$$\max_{r_k,t_l} |\overset{s+1}{\hat{A}_j} - \overset{s}{\hat{A}_j}| \le \varepsilon \max_{r_k,t_l} |\overset{s}{\hat{A}_j}| + \delta,\ j=1-4 \qquad (17)$$

is valid. Here ε, δ are positive parameters which defines the iteration process accuracy.

6. Comparison of computer simulation results with analytical solution

In figure 1 the analytical solution (12) and the corresponding computer simulation results are depicted. We see very good coincidence of both dependencies. In figure 2 the comparison is shown for different values of diffraction and dispersion coefficients: $\tilde{D}_{rj}$ and D_j. We see that with increasing

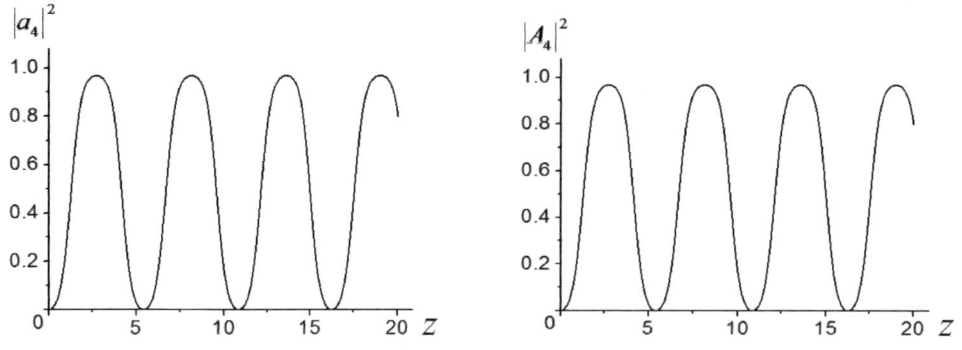

(a) (b)

Figure 1. Analytical solution (a) and computer simulation result (b) corresponding to parameters $\alpha_1 = \alpha_2 = \alpha_3 = \alpha_4 = 1$, $a_{10} = A_{10} = a_{30} = A_{30} = 1$, $a_{20} = A_{20} = 0.2$, $a_{40} = A_{40} = 0$, $D_{j4} = 1\cdot 10^{-12}$, $\tilde{D}_{rj} = 10^{-12}$.

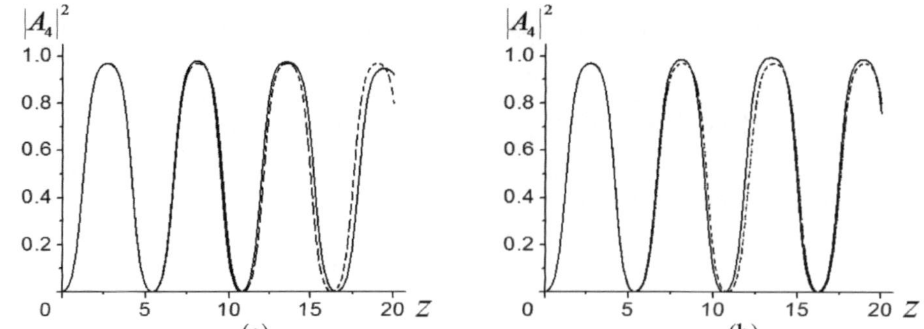

(a) (b)

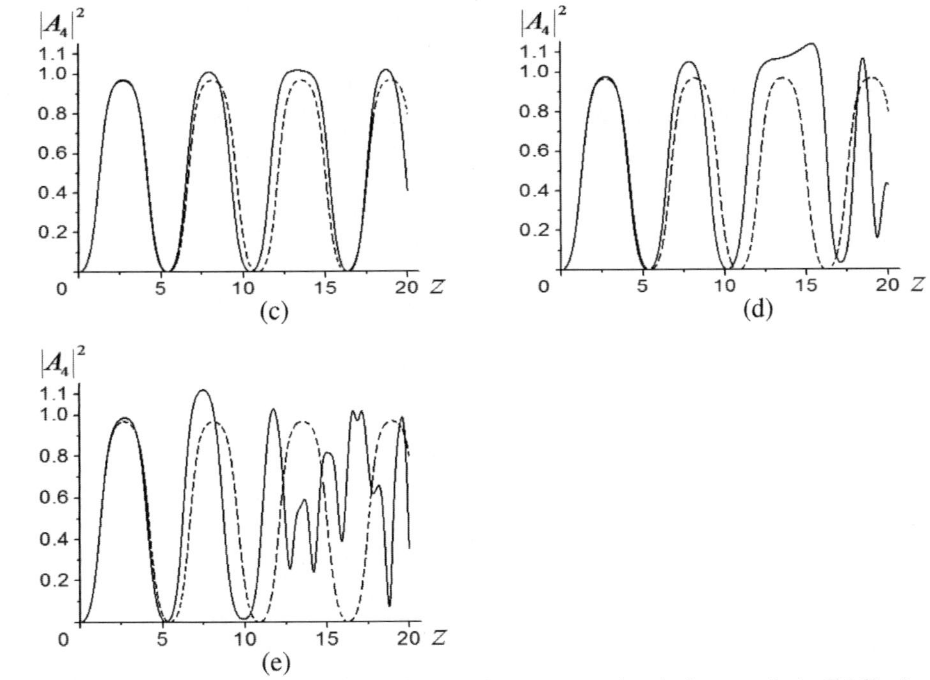

Figure 2. Analytical solution (dash line) and computer simulation result (solid line) corresponding to parameters: $\alpha_1 = \alpha_2 = \alpha_3 = \alpha_4 = 1$, $a_{10} = A_{10} = a_{30} = A_{30} = 1$, $a_{20} = A_{20} = 0.2$, $a_{40} = A_{40} = 0$ and $D_j = \tilde{D}_{rj} = 10^{-6}$ (a), $D_j = \tilde{D}_{rj} = 10^{-5}$ (b), $D_j = \tilde{D}_{rj} = 10^{-5}$ (c), $D_j = \tilde{D}_{rj} = 10^{-5}$ (d), $D_j = \tilde{D}_{rj} = 10^{-4}$ (e).

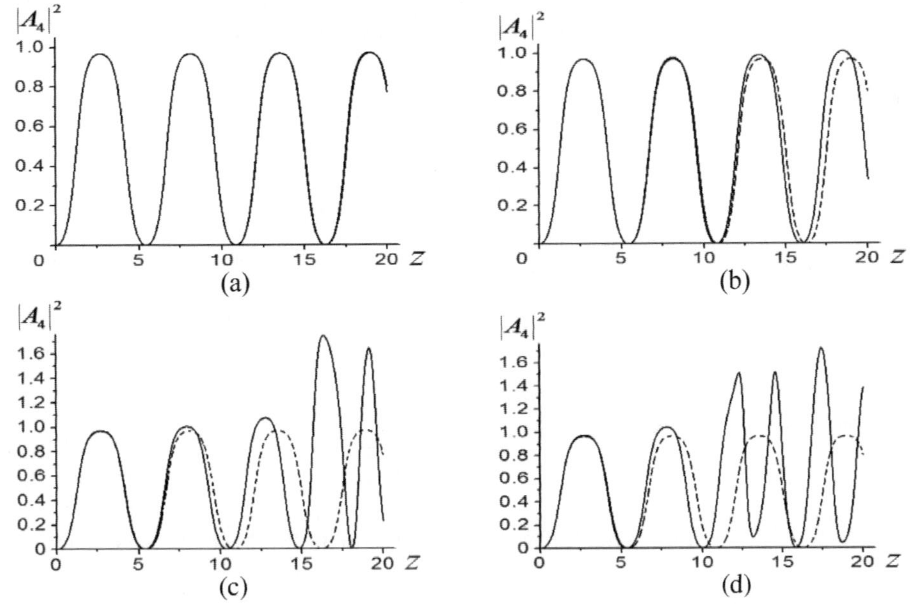

Figure 3. Analytical solution (dash line) and computer simulation result (solid line) corresponding to parameters: $\alpha_1 = \alpha_2 = \alpha_3 = \alpha_4 = 1$, $a_{10} = A_{10} = a_{30} = A_{30} = 1$, $a_{20} = A_{20} = 0.2$, $a_{40} = A_{40} = 0$ and $\tilde{D}_{rj} = 10^{-12}$, $D_j = 10^{-6}$ (a), 10^{-5} (b), $5 \cdot 10^{-5}$ (c), 10^{-4} (d).

values of diffraction of a laser beam and dispersion of a laser pulse the analytical solution deviation from the computer simulation result increases. Nevertheless, coinciding of both curvatures takes place along big distance of the optical beam propagation. To clarify a playing role of a pulse dispersion and a beam diffraction in breaking of the long pulse duration approximation and plane wave approximation we make computer simulation with unchangeable value of the diffraction coefficient with the dispersion coefficient changing and vice versa. The corresponding comparison is shown in figures 3, 4. We see that a second order dispersion influence results in the pulse compression and consequently, the maximal intensity of the pulse increases and this leads to the intensity oscillation period decreasing (see formula (12)).

The beam diffraction also leads to the maximal intensity enhancing. However, this enhancing is not pronounced along a pulse propagation distance under consideration.

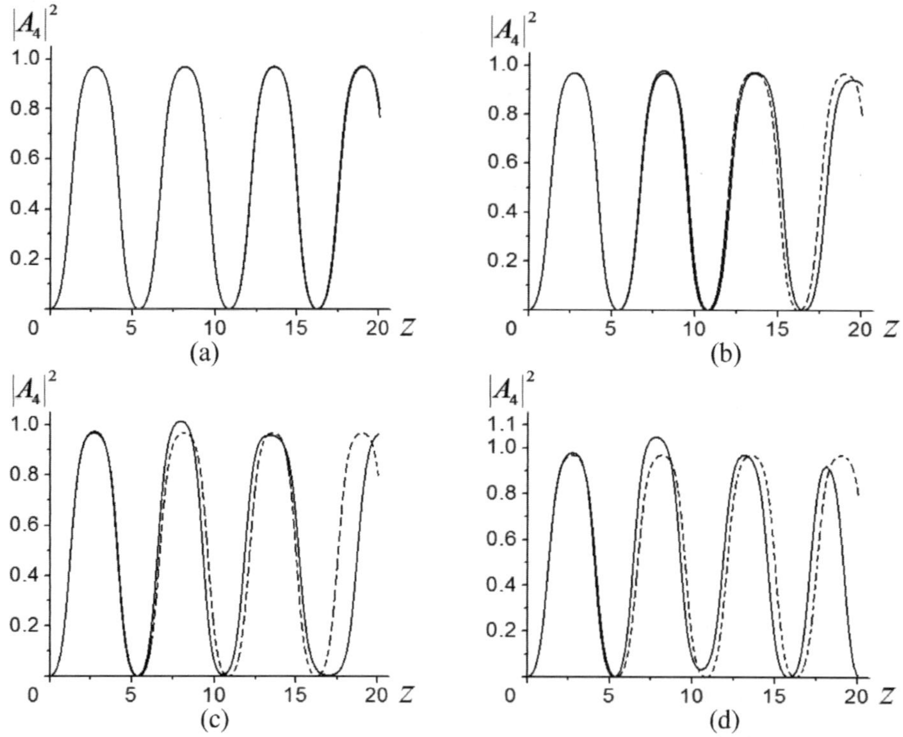

Figure 4. Analytical solution (dash line) and computer simulation result (solid line) corresponding to parameters: $\alpha_1 = \alpha_2 = \alpha_3 = \alpha_4 = 1$, $a_{10} = A_{10} = a_{30} = A_{30} = 1$, $a_{20} = A_{20} = 0.2$, $a_{40} = A_{40} = 0$ and $D_j = 10^{-12}$, $\tilde{D}_{rj} = 10^{-6}$ (a), 10^{-5} (b), $5 \cdot 10^{-5}$ (c), 10^{-4} (d).

7. Conclusions

Thus, in this paper we made a comparison of the analytical solution, obtained in the frame-work of long pulse and plane wave approximation for FWM problem, with computer simulation results, obtained by using the conservative finite-difference scheme proposed for this problem solution.

We consider a case of equality of the pump wave intensities at co-propagating FWM under phase matching condition. We showed that along big distance of a laser pulse propagation both solutions coincide. Main reason of a deviation of the solutions is an influence of second order dispersion. Beam diffraction influences less than pulse dispersion till propagation distance under analysis.

Acknowledgments
The investigation was made using support of the Russian Science Foundation (Grant № 14-21-00081).

References

[1] Abrams R L and Lind R C 1978 Degenerate four-wave mixing in absorbing media *Optics Letters* **2**(4) 94-96

[2] Winful H G and Marburger J H 1980 Hysteresis and optical bistability in degenerate fourwave mixing *Applied Physics Letters* **23** 613-614

[3] Jain R K and Lind R C 1983 Degenerate four-wave mixing in semiconductor-doped glasses *J. Opt. Soc. Am.* **73**(5) 647-653

[4] Slusher R E, Hollberg L W, Yurke B, Mertz J C and Valleys J F 1985 Observation of Squeezed States Generated by Four-Wave Mixing in an Optical Cavity *Physical Review Letters* **55**(22) 2409-2412

[5] Abdullah M N, Shaari S, Ehsan A A, Abidin M N Z and Abidin A R Z 2011 Effect of Photonic Crystal Fibre in Generation Four Wave Mixing *Physics Procedia* **19** 344-351

[6] Wang Z, Liu H, Huang N, Sun Q and Wen J 2011 Influence of spectral broadening on femtosecond wavelength conversion based on four-wave mixing in silicon waveguides *Applied Optics* **50**(28) 5430-5436

[7] Cheng H, Luo Z, Ye C, Huang Y, Liu C and Cai Z 2013 Numerical modeling of mid-infrared fiber optical parametric oscillator based on the degenerated FWM of tellurite photonic crystal fiber *Applied Optics* **52**(3) 525-529

[8] Petersen S R, Alkeskjold T T and Lægsgaard J 2013 Degenerate four wave mixing in large mode area hybrid photonic crystal fibers *Optics Express* **21**(15) 18111-18124

[9] Qin J, Ji Y, Wang H, Wang D, Zhang M and Lu G-W 2015 Multichannel wavelength multicasting for two QPSK signals based on FWM in SOA *Chinese Optics Letters* **13**(1) 010601

[10] Naimi S T, Duill S P and Barry L P 2014 Detailed Investigation of the Pump Phase Noise Tolerance for Wavelength Conversion of 16-QAM Signals Using FWM *J. Opt. Commun. Netw.* **6**(9) 793-800

[11] Lavdas S and Panoiu N C 2015 Comparative analysis of four-wave mixing of optical pulses in slow- and fast-light regimes of a silicon photonic crystal waveguide *Optics Letters* **40**(18) 4233-4236

[12] Xie W, Fsaifes I, Labidi T and Bretenaker F 2015 Investigation of degenerate dual-pump phase sensitive amplifier using multi-wave model *Optics Express* **23**(25) 31896-31907

[13] Liu H, Wang Z, Wen J, Huang N and Sun Q 2012 Efficient femtosecond optical parametric amplification and wavelength conversation in silicon waveguides *ALT Proceedings* **1**

[14] Li J-B, Liang S, Xiao S, He M-D, Kim N-C, Chen L-Q, Wu G-H, Peng Y-X, Luo X-Y and Guo Z-P 2015 Four-wave mixing signal enhancement and optical bistability of a hybrid metal nanoparticle-quantum dot molecule in a nanomechanical resonator *Optics Express* **24**(3) 2360-2369

[15] Grudtsyn Y V, Zubarev I G, Koribut A V, Kuchik I E, Mamaev S B, Mikheev L D, Semjonov S L, Stepanov S G, Trofimov V A and Yalovoi V I 2015 Self-phase modulation in a thin fused silica plate upon interaction with a converging beam of down-chirped femtosecond radiation *Quantum Electronics* **45**(5) 415-420

[16] Duffin T J, Nielsen M P, Diaz F, Palomba S, Maier S A and Oultoni R F 2016 Degenerate four-wave mixing in silicon hybrid plasmonic waveguides *Optics Letters* **41**(1) 155-158

[17] Manna S, Bernard M, Biasil S, Manzano F R, Mancinelli M, Ghulinyan M, Pucker G and Pavesi L 2016 Stimulated degenerate four-wave mixing in Si nanocrystal waveguides *J.Opt* **18** 075801

[18] Yang Y, Ooka Y, Thompson R M, Ward J M and Chormaic S N 2016 Degenerate four-wave mixing in a silica hollow bottle-like microresonator *Optics Letters* **41**(3) 575-578

[19] Yuan J, Zhou G, Xia C, Sang X, Li F, Yu C, Wang K, Yan B, Tam H Y and Wai P K A 2016 Degenerate Four-Wave Mixing-Based Light Source for CARS Microspectroscopy *IEEE Photonics Technology Letters* **28**(7) 575-578

[20] Trofimov V A, Trofimov V V and Levitskiy N V 2011 Up- and down-conversion at three-wave interaction in medium with combined nonlinear response *Proceedings of SPIE* **8071** 807106

[21] Trofimov V A, Kuchik I E and Levitskiy N V 2014 Multi-modes of four-waves mixing at non-collinear interaction of laser beams in medium with cubic nonlinear response *Proceedings of SPIE* **9136** 913619

[22] Trofimov V A and Lysak T M 2009 Highly Efficient SHG of a Sequence of Laser Pulses with a Random Peak Intensity and Duration *Optics and Spectroscopy* **107**(3) 399-406

[23] Lysak T M and Trofimov V A 2001 Bistability and uniqueness of solutions in the problem of second harmonic generation of femtosecond pulses *Computational Mathematics and Mathematical Physics* **41**(8) 1214-1226

[24] Lysak T M and Trofimov V A 2001 Bistable Mode of Second Harmonic Generation by Femtosecond Pulses *Technical Physics* **46**(11) 1401-1406

[25] Trofimov V V 2006 Analytical solution for problem of generation of sum frequency wave in medium with quadratic and cubic nonlinearities *Proceedings of SPIE* **6729** 6729-113

[26] Trofimov V A and Trofimov V V 2007 High effective SHG of femtosecond pulse with ring profile of beam in bulk medium with cubic nonlinear response *Proceedings of SPIE* **6610** 66100R-01- 66100R-8

Soliton self-formation in the medium with nanorods under conditions of the absorption spectrum finite width

Vyacheslav A. Trofimov, Tatiana M. Lysak, Igor E. Kuchik

Lomonosov Moscow State University, Moscow 119992, Russia

E-mail vatro@cs.msu.ru

Abstract. We investigate numerically an influence of the absorption spectrum width and the absorption spectrum central frequency shift on novel type of solitons (chirped soliton) appearence at a femtosecond pulse propagation in a medium with noble nanoparticles. We take into account the TPA of laser radiation by nanorods, and time-dependent nanorod aspect ratio changing due to their melting or reshaping because of laser energy absorption. The chirped solitons are formed due to the trapping of laser radiation by the nanorods reshaping fronts, if a positive or negative phase-amplitude grating is induced by laser radiation. We consider both linear and strong nonlinear dependences of absorption coefficient on the nanorod aspect ratio. We discuss a physical mechanism of the absorption spectrum width influence on the accelerating soliton formation.

1. Introduction

The influence of the noble metal nanoparticle aspect ratio and orientation, and local environment on the nanoparticle optical response has been widely investigated in recent years [1-12] because of nanorods using in many novel technologies. For example, nanoparticles aspect ratio changing because of photo-thermal reshaping provides the ability for five-dimensional recording [1-6]. Other very important question is the physical mechanism for photo-thermal reshaping of nanorods or wires at the temperatures less than the bulk melting [13-16]. Some models are proposed up to now to explain this phenomenon [15,16]

Among a number of problems of laser radiation interaction with a medium containing nanorods, the self-similar mode of laser pulse propagation is of great importance. This is due to the fact that the laser radiation spectrum distortions taking place due to nonlinear refraction can cause false information recording in all-optical data storage devices. Indeed, nanoparticles reshaping due to the laser radiation absorption leads to the pulse spectrum changing caused by the pulse chirping. This results in violation of optimal conditions for information recording, as well as (that is more essential) in the false information recording and reading. That is why the investigation of the laser pulse interaction with the medium, containing nanoparticles, is an actual problem.

In our previous papers [17-20] we investigated a femtosecond pulse propagation in a medium with nanorods under the conditions of nanorod aspect ratio changing and dependence of laser radiation absorption on nanoparticle aspect ratio. In particular, using the density matrix formalism, we derived the equation set which describes the femtosecond pulse propagation in the medium with nanorods reshaping, and obtained some analytical formulas for such propagation [17]. We also demonstrated the possibility for the superluminality effect implementation [21-22] in such media and discussed a physical mechanism of chirped soliton formation and light acceleration. In particular, we showed that the light acceleration takes place for a chirped incident pulse and demonstrated that a nonlinear frequency chirp is crucial for a self-similar mode realization in a medium with nanorods [18]. We analyzed the influence of the relation between the nanorods absorption spectrum bandwidth and laser pulse spectrum bandwidth on the laser pulse spectrum transform [19] and the laser pulse acceleration [20] under its propagation in a medium containing nanorods.

Our attention in this paper is attracted by an influence of the absorption spectrum central frequency on the effects of superliminality and soliton formation at laser pulse propagation in a medium with nanorods under the condition of a narrow bandwidth of the pulse spectrum.

2. Problem statement

We consider a femtosecond laser pulse propagation in a medium with nanorods by taking into account the nanorod aspect ratio changing due to nanorod reshaping because of TPA of optical energy. In the framework of slowly varying envelope of wave packet, this process can be described by the following dimensionless nonlinear equations:

$$\frac{\partial A}{\partial z}+iD\frac{\partial^2 A}{\partial t^2}+f(\varepsilon)\left(\delta_0+i\xi\right)|A|^2\cdot A\cdot\frac{\int_{-\infty}^{+\infty}\delta(\omega)|A_\omega|^2\,d\omega}{\int_{-\infty}^{+\infty}|A_\omega|^2\,d\omega}=0\ , \qquad (1)$$

$$\frac{\partial\varepsilon}{\partial t}=\tilde{\delta}f(\varepsilon)|A|^4\frac{\int_{-\infty}^{+\infty}\delta(\omega)|A_\omega|^2\,d\omega}{\int_{-\infty}^{+\infty}|A_\omega|^2\,d\omega}. \qquad (2)$$

with initial and boundary conditions for complex amplitude and nanorod aspect ratio

$$A(z,0)=A(z,L_t)=0,\quad 0\le z\le L_z,\ A(0,t)=A_0(t),\quad 0\le t\le L_t,\ \varepsilon(z,t=0)=\varepsilon_0,\ 0\le z\le L_z.$$

Above A is dimensionless slowly varying envelope of a wave packet normalized on a square root from the maximum incident pulse intensity ($z=0$). $\varepsilon=a/b$ is a nanorod aspect ratio, where a and b are major and minor axes of nanorods, ε_0 is its initial value. Variable z is a dimensionless longitudinal coordinate, along which the optical radiation propagates; t is a dimensionless time in the system of coordinates moving with a pulse, time is measured in the units of τ_{pulse} - duration of the incident pulse; L_t is a dimensionless time interval, during which the laser pulse interaction with nanorods is analyzed, L_z is a dimensionless length of nonlinear medium. Parameter D characterizes the group velocity dispersion (GVD). Parameters δ_0 and $\tilde{\delta}$ characterize the absorption of laser light and a part of absorbed energy spent on nanorods reshaping, correspondingly. Coefficient ξ characterizes the laser pulse self-action due to the wave packet carrier frequency detuning from the central frequency of the nanorod absorption spectrum. The case of $\xi=0$ corresponds to an optical pulse propagation in a medium with pure amplitude grating. It means an influence only of a laser energy absorption on the laser pulse propagation. In the opposite case ($\xi\neq 0$), the phase grating is also induced by the laser radiation. It should be mentioned, that the positive sign of the parameter ξ (this case is named by us as positive grating) corresponds to pulse compression and the laser pulse decompression occurs at negative sign of this parameter (this case is named by us as negative grating).

The function $f(\varepsilon)$ describes the dependence of TPA process on the nanorod aspect ratio ε. Below we consider two possible types of this dependence:

$$f(\varepsilon) = \varepsilon - 1 \tag{3}$$

$$f(\varepsilon) = \exp(5(\varepsilon - 2)), \quad \varepsilon \le 2.7, \tag{4}$$

These dependences approximate the dependence

$$f(\varepsilon) = \frac{\left(\varepsilon_2 / A^2\right)}{\left(\varepsilon_1 + \frac{1-A}{A}\varepsilon_m\right)^2 + \varepsilon_2^2}, \tag{5}$$

$$A = \left[1 - \frac{\xi Q_1'(\xi)}{Q_1(\xi)}\right]^{-1}, \quad \xi = \left[1 - \left(\frac{1}{\varepsilon}\right)^2\right]^{-1/2}, \quad Q_1(\xi) = \left(\frac{\xi}{2}\right)\ln\left[\frac{\xi+1}{\xi-1}\right] - 1, \quad Q_1'(\xi) = \frac{dQ_1(\xi)}{d\xi}, \tag{6}$$

which follows from Boyd and Shen shape factor expression [23] for the absorption coefficient of gold nanorods, calculated in the dipole approximation [24] for physical parameters $\varepsilon_1 = -22.4$, $\varepsilon_2 = 1.8$ for Au at $\lambda = 800$ and ε_m=3. In Eq.(5)-(6), ε_m is the dielectric permittivity of the ambient medium, $\varepsilon_1 + i\varepsilon_2$ is the complex dielectric permittivity of the nanorods.

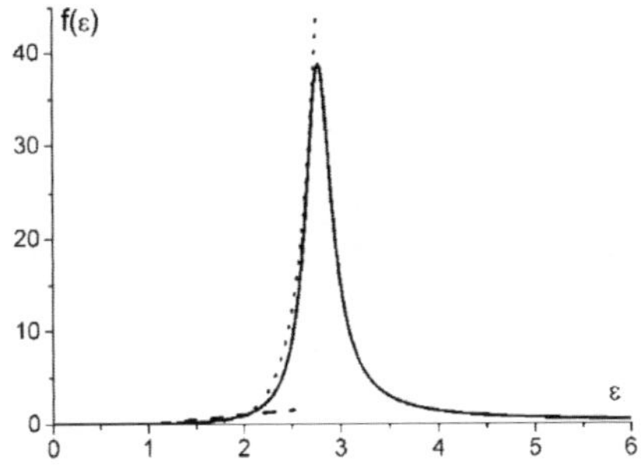

Figure 1. Dependence $f(\varepsilon)$ defined by formulas (5)-(6) for gold nanorods at the falling radiation with $\lambda = 800$ nm (solid line), or by formula (3) (dashed line), or by formula (4) (dotted line).

If the aspect ratio varies from 1 to 2, then a laser radiation interaction with nanorods occurs far from the nonlinear absorption resonance. In this case, a linear dependence (3) adequately approximates dependence (5)-(6) (figure1, dashed line). Strongly non-linear dependence (4) can be used to approximate the dependence (5)-(6) if the aspect ratio changes from 1 to 2.6 (2, dotted line). The expression (4) is more preferable for a laser pulse interaction with nanorods near the nonlinear absorption resonance. Below we consider both dependences and choose initial value of aspect ratio as $\varepsilon_0 = 2$ for the dependence (4) and $\varepsilon_0 = 2.6$ for the dependence (5).

In Eqs. (1)-(2), we also take into account the absorption spectrum bandwidth influence on a laser energy absorption. At writing the spectral absorption coefficient $\delta(\omega)$, we consider the Lorentz law for the absorbance cross-section on light frequency. Hence, it is written as

$$\delta(\omega) = \frac{1}{1 + 4\left(\frac{\omega_0 - \omega}{\Delta\omega}\right)^2} \tag{6}$$

with the absorption spectrum width and the central absorption frequency denoted as $\Delta\omega$ and ω_0, respectively. Obviously, the absorption spectrum central frequency ω_0 does not influence TPA, if the absorption spectrum width is many times larger than the laser pulse spectrum width. Indeed, in this case the integral dependence in (1) and (2) can be omitted because the second item in the denominator of the expression (6) becomes negligible. So, it is possible to choose $\omega_0 = 0$ for the wide absorption spectrum. Below we show that the central absorption frequency ω_0 greatly influences the process of soliton formation if the absorption spectrum is comparable with the pulse spectrum or narrower.

We specify the following complex amplitude of incident pulse

$$A(z=0,t) = A_0(t) = \exp\left(-((t - L_t/2)/\tau)^{m_t}\right), 0 \le t \le L_t$$

τ is a dimensionless pulse duration, m_t=2 describes the Gaussian pulse shape.
We also follow the laser pulse centre position

$$\tau_c(z) = \int_0^{L_t} (t - L_t/2)|A(z,t)|^2 dt \bigg/ \int_0^{L_t} |A(z,t)|^2 dt$$

3. Computer simulation results

Below we present computer simulation results for the incident Gaussian pulse ($m_t = 2, \tau = 1$) propagation at $D = 0.1$ and positive phase-amplitude grating $\xi = 5$, and sufficiently small strength of reshaping $\tilde{\delta} = 5$ for both types (3) and (4) of the absorption coefficient dependence on aspect ratio. Let us mention that the considered value of GVD corresponds to the pulse duration about 500 fs. We consider narrow absorption spectrum with bandwidth $\Delta\omega = 1$, which is less than the pulse spectrum width and three values of the central absorption frequency $\omega_0 = \{0, 1, -1\}$ with respect to the carrier frequency of a wave packet.

Before we discuss the computer simulation results, let us note that the absorption spectrum bandwidth $\Delta\omega$ affects the pulse spectrum changing and the pulse propagation only if it becomes comparable with the initial spectrum width. This is valid for all signs of the phase grating as well as for the pure amplitude grating [19]. Moreover, as it was demonstrated in [20], if the absorption spectrum bandwidth is comparable with the incident pulse spectrum bandwidth or narrower than it, a light acceleration is not so pronounced as in the case of the absorption spectrum that is much wider than the incident pulse spectrum. The pulse splitting also depends on the absorption spectrum bandwidth: it occurs only for a wide spectrum or for the spectrum comparable with the incident pulse spectrum bandwidth. If the absorption spectrum is narrower than the incident pulse spectrum, no pulse splitting occurs, and the pulse propagates as a whole at the velocity corresponding to the pulse front propagation in a linear medium.

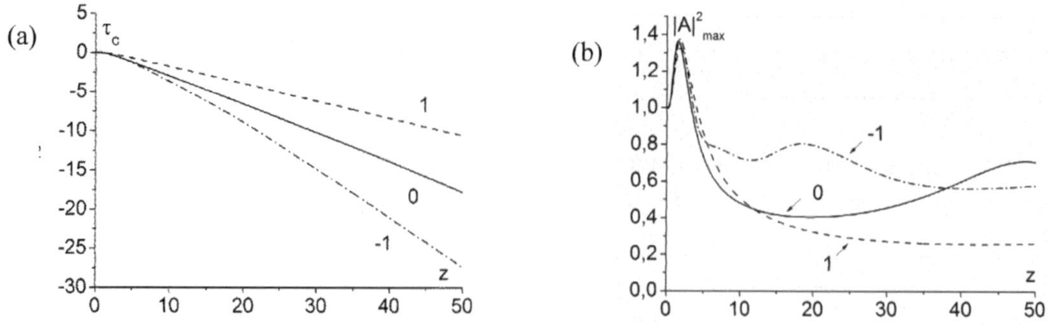

Figure 2. Pulse centre shifting along the z-coordinate (a), maximal intensity evolution (b) for the three values of the central absorption frequency ω_0 (shown in Figure), positive phase amplitude grating $\xi = 5$ and $\delta_0 = 0.005, \tilde{\delta} = 5$, $\Delta\omega = 1$.

3.1. Soliton formation far fromnonlinear absorption resonance

We provide computer simulations for the linear dependence (3) of absorption coefficient on nanorod aspect ratio which is valid far from the nonlinear absorption resonance. We considered a small depletion ($\delta_0 = 0.005$) of laser energy due to TPA at a short distance ($z \approx 1$). It should be mentioned that the considered values of phase-amplitude grating and energy depletion correspond to the dimensionless frequency detuning parameter $\theta = (2\omega_p - \omega_{21})T_\perp = \xi/\delta_0 = 10^3$, which is much bigger than the dimensionless value of $\omega_p T_\perp \approx 236$, corresponding to $\lambda = 800\,\text{nm}$ (we consider $T_\perp \approx 10^{-13}$ s [25]). Nevertheless, this parameter values are very important for theoretical consideration.

As it is well seen in figures 2-4, if the central absorption frequency belongs to low frequency area ($\omega_0 = -1$) then the pulse acceleration enhancing occurs in comparison with its zero-value ($\omega_0 = 0$). Shift of the central absorption frequency to the high frequency area ($\omega_0 = 1$) results in the pulse acceleration decreasing.

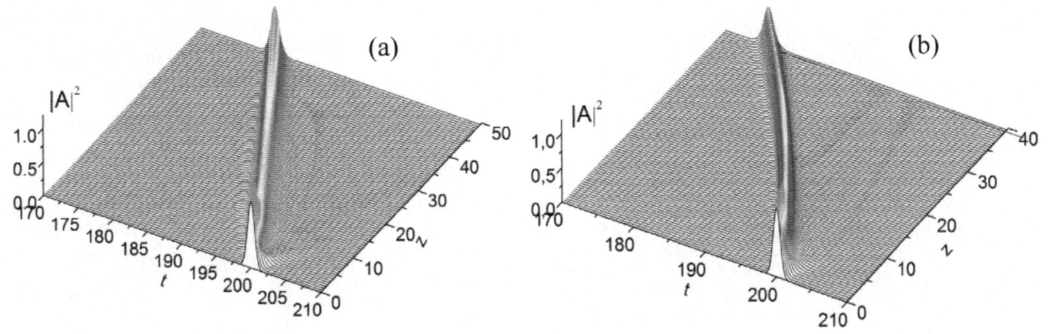

Figure 3. Soliton propagation in the medium with positive phase grating $\xi = 5$ and $\delta_0 = 0.005$, $\tilde{\delta} = 5$ for $\Delta\omega = 1$, $\omega_0 = 0$ (a), -1(b).

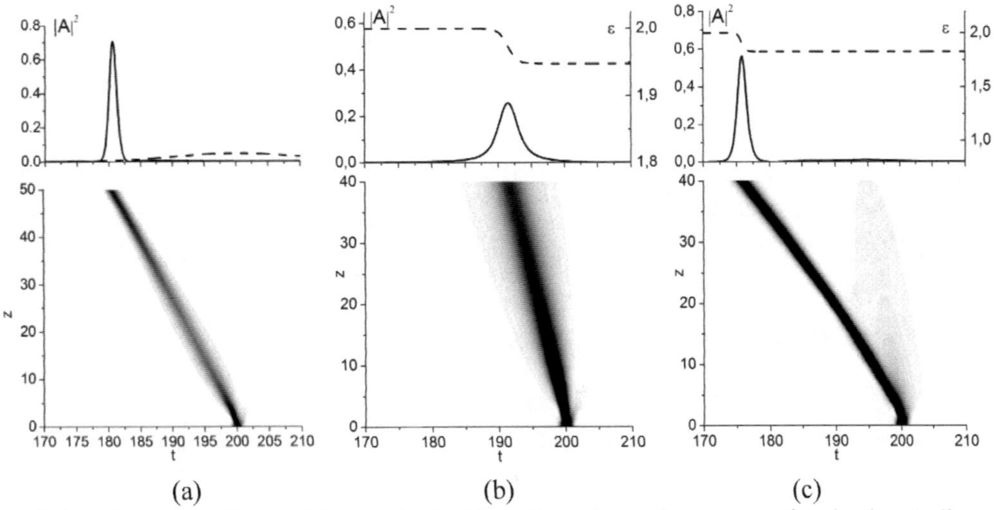

Figure 4. Maximal intensity position at the incident Gaussian pulse propagation in the medium with positive phase grating $\xi = 5$ and $\delta_0 = 0.005$, $\tilde{\delta} = 5$ for $\Delta\omega = 1$, $\omega_0 = 0$ (a),1(b),-1 (c). The solid lines at the top panels show soliton shapes at the exit section. Dashed line shows pulse shape at its propagation in linear medium (a) or aspect ratio distribution (b, c) at the exit section.

IOP Conf. Series: Journal of Physics: Conf. Series **826** (2017) 012019 doi:10.1088/1742-6596/826/1/012019

Note, that in the considered case, the soliton position coincides with the pulse front position at its propagation in a linear medium without nanorods (dashed line in figure 4a), if the central absorption frequency is equal to zero ($\omega_0 = 0$). The pulse compression observed at the initial stage of propagation is due to the positive phase grating action and small laser energy depletion (figure 2b). Moreover, a negative value of the central absorption frequency ($\omega_0 = -1$) results in a larger average maximal intensity of the soliton compared to its positive value (compare lines 1 and -1 in figure 2b).

These laser pulse propagation peculiarities can be explained using the physical mechanism of the soliton acceleration due to the laser radiation trapping by the nanorods reshaping front [17]. The trapping of laser radiation is well-seen in figures 4b, 4c, in which the dashed lines show the nanorod aspect ratio distribution. Indeed, as is well seen in figure 5b, up to the section z=30, the most part of the pulse leaves the absorption area in the frequency range if $\omega_0 = 1$. So, after this section, the pulse

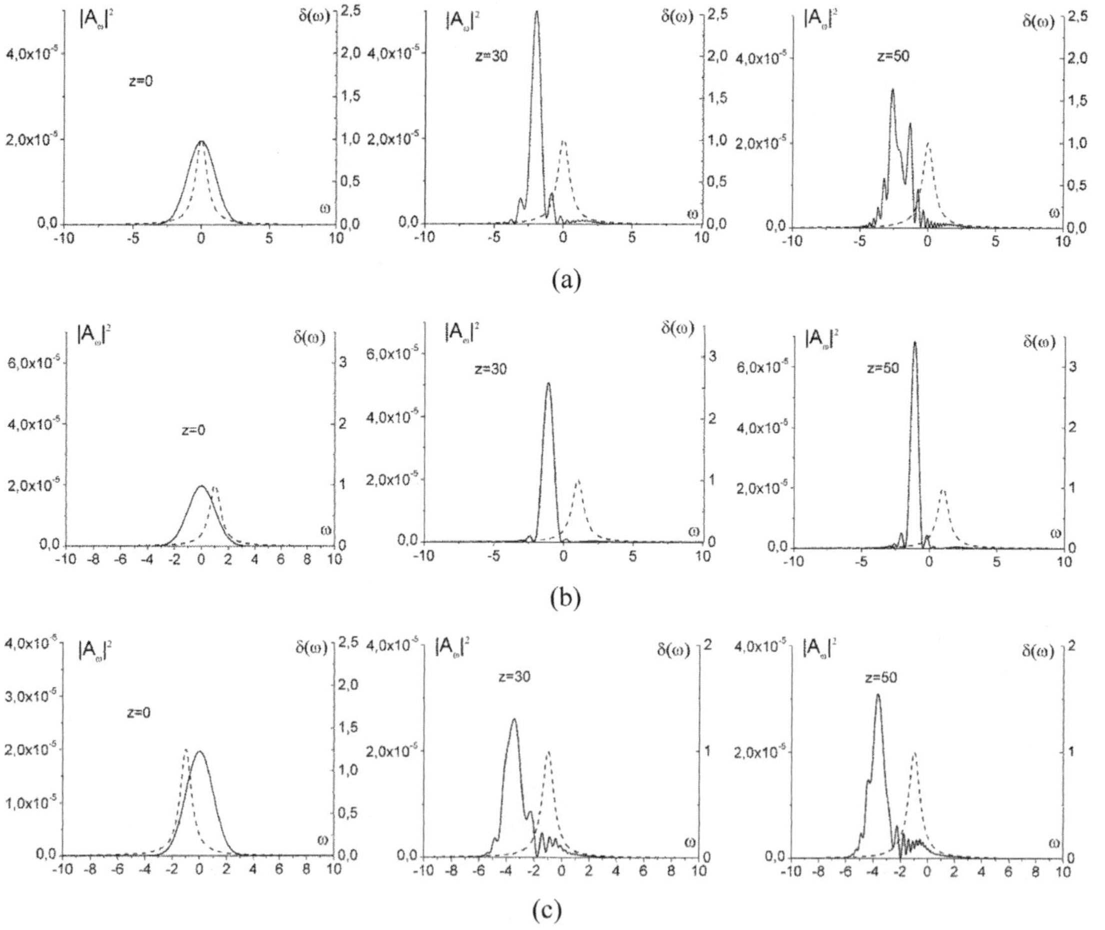

Figure 5. Pulse spectrum at the incident Gaussian pulse propagation in the medium with positive phase grating $\xi = 5$ and $\delta_0 = 0.005$, $\tilde{\delta} = 5$ for $\Delta\omega = 1$ and $\omega_0 = 0$ (a),1 (b), -1(c). Dashed lines show the absorption spectrum.

acceleration stops because the slow waves, which are at the pulse front due to the positive phase grating action, are no longer absorbed. If the central absorption frequency is negative ($\omega_0 = -1$), the pulse remains in the frequency area corresponding to a laser energy absorption for a larger distance of propagation, and therefore its acceleration is greater (figure 5c). As a result, the pulse possesses a higher velocity.

3.2. Soliton formation for strong nonlinear dependence of laser energy absorption

We explore the strong nonlinear dependence (4) of the laser energy absorption on the nanorod aspect ratio for its description near the nonlinear absorption resonance. For example, we choose a maximal value of the absorption coefficient as $\delta_0 = 0.1$. In this case, the dimensionless frequency detuning parameter $\theta = (2\omega_p - \omega_{21})T_\perp = \xi/\delta_0 = 50$ corresponds to the frequency detuning of $\approx 0.2\omega_p$ for $\omega_p \approx 2360$ THz ($\lambda = 800$ nm). This frequency detuning is available in physical experiments.

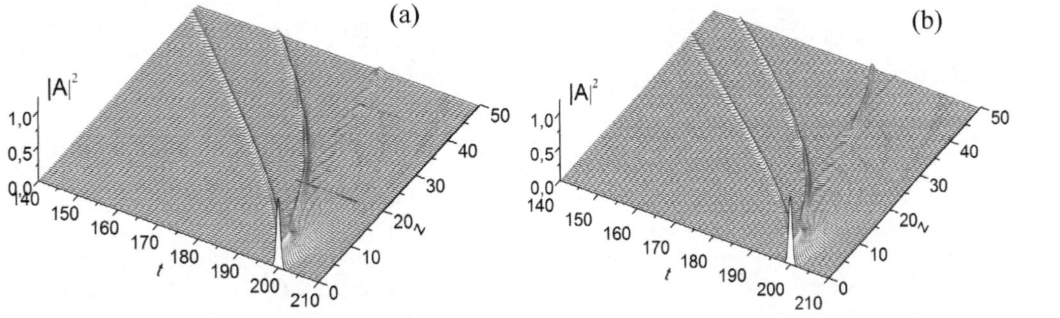

Figure 6. Soliton propagation in the medium with positive phase grating $\xi = 5$ and $\delta_0 = 0.1$, $\tilde{\delta} = 5$ for $\Delta\omega = 1$, $\omega_0 = 0$ (a), -1(b).

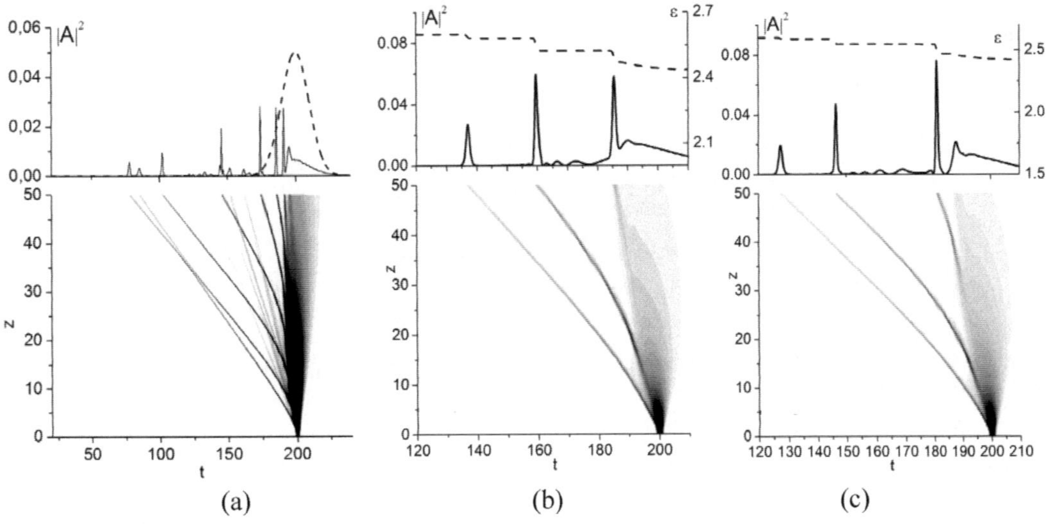

Figure 7. Maximal intensity position at the incident Gaussian pulse propagation in the medium with positive phase grating $\xi = 5$ and $\delta_0 = 0.1$, $\tilde{\delta} = 5$ for $\omega_0 = 0$, $\Delta\omega = 1000$ (a), 1 (b) and $\omega_0 = -1$, $\Delta\omega = 1$ (c) near the nonlinear absorption resonance. The solid lines at the top panels show soliton shapes at the exit section. Dashed line shows pulse shape at its propagation in a linear medium (a) or aspect ratio distribution (b, c) at the exit section.

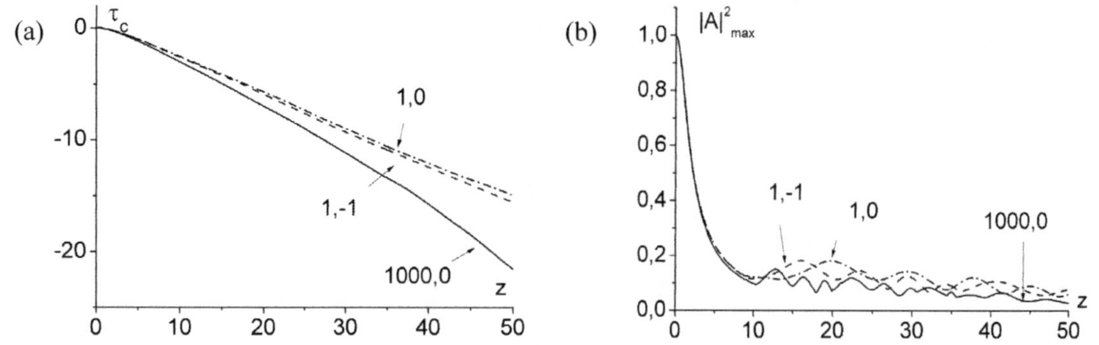

Figure 8. Pulse centre shifting along z-coordinate (a), maximal intensity evolution (b) for positive phase grating $\xi = 5$ and $\delta_0 = 0.1$, $\tilde{\delta} = 5$. The values of $\Delta\omega$ and ω_0 are shown in the figure by the pairs of numbers (the first number is for $\Delta\omega$ and the second is for ω_0).

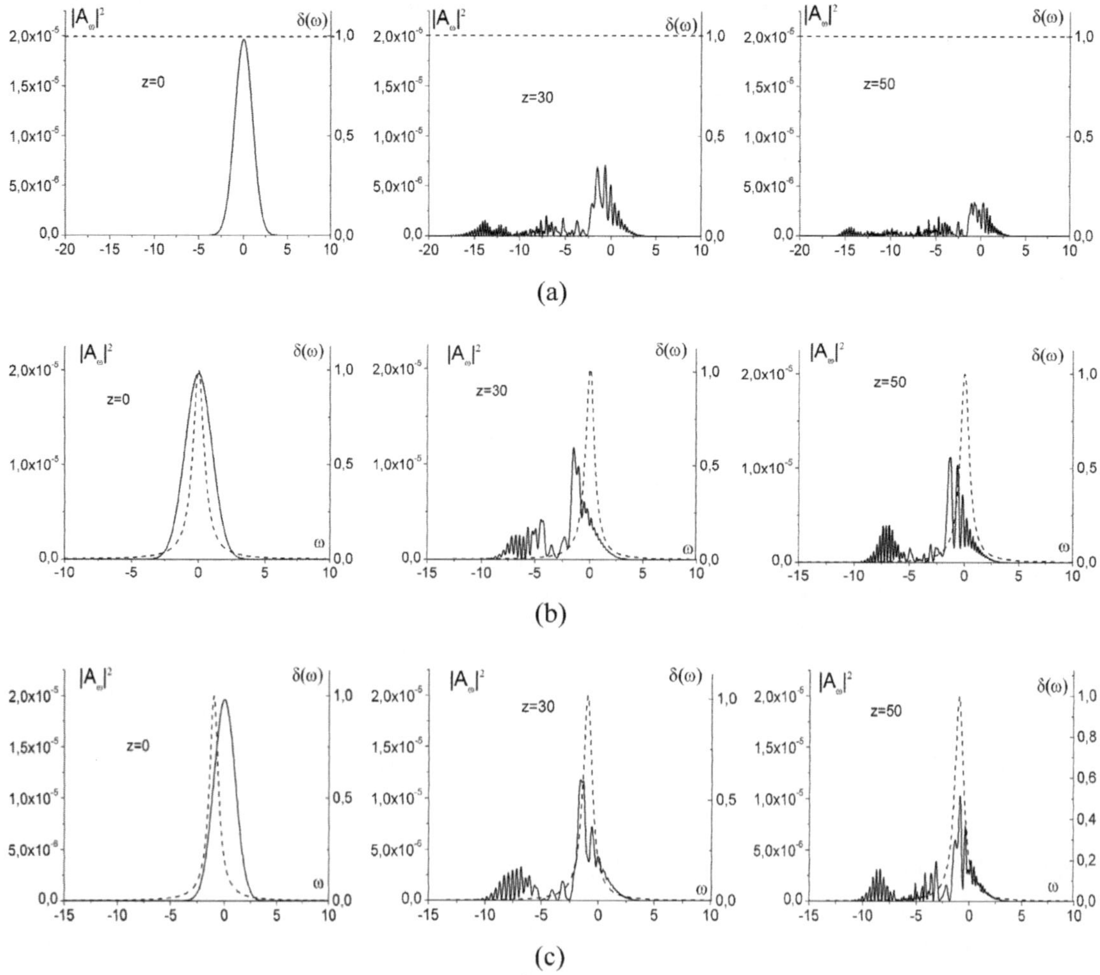

Figure 9. Pulse spectrum at the incident Gaussian pulse propagation in the medium with positive phase grating $\xi = 5$ and $\delta_0 = 0.1$, $\tilde{\delta} = 5$ for $\Delta\omega = 1000$ (a), 1(b,c) and $\omega_0 = 0$ (a,b), -1(c) and near the nonlinear absorption resonance. Dashed lines show the absorption spectrum.

Let us notice that the function $f(\varepsilon)$ in the form (4) instead of (3), does not influence on the dimensionless frequency detuning θ, but this function type strongly changes the optical energy absorption and the strength of induced phase grating, and, thus, decrease the propagation distance at which the effects under consideration can be observed.

In this case, the multiple solitons formation occurs (figures 6, 7). Nevertheless, similar to the previous case of the dependence (3), the absorption spectrum bandwidth influence on the sub-pulses acceleration becomes noticeable only for the absorption spectrum bandwidth which is comparable with the incident pulse spectrum. Also, the acceleration of the sub-pulses decreases and their number decreases with the absorption spectrum bandwidth decreasing (compare figures 7a and 7b; solid and dashed-dotted curves in figure 8a, marked 1000,0 and 1,0, respectively).

The action of the central absorption frequency position in a spectral domain on the sub-pulses accelerating is also similar to the previous case (section 3.2): its negative value ($\omega_0 = -1$) promotes a pulse acceleration in comparison with its zero-value ($\omega_0 = 0$) (figures 6, 7b,c). But this promotion is much less pronounced then in the case of the linear dependence (3) (compare figures 2a and 8a). Indeed, similar to the previous case, the laser energy absorption occurs for the larger propagation distance if $\omega_0 = -1$ (figures 9b,c). Therefore, the acceleration is more pronounced in this case. For the wide absorption spectrum, the sub-pulses do not leave the absorption area in spectral domain, so they achieve higher velocities (figure 9a).

Due to the stronger depletion of light energy and the nonlinear function $f(\varepsilon)$, there is no compression of the light pulse at the initial stage of propagation and the maximum intensities of the sub-pulses at the medium exit are less 0.2 dimensionless units (figure 8b).

4. Conclusions

We investigated the influence of the central absorption frequency on the process of soliton self-formation and light acceleration for the absorption far and close to the nonlinear absorption resonance. This influence occurs mainly if the absorption spectrum bandwidth is compared to the incident pulse spectrum bandwidth.

We showed that the number of self-formed sub-pulses depends on the absorption spectrum bandwidth: it decreases with the absorption spectrum bandwidth decreasing. In particular, if the absorption spectrum is narrower than the initial pulse spectrum, no pulse splitting occurs for the absorption which is far from the nonlinear absorption resonance.

Acceleration of the sub-pulses is the most pronounced if the absorption spectrum is much wider than the incident pulse spectrum. For the narrow absorption spectrum, the shift of its central absorption frequency into the area of lower frequencies promotes acceleration of the sub-pulses, while the shift into the area of higher frequencies results in a lower light velocity.

Acknowledgments

The investigation was made using support of the Russian Science Foundation (Grant № 14-21-00081).

References

[1] Zijlstra P, Chon J W M, and Gu M 2009 Five-dimensional optical recording mediated by surface plasmons in gold nanorods *Nature* **459** 410-413
[2] Ditlbacher H, Krenn J R, Lamprecht B, Leitner A, and Aussenegg F R 2000 Spectrally coded optical data storage by metal nanoparticles *Opt. Lett.* **25** 563–565
[3] Wilson O, Wilson G J, and Mulvaney P 2002 Laser writing in polarized silver nanorod films *Adv. Mater.* **14** 1000–1004
[4] Pérez-Juste J, Rodríguez-González B, Mulvaney P, and Liz-Marzán L M 2005 Optical control and patterning of gold-nanorod-poly(vinyl alcohol) nanocomposite films *Adv. Funct. Mater.*

15 1065–1071

[5] Taylor A B, Kim J and Chon J W M 2012 Detuned surface plasmon resonance scattering of gold nanorods for continuous wave multilayered optical recording and readout *Opt. Express* **20** (5) 5069–5081

[6] Roxworthy B J, Bhuiya A M, Inavalli V V G K, Chen H and Toussaint K C, Jr. 2014 Multifunctional plasmonic film for recording near-field optical intensity *Nano Let.* **14** 4687-4693

[7] Del Fatti N, Christofilos D and Vallée F 2008 Optical response of a single gold nanoparticle *Gold Bulletin* **41**(2) 147-157

[8] Wijers C M J, Chu J-H, Liu J L and Voskoboynikov O 2006 Optical response of layers of embedded semiconductor quantum dots *Phys. Rev. B* **74** 035323

[9] Mohammadi A, Sandoghdar V and Agio M 2008 Gold nanorods and nanospheroids for enhancing spontaneous emission *New Journal of Physics* **10** 105015

[10] Jin R, Jureller J E, Kim H Y and Scherer N F 200 5Correlating second harmonic optical responses of single Ag nanoparticles with morphology *J. Am. Chem. Soc.* **127** 12482-12483

[11] Drachev V P, Buin A K, Nakotte H and Shalaev V M 2004 Size dependent χ^3 for conduction electrons in Ag nanoparticles *Nano Lett.* **4** (8) 1535-1539

[12] Eustis S and El-Sayed M A 2006 Why gold nanoparticles are more precious than pretty gold: Noble metal surface plasmon resonance and its enhancement of the radiative and nonradiative properties of nanocrystals of different shapes *Chemical Society Reviews* **35**, 209-217

[13] Taylor A B, Siddiquee A M and Chon J W M 2014 Below melting point photothermal reshaping of single gold nanorods driven by surface diffusion *ACS Nano* **8** 12071-12079

[14] Goswami G K and Nanda K K 2010 Size-Dependent Melting of Finite-Length Nanowires *J. Phys. Chem. C* **114** (34) 14327–14331

[15] Gan Y and Jiang S 2013 Ultrafast Laser-Induced Premelting and Structural Transformation of Gold Nanorod *J. Appl. Phys.* **113** 073507

[16] Wang Ningyu, Rokhlin S I and Farson D F 2011 Ultrafast laser melting of Au nanoparticles: Atomistic simulations *J Nanopart Res* **13** (10), 4491

[17] Trofimov V A and Lysak T M 2016 Acceleration of light and soliton formation due to nonlinear absorption of femtosecond laser pulse energy by the medium containing nanorods *J. Opt. Soc. Am. B* **33** (1) 62-74

[18] Trofimov VA and Lysak T M 2016 Self-similar mode of laser pulse propagation in a medium containing nanorods *J. Opt.* **18** (1) 025501

[19] Trofimov V A, Lysak T M and Lan Sh 2012 Influence of ellipticity of nanorods on both TPA of femtosecond laser pulse and transformation of pulse spectrum *Proc. of SPIE* **8424** 842435

[20] Trofimov V A and Lysak T M 2016 Influence of nanorod absorption spectrum width on superluminality effect for laser pulse propagation *Proc. of SPIE* **9763** 97631Q

[21] Hache′A and Poirier L 2002 Anomalous dispersion and superluminal group velocity in a coaxial photonic crystal: Theory and experiment *Phys. Rev. E.* **65** (3) 036608

[22] Blonskyi I, Kadan V, Shpotyuk O and Dmitruk I 2009 Manifestations of sub- and superluminality in filamented femtosecond laser pulse in fused silica *Opt. Commun.* **282** (9) 1913–1917

[23] Boyd G T, Yu Z H and Shen Y R 1986 Photoinduced Luminescence From the Noble-Metals and Its Enhancement On Roughened Surfaces *Phys. Rev. B* **33**, (12) 7923-7936

[24] Link S, Mohamed M B and El-Sayed M A 1999 Simulation of the optical absorption spectra of gold nanorods as a function of their aspect ratio and the effect of the medium dielectric constant *J. Phys. Chem. B* **103** (16) 3073-3077

[25] Butylkin V S, Kaplan A E, Khronopulo Yu G and Yakubovich EU 1989 *Resonant nonlinear interactions of light with matter* (Berlin: Springer)

Uncertainty product of an out-of-equilibrium Bose-Einstein condensate

S Klaiman[1], A I Streltsov[1] and O E Alon[2]

[1] Theoretische Chemie, Heidelberg University, Heidelberg, Germany
[2] Department of Physics, University of Haifa at Oranim, Tivon, Israel

E-mail: `ofir@research.haifa.ac.il`

Abstract. The variance and uncertainty product of the position and momentum many-particle operators of structureless bosons interacting by a long-range inter-particle interaction and trapped in a single-well potential are investigated. In the first example, of an out-of-equilibrium interaction-quench scenario, it is found that, despite the system being fully condensed, already when a fraction of a particle is depleted differences with respect to the mean-field quantities emerge. In the second example, of the pathway from condensation to fragmentation of the ground state, we find out that, although the cloud's density broadens while the system's fragments, the position variance actually decreases, the momentum variance increases, and the uncertainty product is not a monotonous function but has a maximum. Implication are briefly discussed.

1. Introduction

Bose-Einstein condensates (BECs) made of ultracold trapped bosonic atoms have become a vast ground to study interacting quantum systems [1-12]. There has been a substantial theoretical interest in BECs, and ample studies have been made to describe their static and dynamic properties using Gross-Pitaevskii theory. The time-dependent Gross-Pitaevskii equation governs a mean-field theory which assumes that all bosons are described by a single time-dependent one-particle function throughout the evolution of the BEC in time. It is generally accepted that Gross-Pitaevskii theory adequately describes the ground state and out-of-equilibrium dynamics of BECs in the limit of an infinite number of particles and at constant interaction parameter (i.e., when the product of the number of particles times the scattering length is kept fixed). Here, mathematically rigorous results exist and show that, under certain conditions, the energy per particle and density per particle of the many-boson system coincide in this limit with the Gross-Piteavekii results and that the system is 100% condensed [13-16].

Whereas a condensate fraction of 100% implies that the number of depleted (non-condensed) particles divided by the total number of particles vanishes in the infinite-particle limit, the former is always non-zero in an interacting many-boson system. This observation has motivated us recently to look at properties of BECs which depend on the number of depleted particles rather than the condensate fraction. In [17, 18] we showed, for the ground state as well as for an out-of-equilibrium BEC, that even in the infinite-particle limit when the BEC is 100% condensed, the variance of a many-particle operator and the uncertainty product of two such operators can differ from those predicted by the Gross-Pitaevskii theory. The existence of many-body effects beyond those predicted by Gross-Pitaevskii theory stems from the necessity of performing the infinite-particle limit only after a many-particle quantum mechanical observable is evaluated and not prior to its evaluation. Unlike the variance of operators of a single particle [19], the variance and uncertainty product of many-particle operators is more involved, also see [20-22] in this context. It has furthermore been shown that the overlap of the exact and Gross-Pitaevskii wave-functions of a trapped BEC in the limit of an infinite number of particles is always smaller than unity and may even become vanishingly small [23].

The purpose of the present work is to build on and go beyond [17, 18] in two directions, first, by studying the variance and uncertainty product of trapped bosons with a long-range inter-particle interaction and, second, when the system is no longer condensed. The structure of the paper is as follows. In Sec. 2 we briefly discuss a general theory for the many-body variance and uncertainty product of an out-of-equilibrium trapped BEC. In Sec. 3 we present two applications, for the breathing dynamics of a trapped BEC (Subsec. 3.1) and for the pathway from condensation to fragmentation of the ground state of trapped interacting bosons (Subsec. 3.2). Concluding remarks are put forward in Sec. 4. Numerical and convergence details are collected in the Appendix.

2. Theoretical Framework

Consider the many-body Hamiltonian of N interacting bosons in a trap $V(\mathbf{r})$,

$$\hat{H}(\mathbf{r}_1,\ldots,\mathbf{r}_N) = \sum_{j=1}^{N}\left[-\frac{1}{2}\frac{\partial^2}{\partial \mathbf{r}_j^2} + \hat{V}(\mathbf{r}_j)\right] + \sum_{j<k}\lambda_0 \hat{W}(\mathbf{r}_j - \mathbf{r}_k). \tag{1}$$

Here, $\hbar = m = 1$, and $\hat{W}(\mathbf{r}_1 - \mathbf{r}_2)$ is the inter-particle interaction with λ_0 its strength.

The system evolves in time according to the time-dependent Schrödinger equation,

$$\hat{H}(\mathbf{r}_1,\ldots,\mathbf{r}_N)\Psi(\mathbf{r}_1,\ldots,\mathbf{r}_N;t) = i\frac{\partial\Psi(\mathbf{r}_1,\ldots,\mathbf{r}_N;t)}{\partial t}, \tag{2}$$

where the wave-function $\Psi(\mathbf{r}_1,\ldots,\mathbf{r}_N;t)$ is normalized to unity. Typically, the system is initially prepared in the ground state of the trap $V(\mathbf{r})$, $\hat{H}(\mathbf{r}_1,\ldots,\mathbf{r}_N)\Phi(\mathbf{r}_1,\ldots,\mathbf{r}_N) = E\Phi(\mathbf{r}_1,\ldots,\mathbf{r}_N)$, where E is the ground-state energy and $\Phi(\mathbf{r}_1,\ldots,\mathbf{r}_N)$ normalized to unity.

In what follows we employ the reduced one-body and two-body density matrices of $\Psi(\mathbf{r}_1,\ldots,\mathbf{r}_N;t)$ [24-27]. The reduced one-body density matrix is given by

$$\begin{aligned}\frac{\rho^{(1)}(\mathbf{r}_1,\mathbf{r}_1';t)}{N} &= \int d\mathbf{r}_2\ldots d\mathbf{r}_N\,\Psi^*(\mathbf{r}_1',\mathbf{r}_2,\ldots,\mathbf{r}_N;t)\Psi(\mathbf{r}_1,\mathbf{r}_2,\ldots,\mathbf{r}_N;t) = \\ &= \sum_j \frac{n_j(t)}{N}\alpha_j(\mathbf{r}_1;t)\alpha_j^*(\mathbf{r}_1';t).\end{aligned} \tag{3}$$

The quantities $\alpha_j(\mathbf{r};t)$ are the so-called natural orbitals and $n_j(t)$ their respective occupations which are time dependent and used to define the degree of condensation in a system of interacting bosons [28]. The density of the system is simply the diagonal of the reduced one-body density matrix, $\rho(\mathbf{r};t) = \rho^{(1)}(\mathbf{r},\mathbf{r};t)$.

We express in what follows quantities using the time-dependent natural orbitals $\alpha_j(\mathbf{r};t)$. The diagonal part of the reduced two-body density matrix is given by

$$\begin{aligned}\frac{\rho^{(2)}(\mathbf{r}_1,\mathbf{r}_2,\mathbf{r}_1,\mathbf{r}_2;t)}{N(N-1)} &= \int d\mathbf{r}_3\ldots d\mathbf{r}_N\,\Psi^*(\mathbf{r}_1,\mathbf{r}_2,\ldots,\mathbf{r}_N;t)\Psi(\mathbf{r}_1,\mathbf{r}_2,\ldots,\mathbf{r}_N;t) = \\ &= \sum_{jpkq}\frac{\rho_{jpkq}(t)}{N(N-1)}\alpha_j^*(\mathbf{r}_1;t)\alpha_p^*(\mathbf{r}_2;t)\alpha_k(\mathbf{r}_1;t)\alpha_q(\mathbf{r}_2;t),\end{aligned} \tag{4}$$

where the matrix elements are $\rho_{jpkq}(t) = \langle\Psi(t)|\hat{b}_j^\dagger\hat{b}_p^\dagger\hat{b}_k\hat{b}_q|\Psi(t)\rangle$, and the creation and annihilation operators are associated with the time-dependent natural orbitals $\alpha_j(\mathbf{r};t)$.

To express the variance and uncertainty product of operators we begin simply with the operator

$$\hat{A} = \sum_{j=1}^{N} \hat{a}(\mathbf{r}_j) \quad (5)$$

of the many-particle system, where $\hat{a}(\mathbf{r})$ is a Hermitian operator. A straightforward calculation gives the average per particle of $\hat{A}$ in the state $|\Psi(t)\rangle$,

$$\frac{1}{N}\langle\Psi(t)|\hat{A}|\Psi(t)\rangle = \int d\mathbf{r}\frac{\rho(\mathbf{r};t)}{N}a(\mathbf{r}), \quad (6)$$

which is directly expressed in terms of the system's density per particle in case of a local operator in coordinate space $\mathbf{r}$. In the case of a local operator in momentum space $\mathbf{p}$ we can write $\frac{1}{N}\langle\Psi(t)|\hat{A}|\Psi(t)\rangle = \int d\mathbf{p}\frac{\rho(\mathbf{p};t)}{N}a(\mathbf{p})$.

To proceed we also need the expectation value of the square of $\hat{A}$,

$$\hat{A}^2 = \sum_{j=1}^{N} \hat{a}^2(\mathbf{r}_j) + \sum_{j<k} 2\hat{a}(\mathbf{r}_j)\hat{a}(\mathbf{r}_k), \quad (7)$$

which is comprised of one-body *and* two-body operators. Expressed in terms of the density per particle and natural orbitals, the time-dependent variance per particle of the operator $\hat{A}$ can be written as a sum of two terms

$$\begin{aligned}
\frac{1}{N}\Delta^2_{\hat{A}}(t) &= \frac{1}{N}\left[\langle\Psi(t)|\hat{A}^2|\Psi(t)\rangle - \langle\Psi(t)|\hat{A}|\Psi(t)\rangle^2\right] \equiv \Delta^2_{\hat{a},density}(t) + \Delta^2_{\hat{a},MB}(t),\\
\Delta^2_{\hat{a},density}(t) &= \int d\mathbf{r}\frac{\rho(\mathbf{r};t)}{N}a^2(\mathbf{r}) - \left[\int d\mathbf{r}\frac{\rho(\mathbf{r};t)}{N}a(\mathbf{r})\right]^2,\\
\Delta^2_{\hat{a},MB}(t) &= \frac{\rho_{1111}(t)}{N}\left[\int d\mathbf{r}|\alpha_1(\mathbf{r};t)|^2a(\mathbf{r})\right]^2 - (N-1)\left[\int d\mathbf{r}\frac{\rho(\mathbf{r};t)}{N}a(\mathbf{r})\right]^2 +\\
&+ \sum_{jpkq\neq 1111}\frac{\rho_{jpkq}(t)}{N}\left[\int d\mathbf{r}\alpha_j^*(\mathbf{r};t)\alpha_k(\mathbf{r};t)a(\mathbf{r})\right]\left[\int d\mathbf{r}\alpha_p^*(\mathbf{r};t)\alpha_q(\mathbf{r};t)a(\mathbf{r})\right]. \quad (8)
\end{aligned}$$

The first term, which is denoted by $\Delta^2_{\hat{a},density}(t)$, describes the variance of $\hat{a}(\mathbf{r})$ resulting *solely* from the density per particle $\frac{\rho(\mathbf{r};t)}{N}$. The second term, which is denoted by $\Delta^2_{\hat{a},MB}(t)$, collects all other contributions to the many-particle variance and is identically zero within Gross-Pitaevskii theory. $\Delta^2_{\hat{a},MB}(t)$ is generally non-zero within a many-body theory.

Finally, taking two many-particle operators,

$$\hat{A} = \sum_{j=1}^{N} \hat{a}(\mathbf{r}_j), \qquad \hat{B} = \sum_{j=1}^{N} \hat{b}(\mathbf{r}_j), \quad (9)$$

and their respective time-dependent variances per particle, $\frac{1}{N}\Delta^2_{\hat{A}}(t)$ and $\frac{1}{N}\Delta^2_{\hat{B}}(t)$, their uncertainty product satisfies the inequality

$$\frac{1}{N}\Delta^2_{\hat{A}}(t)\frac{1}{N}\Delta^2_{\hat{B}}(t) \equiv \Delta^2_{\frac{\hat{A}}{N}}(t)\Delta^2_{\hat{B}}(t) \geq \frac{1}{4}\left|\int d\mathbf{r}\frac{\rho(\mathbf{r};t)}{N}[\hat{a}(\mathbf{r}),\hat{b}(\mathbf{r})]\right|^2, \quad (10)$$

where $[\hat{a}(\mathbf{r}),\hat{b}(\mathbf{r})]$ is the commutator of the two Hermitian operators $\hat{a}(\mathbf{r})$ and $\hat{b}(\mathbf{r})$.

3. Applications

As mentioned above we would like to go beyond [17, 18] in the investigation of the variance and uncertainty product of trapped BECs. To this end, we here study trapped bosons with a long-range interaction and when the system is no longer condensed. We treat structureless bosons with harmonic inter-particle interaction trapped in a single-well anharmonic potential. The time-dependent Schrödinger equation of the trapped BEC has no analytical solution in the present study, see in this respect [23, 29], nor even the variance and uncertainty product can be computed analytically, thus a numerical solution of the out-of-equilibrium dynamics is a must. This will lead to interesting results.

We need a suitable and proved many-body tool to make the calculations. Such a many-body tool is the multiconfigurational time-dependent Hartree for bosons (MCTDHB) method, which has been well documented [30-35], benchmarked [36], and extensively used [37-55] in the literature.

3.1. Breathing dynamics

We consider $N = 100$ and separately $N = 100,000$ trapped bosons in one spatial dimension. The one-body Hamiltonian is $-\frac{1}{2}\frac{\partial^2}{\partial x^2} + \frac{x^4}{4}$ and the inter-particle interaction is harmonic, $\lambda_0 \hat{W}(x_1 - x_2) = -\lambda_0(x_1 - x_2)^2, \lambda_0 > 0$. The system is prepared in the ground state of the trap for the interaction parameter $\Lambda = \lambda_0(N-1) = 0.19$. At $t = 0$ the interaction parameter is suddenly quenched to $\Lambda = 0.38$ and we inquire what the out-of-equilibrium dynamics of the system would be like. Fig. 1 collects the results.

Fig. 1a depicts snapshots of the density per particle, $\frac{\rho(x;t)}{N}$, as a function of time. The Gross-Pitaevskii and many-body results are seen to match very well. The density is seen to perform breathing dynamics [56-58]. Since the interaction is repulsive and at $t = 0$ quenched up, the density first expands at short times. In Fig. 1b the total number of depleted particles outside the condensed mode [$\alpha_1(x;t)$ natural orbital] are shown as a function of time. The systems are essentially fully condensed with only a fraction of a particle depleted. In Fig. 1c the time-dependent many-particle position variance per particle, $\frac{1}{N}\Delta^2_{\hat{X}}(t)$, momentum variance, $\frac{1}{N}\Delta^2_{\hat{P}}(t)$, and uncertainty product, $\frac{1}{N}\Delta^2_{\hat{X}}(t)\frac{1}{N}\Delta^2_{\hat{P}}(t) \equiv \Delta^2_{\hat{X}_{CM}}(t)\Delta^2_{\hat{P}_{CM}}(t)$ are shown. Note the opposite behavior of the variance at short times when computed at the many-body and Gross-Pitaevskii level. Despite the expansion of the cloud (at short times), the time-dependent position variance increases and momentum variance decreases, implying that $\Delta^2_{\hat{x},MB}(t)$ and $\Delta^2_{\hat{p},MB}(t)$ are opposite in sign with respect to and dominate $\Delta^2_{\hat{x},density}(t)$ and $\Delta^2_{\hat{p},density}(t)$. This is an interesting time-dependent many-body effect.

3.2. Pathway from condensation to fragmentation

We next study the pathway from condensation to fragmentation of the ground state [59-63]. Fragmentation of BECs has drawn much attention, see, e.g., [64-71]. In particular for structureless bosons with a long-range interaction in a single-trap, the ground state has been shown to become fragmented when increasing the inter-particle repulsion [58, 72-75]. Fig. 2 depicts the results for $N = 20$ bosons. The one-body Hamiltonian is again $-\frac{1}{2}\frac{\partial^2}{\partial x^2} + \frac{x^4}{4}$ and the inter-particle interaction is harmonic, $\lambda_0 \hat{W}(x_1 - x_2) = -\lambda_0(x_1 - x_2)^2, \lambda_0 > 0$. Fig. 2a shows snapshots of the density per particle, $\frac{\rho(x;t)}{N}$, as a function of the interaction strength λ_0. As the interaction is increased the density broadens and splits into two parts; Side by side, the ground state fragments [58, 72-75], see Fig. 2b. The ground state essentially evolves from 100% condensed to 50%-50% fragmented. Finally, Fig. 2c displays the many-particle position variance per particle, $\frac{1}{N}\Delta^2_{\hat{X}}$, momentum variance, $\frac{1}{N}\Delta^2_{\hat{P}}$, and uncertainty product, $\frac{1}{N}\Delta^2_{\hat{X}}\frac{1}{N}\Delta^2_{\hat{P}} \equiv \Delta^2_{\hat{X}_{CM}}\Delta^2_{\hat{P}_{CM}}$ of the ground state as a function of the interaction strength. We find

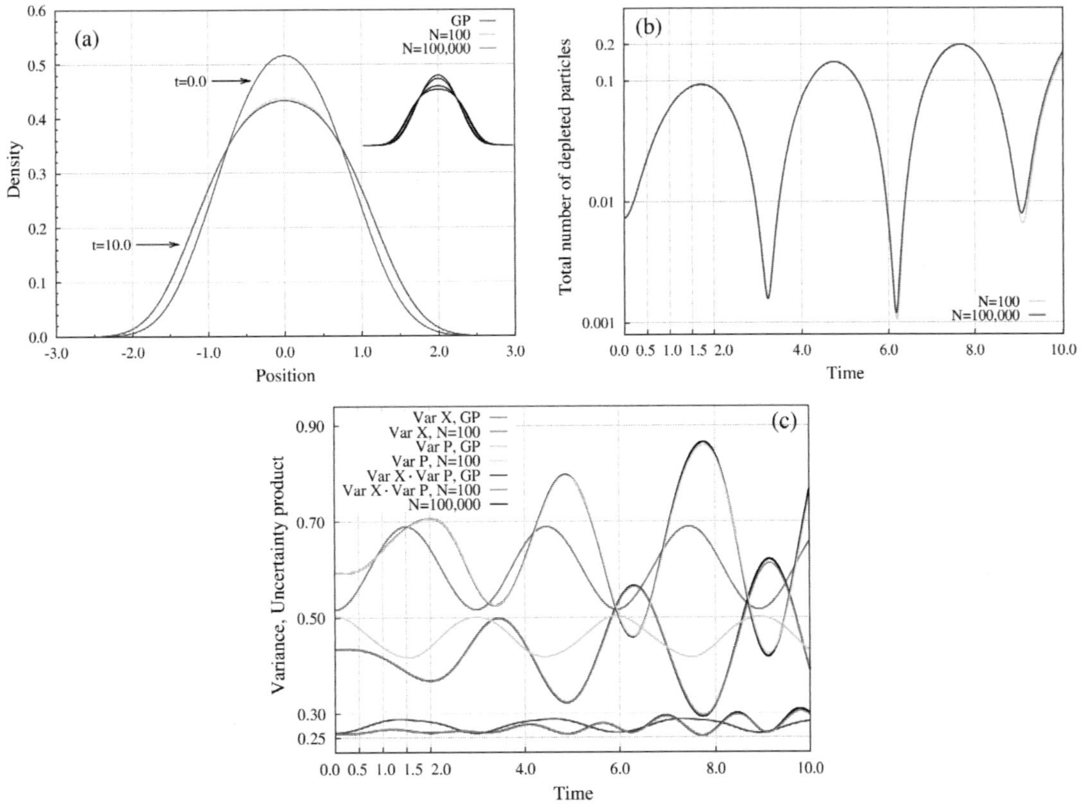

Figure 1. Breathing dynamics following an interaction quench. Shown and compared are many-body results for $N = 100$ (using $M = 4$ time-adaptive orbitals) and $N = 100,000$ (using $M = 2$ time-adaptive orbitals) bosons, and the mean-field results (equivalent to $M = 1$ time-adaptive orbitals). The one-body Hamiltonian is $-\frac{1}{2}\frac{\partial^2}{\partial x^2} + \frac{x^4}{4}$ and the inter-particle interaction is harmonic, $\lambda_0 \hat{W}(x_1 - x_2) = -\lambda_0(x_1 - x_2)^2, \lambda_0 > 0$. (a) Snapshots of the density per particle, $\frac{\rho(x;t)}{N}$, as a function of time following a sudden increase of the interaction parameter from $\Lambda = \lambda_0(N-1) = 0.19$ to 0.38 at $t = 0$. The inset shows the density per particle at instances, from top to bottom, $t = 0.0$, 0.5, 1.0, and 1.5 for $N = 100$ bosons. (b) Total number of depleted particles outside the condensed mode as a function of time. Note the values on the y axis. (c) Time-dependent many-particle position variance per particle, $\frac{1}{N}\Delta^2_{\hat{X}}(t)$, momentum variance, $\frac{1}{N}\Delta^2_{\hat{P}}(t)$, and uncertainty product, $\frac{1}{N}\Delta^2_{\hat{X}}(t)\frac{1}{N}\Delta^2_{\hat{P}}(t) \equiv \Delta^2_{\hat{X}_{CM}}(t)\Delta^2_{\hat{P}_{CM}}(t)$. Note the opposite behavior of the variance at short times when computed at the many-body and mean-field level. See the text for more details. The quantities shown are dimensionless.

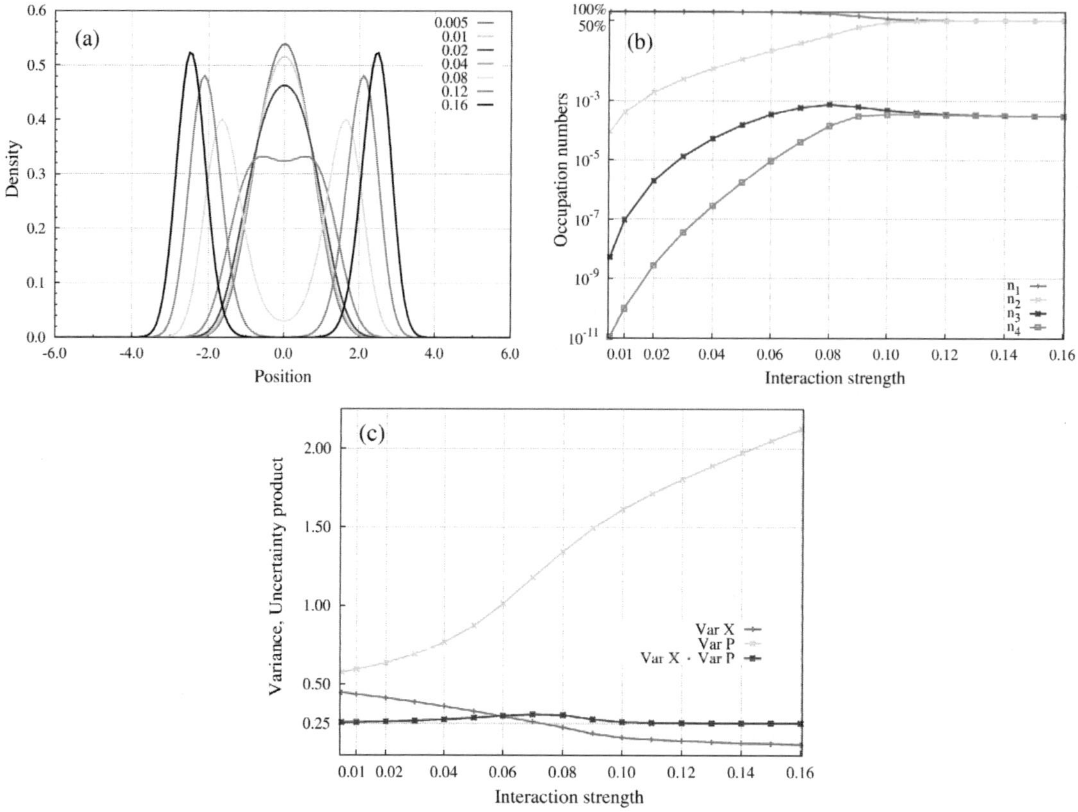

Figure 2. Pathway from condensation to fragmentation of the ground state. Shown are many-body results for $N = 20$ bosons (using $M = 4$ time-adaptive orbitals). The one-body Hamiltonian is $-\frac{1}{2}\frac{\partial^2}{\partial x^2} + \frac{x^4}{4}$ and the inter-particle interaction is harmonic, $\lambda_0 \hat{W}(x_1 - x_2) = -\lambda_0(x_1 - x_2)^2, \lambda_0 > 0$. (a) Snapshots of the density per particle, $\frac{\rho(x;t)}{N}$, as a function of the interaction strength $\lambda_0 = 0.005, \ldots, 0.16$. As the interaction is increased the density broadens and splits into two parts. (b) The four largest occupation numbers as a function of the interaction strength λ_0 (points' symbols are data; smooth curves are to guide the eye). The ground state essentially evolves from 100% condensed to 50%–50% fragmented. (c) Many-particle position variance per particle, $\frac{1}{N}\Delta^2_{\hat{X}}$, momentum variance, $\frac{1}{N}\Delta^2_{\hat{P}}$, and uncertainty product, $\frac{1}{N}\Delta^2_{\hat{X}}\frac{1}{N}\Delta^2_{\hat{P}} \equiv \Delta^2_{\hat{X}_{CM}}\Delta^2_{\hat{P}_{CM}}$. Note that, although the cloud's density broadens while the system's fragments, the position variance actually decreases whereas the momentum variance increases. See the text for more details. The quantities shown are dimensionless.

that, although the cloud's density broadens while the system's fragments, the position variance decreases and the momentum variance increases, unlike from what one would expect by just examining the density. When the variances are combined, the uncertainty product exhibits a maximum along the pathway from condensation to fragmentation. This is an interesting static many-body effect.

4. Summary and Conclusions

We have studied in the present work the variance and uncertainty product of the position and momentum many-particle operators of structureless bosons interacting by a long-range inter-particle interaction and trapped in an anharmonic single-well potential. There is no analytical solution to the many-particle Schrödinger equation of this system, not even to the variance and uncertainty product themselves, which makes a numerical solution of the out-of-equilibrium dynamics a must. In the out-of-equilibrium interaction-quench scenario, we have found that, despite the system being fully condensed, already when a fraction of a particle is depleted differences with respect to the mean-field quantities arise. In the static pathway from condensation to fragmentation scenario, we found out that, although the cloud's density broadens while the system's fragments, the position variance decreases, the momentum variance increases, and the uncertainty product exhibits a maximum. Both scenarios suggest a richness of effects emanating from the many-body term of the variance and uncertainty product in interacting trapped many-boson systems. Such many-body effects need not coincide with the information that can be extracted based on the system's density alone.

Acknowledgments

This research was supported by the Israel Science Foundation (Grant No. 600/15). Partial financial support by the Deutsche Forschungsgemeinschaft (DFG) is acknowledged. We thank Lorenz Cederbaum for discussions. Computation time on the Cray XC40 system Hazelhen at the High Performance Computing Center Stuttgart (HLRS) is gratefully acknowledged.

Appendix A. Details and convergence of the numerical computations

The multiconfigurational time-dependent Hartree for bosons (MCTDHB) method [30-35] is used in the present work to compute the time-dependent and ground-state properties of trapped bosons interacting by a long-range inter-particle interaction. To obtain the ground state we propagate the MCTDHB equations of motion in imaginary time [36, 63]. For the computations the many-body Hamiltonian is represented by 256 exponential discrete-variable-representation grid points (using a Fast-Fourier Transform routine) in a box of size $[-10, 10)$. We use the numerical implementation in the software packages [76, 77]. Convergence of the variance and uncertainty product with increasing number M of time-adaptive orbitals is demonstrated for $N = 100$ bosons in Fig. A1a for the out-of-equilibrium dynamics [18] and for $N = 20$ bosons in Fig. A1b for the ground state [17], also see in this context [78]. It is found that, respectively, the results with $M = 2$ and $M = 4$ orbitals are converged.

References

[1] Anderson M H, Ensher J R, Matthews M R, Wieman C E and Cornell E A 1995 *Science* **269** 198
[2] Bradley C C, Sackett C A, Tollett J J and Hulet R G 1995 *Phys. Rev. Lett.* **75** 1687
[3] Davis K B, Mewes M O, Andrews M R, van Druten N J, Durfee D S, Kurn D M and Ketterle W 1995 *Phys. Rev. Lett.* **75** 3969
[4] Cornell E A and Wieman C E 2002 *Rev. Mod. Phys.* **74** 875
[5] Ketterle W 2002 *Rev. Mod. Phys.* **74** 1131
[6] Dalfovo F, Giorgini S, Pitaevskii L P and Stringari S 1999 *Rev. Mod. Phys.* **71** 463
[7] Leggett A J 2001 *Rev. Mod. Phys.* **73** 307
[8] Bloch I, Dalibard J and Zwerger W 2008 *Rev. Mod. Phys.* **80** 885

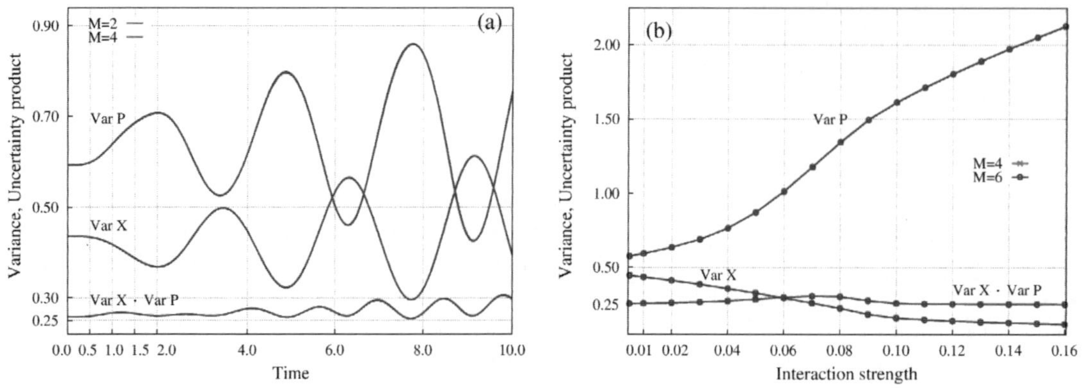

Figure A1. Convergence of the many-particle position variance per particle, $\frac{1}{N}\Delta^2_{\hat{X}}(t)$, momentum variance, $\frac{1}{N}\Delta^2_{\hat{P}}(t)$, and uncertainty product, $\frac{1}{N}\Delta^2_{\hat{X}}(t)\frac{1}{N}\Delta^2_{\hat{P}}(t) \equiv \Delta^2_{\hat{X}_{CM}}(t)\Delta^2_{\hat{P}_{CM}}(t)$ with the number of time-adaptive orbitals M used in the MCTDHB computations for the systems consisting of $N = 100$ and $N = 20$ bosons discussed in the main text. (a) The out-of-equilibrium interaction-quench breathing dynamics in Sec. 3.1 (propagation of the MCTDHB equations of motion in real time). It is found that the results with $M = 2$ and $M = 4$ orbitals lie atop each other. (b) The pathway from condensation to fragmentation of the ground state in Sec. 3.2 (propagation of the MCTDHB equations of motion in imaginary time). It is found that the results with $M = 4$ and $M = 6$ orbitals lie atop each other. The quantities shown are dimensionless.

[9] Yukalov V I 2011 *Phys. Part. Nucl.* **42** 460
[10] Pitaevskii L and Stringari S 2003 *Bose-Einstein Condensation* (New York: Oxford University Press)
[11] Leggett A J 2008 *Quantum Liquids: Bose Condensation and Cooper Pairing in Condensed-Matter Systems* (New York: Oxford University Press)
[12] Pethick C J and Smith H 2008 *Bose-Einstein Condensation in Dilute Gases* 2nd ed (New York: Cambridge University Press)
[13] Lieb E H, Seiringer R and Yngvason J 2000 *Phys. Rev. A* **61** 043602
[14] Lieb E H and Seiringer R 2002 *Phys. Rev. Lett.* **88** 170409
[15] Erdős L, Schlein B and Yau H T 2007 *Phys. Rev. Lett.* **98** 040404
[16] Erdős L, Schlein B and Yau H T 2007 *Invent. Math.* **167** 515
[17] Klaiman S and Alon O E 2015 *Phys. Rev. A* **91** 063613
[18] Klaiman S, Streltsov A I and Alon O E 2016 *Phys. Rev. A* **93** 023605
[19] Cohen-Tannoudji C, Diu B and F. Laloë F 1977 *Quantum Mechanics vol 1* (New York: Wiley)
[20] Vaughan T, Drummond P and Leuchs G 2007 *Phys. Rev. A* **75** 033617
[21] Chen Y J, Pabst S, Li Z, Vendrell O and Santra R 2014 *Phys. Rev. A* **89** 052113
[22] Klaiman S, Streltsov A I and Alon O E 2016 *Chem. Phys.* http://dx.doi.org/10.1016/j.chemphys.2016.07.011
[23] Klaiman S and Cederbaum L S 2016 arXiv:1609.05895v1 [cond-mat.quant-gas]
[24] Löwdin P O 1955 *Phys. Rev.* **97** 1474
[25] Coleman A J and Yukalov V I 2000 *Reduced Density Matrices: Coulson's Challenge (Lectures Notes in Chemistry vol 72)* (Berlin: Springer)
[26] Mazziotti D A (ed) 2007 *Reduced-Density-Matrix Mechanics: with Application to Many-Electron Atoms and Molecules (Advances in Chemical Physics vol 134)* (New York: Wiley)
[27] Sakmann K, Streltsov A I, Alon O E and Cederbaum L S 2008 *Phys. Rev. A* **78** 023615
[28] Penrose O and Onsager L 1956 *Phys. Rev.* **104** 576
[29] Cohen L and Lee C 1985 *J. Math. Phys.* **26** 3105
[30] Streltsov A I, Alon O E and Cederbaum L S 2007 *Phys. Rev. Lett.* **99** 030402
[31] Alon O E, Streltsov A I and Cederbaum L S 2008 *Phys. Rev. A* **77** 033613

[32] Meyer H D, Gatti F and Worth G A (ed) 2009 *Multidimensional Quantum Dynamics: MCTDH Theory and Applications* (Weinheim: Wiley-VCH)
[33] Sakmann K 2011 *Many-Body Schrödinger Dynamics of Bose-Einstein Condensates (Springer Theses)* (Heidelberg: Springer)
[34] Proukakis N P, Gardiner S A, Davis M J and Szymanska M H (ed) 2013 *Quantum Gases: Finite Temperature and Non-Equilibrium Dynamics (Cold Atoms Series vol 1)* (London: Imperial College Press)
[35] Lode A U J 2015 *Tunneling Dynamics in Open Ultracold Bosonic Systems (Springer Theses)* (Heidelberg: Springer)
[36] Lode A U J, Sakmann K, Alon O E, Cederbaum L S and Streltsov A I 2012 *Phys. Rev. A* **86** 063606
[37] Sakmann K, Streltsov A I, Alon O E and Cederbaum L S 2009 *Phys. Rev. Lett.* **103** 220601
[38] Grond J, Schmiedmayer J and Hohenester U 2009 *Phys. Rev. A* **79** 021603(R)
[39] Grond J, Betz T, Hohenester U, Mauser N J, Schmiedmayer J and Schumm T 2011 *New J. Phys.* **13** 065026
[40] Heimsoth H, Hochstuhl D, Creffield C E, Carr L D and Sols F 2013 *New J. Phys.* **15** 103006
[41] Klaiman S, Lode A U J, Streltsov A I, Cederbaum L S and Alon O E 2014 *Phys. Rev. A* **90** 043620
[42] Březinová I, Burgdörfer J, Lode A U J, Streltsov A I, Cederbaum L S, Alon O E, Collins L A and Schneider B I 2014 *J. Phys.: Conf. Ser.* **488** 012032
[43] Mistakidis S I, Cao L and Schmelcher P 2015 *Phys. Rev. A* **91** 033611
[44] Krönke S and Schmelcher P 2015 *Phys. Rev. A* **91** 053614
[45] Fischer U R, Lode U J L and Chatterjee B 2015 *Phys. Rev. A* **91** 063621
[46] Krönke S and Schmelcher P 2015 *Phys. Rev. A* **92** 023631
[47] Beinke R, Klaiman S, Cederbaum L S, Streltsov A I and Alon O E 2015 *Phys. Rev. A* **92** 043627
[48] Mistakidis S I, Wulf T, Negretti A and Schmelcher P 2015 *J. Phys. B* **48** 244004
[49] Brouzos I, Streltsov A I, Negretti A, Said R S, Caneva T, Montangero S and Calarco T 2015 *Phys. Rev. A* **92** 062110
[50] Alon O E 2015 *J. Phys.: Conf. Ser.* **594** 012039
[51] Sakmann K and Kasevich M 2016 *Nat. Phys.* **12** 451
[52] Klaiman S and Alon O E 2016 *J. Phys.: Conf. Ser.* **691** 012015
[53] Lode A U J 2016 *Phys. Rev. A* **93** 063601
[54] Dutta S and Basu S 2016 *J. Phys.: Conf. Ser.* **759** 012036
[55] Weiner S E, Tsatsos M C, Cederbaum L S and Lode A U J 2014 arXiv:1409.7670v3 [cond-mat.quant-gas]
[56] Bauch S, Balzer K, Henning C and Bonitz M 2009 *Phys. Rev. B* **80** 054515
[57] Schmitz R, Krönke S, Cao L and Schmelcher P 2013 *Phys. Rev. A* **88** 043601
[58] Streltsova O I, Alon O E, Cederbaum L S and Streltsov A I 2014 *Phys. Rev. A* **89** 061602(R)
[59] Spekkens R W and Sipe J E 1999 *Phys. Rev. A* **59** 3868
[60] Streltsov A I, Cederbaum L S and Moiseyev N 2004 *Phys. Rev. A* **70** 053607
[61] Alon O E and Cederbaum L S 2005 *Phys. Rev. Lett.* **95** 140402
[62] Zöllner S, Meyer H D and Schmelcher P 2006 *Phys. Rev. A* **74** 063611
[63] Streltsov A I, Alon O E and Cederbaum L S 2006 *Phys. Rev. A* **73** 063626
[64] Nozières P and Saint James D 1982 *J. Phys. (Paris)* **43** 1133
[65] Nozières P 1996 *Bose-Einstein Condensation* ed Griffin A, Snoke D W and Stringari S (Cambridge: Cambridge University Press)
[66] Mueller E J, Ho T L, Ueda M and Baym G 2006 *Phys. Rev. A* **74** 033612
[67] Fischer U R and Bader P 2010 *Phys. Rev. A* **82** 013607
[68] Zhou Q and Cui X 2013 *Phys. Rev. Lett.* **110** 140407
[69] Kawaguchi Y 2014 *Phys. Rev. A* **89** 033627
[70] Song S W, Zhang Y C, Zhao H, Wang X and Liu W M 2014 *Phys. Rev. A* **89** 063613
[71] Jen H H and Yip S K 2015 *Phys. Rev. A* **91** 063603
[72] Bader P and Fischer U R 2009 *Phys. Rev. Lett.* **103** 060402
[73] Streltsov A I 2013 *Phys. Rev. A* **88** 041602(R)
[74] Kang M K and Fischer U R 2014 *Phys. Rev. Lett.* **113** 140404
[75] Fischer U R and Kang M K 2015 *Phys. Rev. Lett.* **115** 260404
[76] Streltsov A I and Streltsova O I 2015 MCTDHB-Lab http://www.mctdhb-lab.com, version 1.5
[77] Streltsov A I, Cederbaum L S, Alon O E, Sakmann K, Lode A U J, Grond J, Streltsova O I and Klaiman S 2006-Present The Multiconfigurational Time-Dependent Hartree for Bosons Package http://mctdhb.org, version 3.x, Heidelberg
[78] Cosme J G, Weiss C and Brand J 2016 *Phys. Rev. A* **94** 043603

Entanglement production by evolution operator

V.I. Yukalov[1] **and E.P. Yukalova**[2]

[1]Bogolubov Laboratory of Theoretical Physics, Joint Institute for Nuclear Research, Dubna 141980, Russia
[2]Laboratory of Information Technologies, Joint Institute for Nuclear Research, Dubna 141980, Russia

E-mail: yukalov@theor.jinr.ru

Abstract. Entanglement production, generated by an evolution operator, is considered. The measure of entanglement production, introduced earlier by one of the authors, is employed. As an illustration, a two-qubit register is studied and the corresponding measure of evolutional entanglement production is calculated. Such two-qubit registers can be realized by atomic systems in a double well or by trapped atoms with two coherent modes.

1. Entangled and disentangled states

Entanglement is a pivotal quantum property that can give a principal advantage for the use of quantum processes, as compared to classical ones, in such fields as quantum information processing, quantum computing, and the functioning of artificial quantum intelligence [1–7].

The states, represented by wave functions or state vectors, are distinguished onto non-entangled and entangled, as explained below.

One considers a composite system characterized by a Hilbert space

$$\mathcal{H} = \bigotimes_{i=1}^{N} \mathcal{H}_i \; , \qquad (1)$$

which is a tensor product of partial Hilbert spaces

$$\mathcal{H}_i = {\rm span}\{ |n_i\rangle \} \qquad (2)$$

that are closed linear envelopes over the related bases. So that the total Hilbert space can be written as

$$\mathcal{H} = {\rm span} \left\{ \bigotimes_{i=1}^{N} |n_i\rangle \right\} . \qquad (3)$$

Disentangled states have the form of the products

$$\varphi_{dis} = \bigotimes_{j=1}^{N} \varphi_j \qquad (4)$$

of the partial states

$$\varphi_j = \sum_{n_j} c_{n_j} |n_j\rangle \in \mathcal{H}_j \tag{5}$$

pertaining to the partial Hilbert spaces. Respectively, a disentangled state is of the form

$$\varphi_{dis} = \bigotimes_{j=1}^{N} \sum_{n_j} c_{n_j} |n_j\rangle \ . \tag{6}$$

All disentangled states of the space $\mathcal{H}$ compose a *disentangled set*

$$\mathcal{D} \equiv \{\forall \varphi_{dis} \in \mathcal{H}\} \ . \tag{7}$$

Entangled states of the space $\mathcal{H}$ can be written as

$$\varphi_{ent} = \sum_{\{n_i\}} c_{n_1 n_2 \ldots n_N} |\, n_1 n_2 \ldots n_N\rangle \ , \tag{8}$$

where at least one of the coefficients cannot be represented as a product of c_{n_i}, so that at least for one coefficient

$$c_{n_1 n_2 \ldots n_N} \neq \prod_{i=1}^{N} c_{n_i} \ . \tag{9}$$

The collection of all entangled states forms an *entangled set*

$$\mathcal{H} \setminus \mathcal{D} \equiv \{\varphi_{ent} \in \mathcal{H},\ \varphi_{ent} \notin \mathcal{D}\} \ . \tag{10}$$

Thus the total Hilbert space $\mathcal{H}$ consists of two sets, a disentangled set $\mathcal{D}$ and an entangled set $\mathcal{H} \setminus \mathcal{D}$.

2. Measure of entanglement production

One should distinguish entangled states and entanglement production by quantum operations. Entanglement of states can be induced by operators acting on the states of a Hilbert space. So, entangled states characterize a static feature of a considered set, while entanglement production describes the operational feature of an operator.

Let $\hat{A}$ be a non-singular trace-class operator on $\mathcal{H}$, such that

$$0 \neq |\mathrm{Tr}_{\mathcal{H}} \hat{A}| < \infty \ . \tag{11}$$

The action of the operator on a disentangled state $\varphi_{dis} \in \mathcal{D}$ can result either in a different disentangled state $\varphi'_{dis} \in \mathcal{D}$ or in an entangled state $\varphi_{ent} \in \mathcal{H} \setminus \mathcal{D}$. How would it be possible to quantify the entangling ability of an operator?

For two given states $\varphi_{dis} \in \mathcal{D}$ and $\varphi_{ent} \in \mathcal{H} \setminus \mathcal{D}$, one could define a kind of the entanglement probability

$$p_{ent}\left(\hat{A}\varphi_{dis} \rightarrow \varphi_{ent}\right) \propto |(\varphi_{ent}|\hat{A}\varphi_{dis})|^2 \ , \tag{12}$$

being appropriately normalized. However, this would be not a general property of the operator on the whole Hilbert space, but merely its property with regard to the two given particular states.

The global ability of an operator of producing entanglement in the total Hilbert space $\mathcal{H}$, when

$$\hat{A}\mathcal{D} \rightarrow \mathcal{H} \setminus \mathcal{D} \ , \tag{13}$$

can be quantified as follows.

The entangling ability of the operator $\hat{A}$ can be understood by comparing its action on the given Hilbert space with the action of its non-entangling counterpart $\hat{A}^{\otimes}$ that does not entangle the states,

$$\hat{A}^{\otimes}\mathcal{D} \rightarrow \mathcal{D} , \tag{14}$$

being defined as

$$\hat{A}^{\otimes} \equiv \frac{\bigotimes_{i=1}^{N} \hat{A}_i}{(\mathrm{Tr}_{\mathcal{H}}\hat{A})^{N-1}} , \tag{15}$$

where the partially traced operators

$$\hat{A}_i \equiv \mathrm{Tr}_{\mathcal{H}/\mathcal{H}_i}\hat{A} \tag{16}$$

are obtained by tracing out the operator $\hat{A}$ over all partial Hilbert spaces, composing $\mathcal{H}$, except one space $\mathcal{H}_i$. The denominator in the non-entangling counterpart is such that the normalization

$$\mathrm{Tr}_{\mathcal{H}}\hat{A}^{\otimes} = \mathrm{Tr}_{\mathcal{H}}\hat{A} \tag{17}$$

be preserved.

The measure of entanglement production, for an operator $\hat{A}$, is defined [8, 9] as

$$\varepsilon(\hat{A}) \equiv \log \frac{||\hat{A}||_p}{||\hat{A}^{\otimes}||_p} . \tag{18}$$

The logarithm can be taken with respect to any base. The norm here can be accepted as a Schatten p-norm

$$||\hat{A}||_p \equiv \left(\mathrm{Tr}_{\mathcal{H}} |\, \hat{A} \,|^p\right)^{1/p} ,$$

in which

$$p \in [1, \infty) , \qquad |\, \hat{A} \,| \equiv \sqrt{\hat{A}^+\hat{A}} .$$

The Schatten norms are isometrically invariant, and therefore unitary invariant, such that

$$||\hat{A}||_p = ||U_1\hat{A}U_2||_p$$

for any linear isometries, hence, for any unitary transformations U_1 and U_2. Thence, the Schatten norms are independent of the basis used when taking the trace operation.

It is convenient to employ the Hilbert-Schmidt, or Frobenius, norm

$$||\hat{A}||_2 = \sqrt{\mathrm{Tr}_{\mathcal{H}}(\hat{A}^+\hat{A})} , \tag{19}$$

which is the Schatten 2-norm.

The standard operator norm corresponds to the Schatten ∞-norm

$$||\hat{A}||_{\infty} \equiv \sup_{\varphi \in \mathcal{H}} \frac{||\hat{A}\varphi||}{||\varphi||} \qquad (\varphi \neq 0) , \tag{20}$$

denoted in what follows as

$$||\hat{A}|| \equiv ||\hat{A}||_{\infty} .$$

Keeping in mind an orthonormal basis $\{e_\alpha\}$ in $\mathcal{H}$, we can write

$$\mathrm{Tr}_{\mathcal{H}}(\hat{A}^+\hat{A}) = \sum_\alpha (e_\alpha, \hat{A}^+\hat{A}e_\alpha) = \sum_\alpha (\hat{A}e_\alpha, \hat{A}e_\alpha) = \sum_\alpha ||\hat{A}e_\alpha||^2 \; .$$

Therefore the Hilbert-Schmidt norm can be defined as

$$||\hat{A}||_2^2 = \sum_\alpha ||\hat{A}e_\alpha||^2 \; .$$

Quantity (18) enjoys all necessary properties required for being a measure. Thus, it is semi-positive. Because of the importance of this property, we prove it below.

Theorem. For trace-class operators $\hat{A} : \mathcal{H} \to \mathcal{H}$ and $\hat{A}^\otimes : \mathcal{D} \to \mathcal{D}$, defined in (15), the measure of entanglement production (18) is semi-positive,

$$\varepsilon(\hat{A}) \geq 0 \; . \tag{21}$$

Proof. The operators, enjoying finite norm (19), are Hilbert-Schmidt operators. Note that the trace-class operators are also Hilbert-Schmidt operators. A family of the operators $\hat{A} : \mathcal{H} \to \mathcal{H}$, supplemented by norm (19), forms a Banach space

$$\mathcal{B}(\mathcal{H}) \equiv \{\hat{A} : \; \mathcal{H} \to \mathcal{H}, ||\hat{A}||_2\} \; , \tag{22}$$

that is a normed, complete, linear space. Respectively, a family of the operators $\hat{A}^\otimes : \mathcal{D} \to \mathcal{D}$, complimented by norm (19), forms a Banach space

$$\mathcal{B}_\otimes(\mathcal{D}) \equiv \{\hat{A}^\otimes : \; \mathcal{D} \to \mathcal{D}, ||\hat{A}^\otimes||_2\} \; , \tag{23}$$

which is a subspace of space (22),

$$\mathcal{B}_\otimes(\mathcal{D}) \subset \mathcal{B}(\mathcal{H}) \; .$$

It is possible to introduce a projection $P_\otimes$ in the space of bounded linear operators [10], such that

$$\mathcal{P}_\otimes \mathcal{B}(\mathcal{H}) = \mathcal{B}_\otimes(\mathcal{D}) \; , \tag{24}$$

with the standard properties of idempotence and self-adjointness,

$$\mathcal{P}_\otimes^2 = \mathcal{P}_\otimes \; , \qquad \mathcal{P}_\otimes^+ = \mathcal{P}_\otimes \; .$$

The operator norm of the projector reads as

$$||\mathcal{P}_\otimes|| = \sup_{\hat{A}\in\mathcal{B}(\mathcal{H})} \frac{||\mathcal{P}_\otimes \hat{A}||}{||\hat{A}||} \qquad (\hat{A} \neq 0) \; ,$$

where in the right-hand side the operator norms are given by (20). Then

$$||\mathcal{P}_\otimes|| = \sup_{\hat{A}\in\mathcal{B}(\mathcal{H})} \; \sup_{\varphi\in\mathcal{H}} \frac{||\mathcal{P}_\otimes \hat{A}\varphi||}{||\hat{A}\varphi||} \qquad (\hat{A} \neq 0 \; , \; \varphi \neq 0) \; ,$$

expressed through the vector norms. Since $\hat{A}\varphi \in \mathcal{H}$, the projector norm takes the form

$$||\mathcal{P}_\otimes|| = \sup_{\hat{A}\varphi\in\mathcal{H}} \frac{||\mathcal{P}_\otimes \hat{A}\varphi||}{||\hat{A}\varphi||} \qquad (\hat{A}\varphi \neq 0) \; .$$

The non-entangling counterpart (15) is obtained from the initial operator $\hat{A}$ by means of the transformation

$$\mathcal{P}_\otimes \hat{A} = \hat{A}^\otimes \in \mathcal{B}_\otimes(\mathcal{D}) \subset \mathcal{B}(\mathcal{H})$$

preserving normalization (17). Therefore

$$||\mathcal{P}_\otimes|| = 1 \; .$$

Keeping in mind that

$$||\hat{A}^\otimes||_2^2 = ||\mathcal{P}_\otimes \hat{A}||_2^2 = \sum_\alpha ||\mathcal{P}_\otimes \hat{A} e_\alpha||^2 \; ,$$

we have

$$||\mathcal{P}_\otimes \hat{A} e_\alpha|| \leq ||\mathcal{P}_\otimes|| \; ||\hat{A} e_\alpha|| \; .$$

Hence

$$\sum_\alpha ||\mathcal{P}_\otimes \hat{A} e_\alpha||^2 \leq \sum_\alpha ||\hat{A} e_\alpha||^2 = ||\hat{A}||_2^2 \; ,$$

from where

$$||\hat{A}^\otimes||_2 \leq ||\hat{A}||_2 \; , \tag{25}$$

which implies that measure (18) is semi-positive, satisfying (21). □

Employing the properties of the Hilbert-Schmidt norm [11, 12], it is straightforward to prove that measure (18) satisfies all other conditions required for being a measure [9]. Thus, it is continuous with respect to the norm convergence, it is zero for non-entangling operators, it is additive with respect to copies, and invariant under local unitary operations.

3. Evolutional entanglement production

Entanglement production has been considered for density matrices of two-mode Bose systems of trapped atoms and multimode Bose systems in optical lattices [13–20]. Also, it has been studied for pseudospin entanglement production of radiating systems [21]. Here we investigate the entanglement production by the evolution operators [22].

The evolution operator of a closed quantum system,

$$\hat{U}(t) = e^{-iHt} \; , \tag{26}$$

characterizes the evolution of quantum states in time, generated by a Hamiltonian H. Even when an initial state $\varphi(0)$ would be not entangled, it can become entangled with time yielding an entangled state

$$\varphi(t) = \hat{U}(t)\varphi(0) \; . \tag{27}$$

The entangling properties of the evolution operator are described by the entanglement-production measure

$$\varepsilon(\hat{U}(t)) = \log \frac{||\hat{U}(t)||_2}{||\hat{U}^\otimes(t)||_2} \equiv \varepsilon(t) \; . \tag{28}$$

Since evolution operators are unitary, so that

$$\hat{U}^+(t)\hat{U}(t) = \hat{1}_\mathcal{H} \; ,$$

with the right-hand side being the identity operator in $\mathcal{H}$, the related norm equals the dimensionality of the Hilbert space $\mathcal{H}$,

$$||\hat{U}(t)||_2^2 = \mathrm{Tr}_\mathcal{H} \hat{1}_\mathcal{H} = \dim \mathcal{H} \; . \tag{29}$$

The calculation of the norm for the nonentangling counterpart $||\hat{U}^{\otimes}||_2$ is more involved. In the limit of short times, we have

$$||\hat{U}^{\otimes}(t)||_2^2 \simeq \dim\mathcal{H} - \mu t^2 \; ,$$

with a constant μ depending on the system Hamiltonian. As a result, the short-time behavior of the entanglement-production measure is

$$\varepsilon(t) \simeq \frac{1}{2}\,\mu t^2 \qquad (t \to 0) \; . \tag{30}$$

4. Entanglement for two-qubit registers

Two-qubit registers are widely used for quantum information processing [1–7]. Such registers can be realized with the help of cold atoms, trapped ions, spin systems, and photons in cavities [19, 23–30]. Below we show how the entanglement-production measure for arbitrary time can be explicitly calculated for two-qubit registers.

In the case of a two-qubit register, a disentangled state is a product

$$\varphi_{dis} = \varphi_1 \bigotimes \varphi_2 \; ,$$

in which

$$\varphi_1 = a_1|\uparrow\rangle + a_2|\downarrow\rangle \; , \qquad \varphi_2 = b_1|\uparrow\rangle + b_2|\downarrow\rangle \; .$$

And an entangled state reads as

$$\varphi_{ent} = c_1|\uparrow\uparrow\rangle + c_2|\uparrow\downarrow\rangle + c_3|\downarrow\uparrow\rangle + c_4|\downarrow\downarrow\rangle \; ,$$

where at least one of the coefficients c_j is not a product of the type $a_j b_j$.

The two-qubit Hamiltonian has the form

$$H = H_0 + H_{int} \; , \tag{31}$$

consisting of a free term

$$H_0 = -h\left(S_i^z \bigotimes \hat{1}_2 \; + \; \hat{1}_1 \bigotimes S_2^z\right) \; , \tag{32}$$

where S_j^z are spin operators, while h is an external field, and the interaction term is

$$H_{int} = 2J S_1^z \bigotimes S_2^z \; . \tag{33}$$

The Hamiltonian is of a two-site Ising type with the interaction strength J.

Calculating the entanglement-production measure, as explained in the previous sections, we obtain

$$\varepsilon(t) = \frac{1}{2}\,\log\,\frac{1+\cos^2(ht)+2\cos(ht)\cos(Jt)}{[1+\cos(ht)\cos(Jt)]^2} \; . \tag{34}$$

At the initial time, as it should be, no entanglement is yet produced by the evolution operator,

$$\lim_{t\to 0}\varepsilon(t) = \varepsilon(0) = 0 \; .$$

Also, notice the importance of interactions, without which no entanglement could be produced,

$$\lim_{J\to 0}\varepsilon(t) = 0 \qquad (t \geq 0) \; .$$

The measure is invariant with respect to the inversion of the signs in h and J. At short times, the measure behaves as

$$\varepsilon(t) \simeq \frac{J^2}{8}\, t^2 + \frac{J^2(J^2 - 12h^2)}{192}\, t^4 \; , \tag{35}$$

in agreement with formula (30).

The temporal behavior of the measure is either periodic, when the ratio h/J is rational, or quasi-periodic, if this ratio is irrational. This is illustrated by Figs. 1 and 2, where time is measured in units of $1/J$.

5. Conclusion and possible extensions

There exist two types of entanglement measures that characterize either the entanglement of given states or the amount of entanglement produced by quantum operations. Here the latter notion is considered quantified by a measure of entanglement production. The theorem on the semi-positiveness of this measure is proved. The entanglement produced by evolution operators is analyzed. Detailed calculations are accomplished for a two-qubit register modeled by an Ising-type Hamiltonian. The measure of entanglement production for such a case is either periodic or quasi-periodic in time, depending on the relation between the values of an external field and pseudospin interactions.

It is straightforward to notice that in the same way, as has been done in the present paper, one can define *thermal entanglement production* by statistical operators of equilibrium systems. Let $\hat{\rho}$ be a statistical operator. Since this operator is trace-normalized to one, its non-entangling, or distilled, counterpart reads as

$$\hat{\rho}_\otimes = \bigotimes_{i=1}^N \hat{\rho}_i \; ,$$

with the partial statistical operators

$$\hat{\rho}_i \equiv {\rm Tr}_{\mathcal{H}/\mathcal{H}_i} \hat{\rho} \; .$$

The measure of entanglement production by the statistical operator is

$$\varepsilon(\hat{\rho}) = \log \frac{||\hat{\rho}||_2}{||\hat{\rho}_\otimes||_2} \; .$$

In equilibrium, the statistical operator of a system with a Hamiltonian H has the form

$$\hat{\rho} = \frac{1}{Z}\, e^{-\beta H} \; , \qquad Z = {\rm Tr}_{\mathcal{H}} e^{-\beta H} \; ,$$

where $\beta = 1/T$ is inverse temperature. It easy to notice that the exponential here can be represented through the evolution operator in imaginary time as

$$e^{-\beta H} = \hat{U}(-i\beta) \; .$$

Hence we have

$$\hat{\rho} = \frac{1}{Z}\, \hat{U}(-i\beta) \; , \qquad Z = {\rm Tr}_{\mathcal{H}} \hat{U}(-i\beta) \; .$$

With the partial term

$$\hat{\rho}_i = \frac{1}{Z}\, {\rm Tr}_{\mathcal{H}/\mathcal{H}_i} \hat{U}(-i\beta) \; ,$$

the distilled operator becomes

$$\hat{\rho}_{\otimes} = \frac{1}{Z^N} \bigotimes_{i=1}^{N} \mathrm{Tr}_{\mathcal{H}/\mathcal{H}_i} \hat{U}(-i\beta) \; .$$

Thus we come to the measure

$$\varepsilon(\hat{\rho}) = \frac{1}{2} \log \frac{Z^{2N-2} \mathrm{Tr}_{\mathcal{H}} \hat{U}(-2i\beta)}{\prod_{i=1}^{N} \mathrm{Tr}_{\mathcal{H}_i} [\mathrm{Tr}_{\mathcal{H}/\mathcal{H}_i} \hat{U}(-i\beta)]^2} \; .$$

The calculational procedure for this measure, as is seen, reduces to that for the evolution operator. A detailed investigation of the thermal entanglement production, based on the above measure, will be treated in a separate paper.

Acknowledgement
Financial support from the Russian Foundation for Basic Research (grant # 14-02-00723) is appreciated.

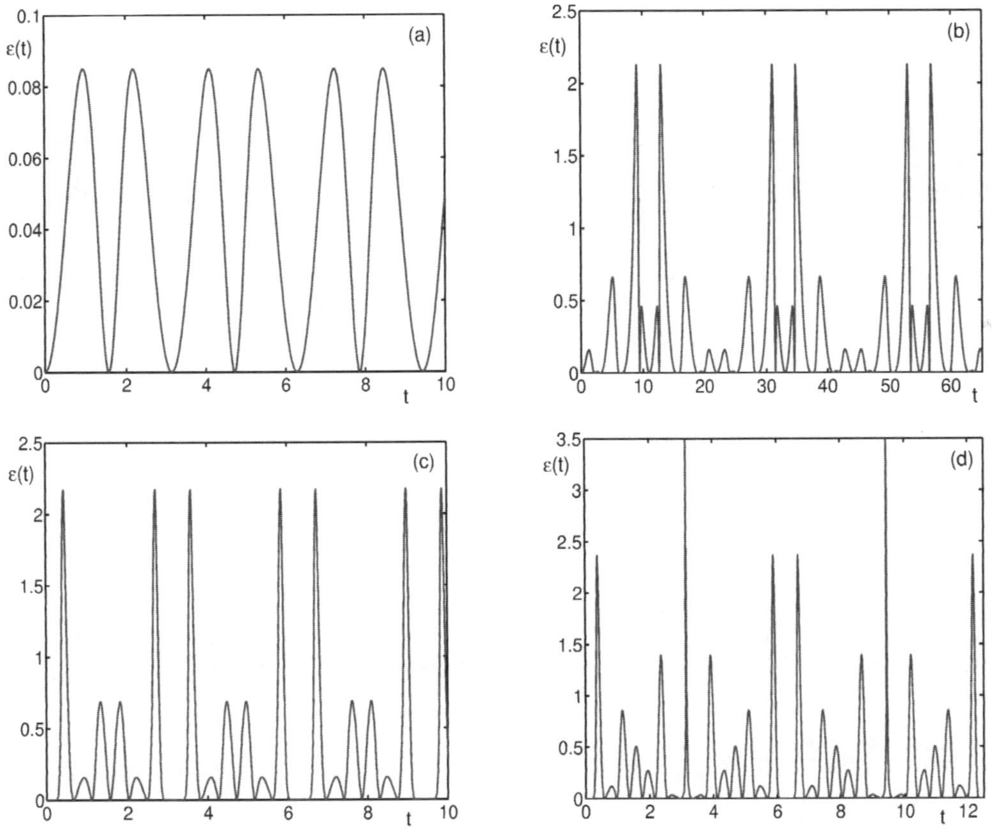

Figure 1. The entanglement-production measure, for the case of periodic evolution, as a function of time measured in units of $1/J$, for different fields: (a) $h/J = 1$ (the period is π); (b) $h/J = 5/7$ (the period is 7π); (c) $h/J = 7$ (the period is π); (d) $h/J = 8$ (the period is 2π).

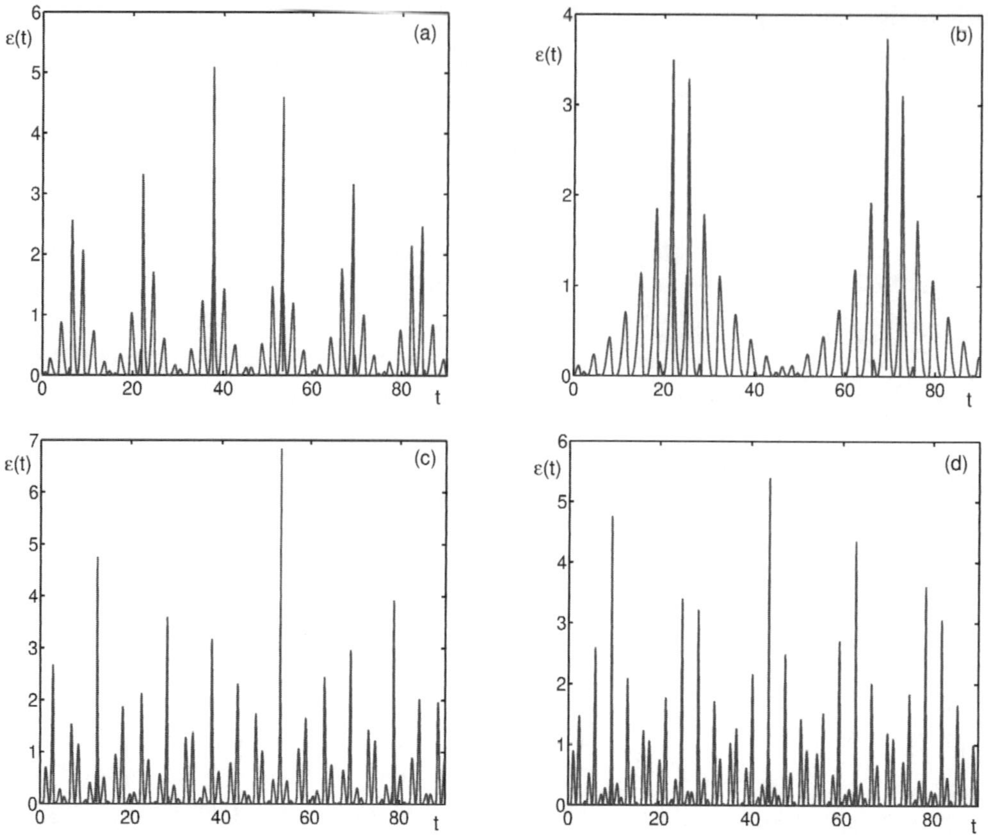

Figure 2. The measure of evolutional entanglement production, illustrating quasi-periodic behavior, for different fields: (a) $h/J = \sqrt{2}$; (b) $h/J = \sqrt{3}/2$; (c) $h/J = \sqrt{5}$; (d) $h/J = \sqrt{7}$.

[1] Williams C P and Clearwater S H 1998, *Explorations in Quantum Computing* (New York: Springer)
[2] Nielsen M A and Chuang I L 2000 *Quantum Computation and Quantum Information* (New York: Cambridge University)
[3] Vedral V 2002 *Rev. Mod. Phys.* **74** 197
[4] Keyl M 2002 *Phys. Rep.* **369** 431
[5] Yukalov V I and Sornette D 2009 *Entropy* **11** 1073
[6] Wilde M 2013, *Quantum Information Theory* (Cambridge: Cambridge University)
[7] Yukalov V I and Sornette D 2013 *Laser Phys.* **23** 105502
[8] Yukalov V I 2003 *Phys. Rev. Lett.* **90** 167905
[9] Yukalov V I 2003 *Phys. Rev. A* **68** 022109
[10] Kuo T H 1974 *Pacif. J. Math.* **52** 475
[11] Reed M and Simon B 1972 *Methods of Modern Mathematical Physics* Vol 1 (New York: Academic)
[12] Horn R A and Johnson C R 1991 *Topics in Matrix Analysis* (New York, Canbridge University)
[13] Yukalov V I 2003 *Mod. Phys. Lett. B* **17** 95
[14] Yukalov V I, Yukalova E P, and Bagnato V S 2003 *Laser Phys.* **13** 551
[15] Yukalov V I, Yukalova E P, and Bagnato V S 2003 *Laser Phys.* **13** 861
[16] Yukalov V I and Yukalova E P 2006 *Laser Phys.* **16** 354
[17] Yukalov V I and Yukalova E P 2006 *Phys. Rev. A* **73** 022335
[18] Yukalov V I and Yukalova E P 2008 *J. Phys. Conf. Ser.* **104** 012003
[19] Yukalov V I 2009 *Laser Phys.* **19** 1
[20] Yukalov V I, Yukalova E P, and Sornette D 2014 *J. Phys. Conf. Ser.* **497** 012034
[21] Yukalov V I 2004 *Laser Phys.* **14** 1403
[22] Yukalov V I and Yukalova E P 2015 *Phys. Rev. A* **92** 052121
[23] Gallagher T F 1994 *Rydberg Atoms* (Cambridge: Cambridge University)

[24] DiVincenzo D P 2000 *Fortschr. Phys.* **48** 771
[25] Raithel G and Morrow N 2006 *Adv. At. Mol. Opt. Phys.* **53** 187
[26] Choi J H, Knuffman B, Leibisch T C, Reinhard A, and Raithel G 2007 *Adv. At. Mol. Opt. Phys.* **54** 131
[27] Gallagher T F and Pillet P 2008 *Adv. At. Mol. Opt. Phys.* **56** 161
[28] Saffman M, Walker T G, and Mølmer K 2010 *Rev. Mod. Phys.* **82** 2313
[29] Buluta I, Ashhab S, and Nori F. 2011 *Rep. Prog. Phys.* **74** 104401
[30] Murray C and Pohl T 2016 *Adv. At. Mol. Opt. Phys.* **65** 321

IOP Conf. Series: Journal of Physics: Conf. Series **826** (2017) 012022 doi:10.1088/1742-6596/826/1/012022

Cavity resonances dominating the photon statistics in the non-equilibrium steady state

Felix Rüting

E-mail: felix.ruting@uam.es
Departamento de Física Teórica de la Materia Condensada, Universidad Autónoma de Madrid, Madrid 28049, Spain
Institut für Physik, Carl von Ossietzky Universität, D-26111 Oldenburg, Germany

Christoph Weiss

E-mail: christoph.weiss@durham.ac.uk
Joint Quantum Centre (JQC) Durham–Newcastle, Department of Physics, Durham University, Durham DH1 3LE, United Kingdom

Abstract. The non-equilibrium-steady state (NEST) for photons in a cavity is investigated theoretically. The NEST is caused by different parts of the cavity being at distinct temperatures or by temperature gradients. By using a rate equation based on the Lindblad equation, we derive an analytic expression for the steady-state distribution of the photon spectrum. We predict differences between the non-equilibrium steady state and a fit to the black-body spectrum calculated via Planck's law with an effective temperature. For two bodies of similar size at two temperatures which differ by a factor of two, the difference would be more than 10%. We also show that cavity resonances have a particularly large influence on the resulting non-equilibrium steady state of the photons. The investigation of thermal spectra in the presence of more than one temperature can be important for high-precision atomic clocks.

1. Introduction

Planck's law describes radiation from a black body in a cavity in thermal equilibrium [1]. It is an important contribution to the development of quantum mechanics; the law plays a role in fields as diverge as photonic crystals [2] and cosmology [3, 4]. The cavity involved in derivations of Planck's law is used in many quantum optics experiments (see, e.g., Refs. [5–8] and references therein). Theoretical investigations include quenching and spontaneous emission [9], entanglement distribution among distant nodes in a quantum network [10], quantum optics with surface plasmons [11] and quantum phase transitions of light [12].

The focus of the present paper lies on the photon-distribution for cases where more than one temperature is involved. For such situations, the non-equilibrium steady state, NEST, is investigated. Such a distribution could be measured experimentally in state-of-the-art quantum optics experiments. However, its applications would not be restricted to quantum optics.

Cavities bordered by bodies at different temperatures appear quite natural, e.g., by studying the temperature dependence of the Casimir-interaction [13–21] or the near-field radiative heat transfer [22–33]. Indeed, this transfer between bodies separated by distances even below

1 nm and with temperature differences up to some 100 K have been investigated in several experiments [26–32, 34], recently. Altfeder *et al.*, for example, have studied the heat transfer between a STM tip and a sample with a separation of only a few angstrom and with a maximal temperature difference of about 200 K [32]. Therefore a theoretical description of these experiments requires the understanding of the photon field in non-equilibrium states and different geometries.

Another important near-field effect is the Casimir force. For metal bodies in thermal equilibrium separated by a distance well below the thermal wavelength the effect of finite temperature is only a relative small correction [13], but recent progress in the measurement of this force allows one to study even this correction both for two bulk bodies and for a bulk body and a gas-phase atom [18–21]. The Casimir force was also investigated theoretically for two parallel plates at different temperatures [16, 17]. However, these studies are limited to the case of two parallel plates since the calculation requires the knowledge of the fluctuating fields between the bodies and for the case of a plane cavity these fields were calculated by Dorofeyev *et al.* [35].

Finally, for optical atomic clocks [36–39], high-precision experiments have reached a regime where the accuracy can be limited by the black-body radiation [39–41]. Recently, accuracy and stability on the 10^{-18} level was reported in an optical lattice clock for which a temperature gradient near the lattice-confined atoms was observed [39]. Thus, calculating deviations from the black-body spectrum due to the influence of more than one temperature can be important for high-precision atomic clocks.

Thus, knowing the photon statistic in a NEST is an important step for determine other field-related quantities. For example, the local energy density is given by product of the local density of states [42] and the mean energy of the mode in the actual state of the system.

The paper is organized as follows: In Sec. 2 we introduce the Lindblad equation. Section 3 solves this master equation for two temperatures as has been realized, e.g. for two parallel plates at distinct temperatures; these results can easily be generalized to the case of more than two temperatures. Section 4 shows that there are significant deviations of the non-equilibrium steady state from the equilibrium distribution. The paper ends with a conclusion in Sec. 5.

2. Rate equation

Even though classical thermal fields typically show a broad spectrum, the thermal near field is mostly dominated by contributions of surface modes of the nearby bodies. These contributions lead to a thermal spectral with only a few well-defined peaks [43, 44]. From this perspective the thermal near field resembles more an electromagnetic field within a cavity than the thermal far field.

The Lindblad equation [45, 46] for the cavity or near field can be found in the appendix. A similar approach has been used by Biehs and Agarwal for describing the heat transfer between nanosystems [47]. For the purpose of the present paper, it is sufficient to use the simplified version. Rather than having to use the complete density matrix in the Lindblad equation [45, 46] [cf. Eq. (A.1)], one can use a rate equation for the probabilities $p_n(t)$ to find $n = 0, 1, 2, \ldots$ photons in mode ω in the cavity at time t:

$$\frac{d}{dt}p_n(t) = \kappa_-(n+1)p_{n+1}(t) + \kappa_+ n p_{n-1}(t) - [\kappa_- n + \kappa_+(n+1)]\, p_n(t)\,. \quad (1)$$

The non-negative coefficients $\kappa_\pm$ correspond to the temperature-dependent rates at which photons are emitted into the cavity, κ_+, or absorbed at its boundaries, κ_-. The temperature dependence can be split into a strong temperature dependence included at the beginning of Sec. 3 and an additional, material-dependent temperature dependence discussed in Sec. 3.3.

The energy of n photons in the mode characterized by the frequency

$$\nu \equiv \omega/(2\pi) \tag{2}$$

is given by

$$E = \hbar\omega\left(n + \frac{1}{2}\right) . \tag{3}$$

The derivation of the non-equilibrium steady state will include the equilibrium distribution

$$\langle n \rangle = \frac{1}{\exp(\hbar\omega\beta) - 1} , \quad \beta \equiv \frac{1}{k_{\mathrm{B}}T} \tag{4}$$

as a special case if all temperatures are equal to T (k_{B} is the Boltzmann constant).

3. Non-equilibrium steady state (NEST)

3.1. Two temperatures

We start with the situation that there are two distinct temperatures

$$\beta_j = \frac{1}{k_{\mathrm{B}}T_j} , \quad j = 1, 2 . \tag{5}$$

involved; at the end of this section, the results will be generalized to the situation with more than two temperatures. We now take Eq. (1) and replace

$$\kappa_- \to \sum_{j=1}^{2} \kappa_{-,j} \quad \text{and} \quad \kappa_+ \to \sum_{j=1}^{2} \kappa_{+,j} . \tag{6}$$

The temperatures enter into the master equation via the Boltzmann law (cf. Ref. [45]) based on the assumption of local thermal equilibrium in each of the bodies:

$$\kappa_{+,j} = \kappa_{-,j} \exp\left(-\hbar\omega\beta_j\right) , \quad j = 1, 2 . \tag{7}$$

As the next step, we introduce a function which, at the present stage of the derivation has no physical meaning. The physical meaning [partially already indicated by our knowledge of Eq. (4)] will become clear at the end of our calculation. The function reads:

$$F(\beta) \equiv \frac{1}{\exp(\hbar\omega\beta) - 1} . \tag{8}$$

With this definition, we have:

$$\frac{\kappa_{+,j}}{\kappa_{-,j}} = \frac{F(\beta_j)}{1 + F(\beta_j)} \tag{9}$$

and we can define two rates κ_j with:

$$\kappa_{-,j} = \kappa_j[1 + F(\beta_j)] \tag{10}$$

and

$$\kappa_{+,j} = \kappa_j F(\beta_j) \tag{11}$$

where

$$\kappa_j > 0 \tag{12}$$

IOP Conf. Series: Journal of Physics: Conf. Series **826** (2017) 012022 doi:10.1088/1742-6596/826/1/012022

might still be temperature-dependent (cf. Sec. 3.3).

The condition for the non-equilibrium steady state reads

$$\frac{d}{dt}p_n(t) = 0 \quad \text{for all } n \tag{13}$$

which leads to the two (equivalent) conditions

$$(\kappa_{+,1} + \kappa_{+,2})\, np_{n-1}(t) = (\kappa_{-,1} + \kappa_{-,2})\, np_n(t) \tag{14}$$

and

$$(\kappa_{+,1} + \kappa_{+,2})\,(n+1)p_n(t) = (\kappa_{-,1} + \kappa_{-,2})\,(n+1)p_{n+1}(t). \tag{15}$$

Taking any of these equations and using that $\kappa_{-,j}$ is positive [see Eqs. (8), (10) and (12)] while $\kappa_{+,j}$ is positive for positive temperatures [Eqs. (8), (11) and (12)]:

$$p_n(t) = bp_{n-1}(t) \tag{16}$$

$$b \equiv \frac{\kappa_{+,1} + \kappa_{+,2}}{\kappa_{-,1} + \kappa_{-,2}} \quad \text{with} \tag{17}$$

$$0 < b < 1 \quad \text{for } T_1 + T_2 > 0\,. \tag{18}$$

The case that both temperatures are zero can be discarded.

The fact that now $p_n \propto b^n$ allows to calculate

$$\langle n \rangle = \sum np_n(t) = \frac{\sum_{n=0}^{\infty} nb^n}{\sum_{n=0}^{\infty} b^n} = \frac{1}{\frac{1}{b} - 1}\,. \tag{19}$$

Using

$$\frac{1}{b} = \frac{\kappa_1\,(1 + F(\beta_1)) + \kappa_2\,(1 + F(\beta_2))}{\kappa_1 F(\beta_1) + \kappa_2 F(\beta_2)} = \frac{\kappa_1 + \kappa_2}{\kappa_1 F(\beta_1) + \kappa_2 F(\beta_2)} + 1 \tag{20}$$

which can be inserted in Eq. (19)

$$\langle n \rangle_{\text{NEST}} = \frac{\kappa_1 F(\beta_1) + \kappa_2 F(\beta_2)}{\kappa_1 + \kappa_2}\,, \tag{21}$$

where NEST refers to the non-equilibrium steady state. A relevant special case is $T_1 = T_2 = T$ for which because of $F(\beta_1) = F(\beta_2)$ we find the expected result [45]

$$\langle n \rangle_{\text{equil.}} = \frac{1}{\exp\left(\frac{\hbar\omega}{k_B T}\right) - 1}\,. \tag{22}$$

For $\kappa_1 = \kappa_2$ we find:

$$\langle n \rangle_{\text{NEST}} = \frac{1}{2}\left(\frac{1}{\exp\left(\frac{\hbar\omega}{k_B T_1}\right) - 1} + \frac{1}{\exp\left(\frac{\hbar\omega}{k_B T_2}\right) - 1}\right). \tag{23}$$

This equation and its more general versions [see Eqs. (21) and (24)] is the first main results of the present paper.

3.2. Several temperatures
Extend the calculation of Sec. 3.1 to the case of several temperatures leads to:

$$\langle n\rangle_{\mathrm{NEST}} = \frac{\sum_j \kappa_j F(\beta_j)}{\sum_j \kappa_j}\,, \tag{24}$$

or, in general:

$$\langle n\rangle_{\mathrm{NEST}} = \frac{\int d^2r\kappa(\vec{r})F(\vec{r})}{\int d^2r\kappa(\vec{r})}\,. \tag{25}$$

where the integral extends over the surface surrounding the cavity.

Equation (24) could also play a role for photon condensates [48–50] — for which contrary to atomic Bose-Einstein condensates grand-canonical statistics can be relevant [48, 51, 52].

3.3. Physical meaning of the constants
What remains to be done is to find a physical interpretation of the parameters $\kappa_\pm$. So far, they seem to be purely phenomenological constants. However, similar to the equilibrium case discuss, e.g., in Ref. [45], they are related to the time-scales at which the non-equilibrium steady state is reached:

$$\frac{d}{dt}\langle n\rangle(t) = \sum_{n=0}^{\infty} n\frac{d}{dt}p_n(t). \tag{26}$$

Using the general expression (24) combined with the master equation (1), we find

$$\frac{d}{dt}\langle n\rangle(t) = -\left(\sum_j \kappa_j\right)\left[\langle n\rangle(t) - \langle n\rangle_{\mathrm{NEST}}\right]\,. \tag{27}$$

For a single κ, this time-scale can easily be related to a dimensionless mode quality factor Q of a cavity via $\kappa = \omega/Q$ [45].

Because of material-properties, the constants κ_j might be temperature dependent.

4. NEST-equivalent of Planck's law
Because of Eq. (3), the average energy in the mode ω is given by

$$\langle E\rangle = \hbar\omega\left(\langle n\rangle_{\mathrm{NEST}} + \frac{1}{2}\right)\,. \tag{28}$$

However, the zero point energy will not be relevant in our derivation. If one starts, e.g., with the zero-photon situation and lets the system approach to the NEST, it is clear that no energy is transferred into the vacuum state. For the purpose of our calculation we can thus use:

$$\langle E\rangle = \hbar\omega\langle n\rangle_{\mathrm{NEST}} \tag{29}$$

While for near-field effects more complicated densities of state are relevant, to derive the NEST-equivalent of the Planck's law [1],

$$I(\omega,T) = \frac{\hbar\omega^3}{\pi^2c^3}\frac{1}{\exp(\hbar\omega\beta)-1}, \tag{30}$$

we use the vacuum density [1,42] in the following

$$D = \frac{\omega^2}{\pi^2 c^3} \, . \tag{31}$$

Thus,

$$I_{\rm NEST}(\omega, \{T_j\}) = \frac{\hbar\omega^3}{\pi^2 c^3} \langle n \rangle_{\rm NEST}, \tag{32}$$

or, for the two-temperature case with $\kappa_1 = \kappa_2$:

$$I_{\rm NEST}(\omega, \{T_1, T_2\}) = \frac{\hbar\omega^3}{2\pi^2 c^3} \left(\frac{1}{\exp\left(\frac{\hbar\omega}{k_B T_1}\right) - 1} + \frac{1}{\exp\left(\frac{\hbar\omega}{k_B T_2}\right) - 1} \right) \tag{33}$$

In an experiment, a natural way to approach the non-equilibrium steady state data would be to try and fit the equilibrium distribution (30) using the temperature as a fitting parameter. For the prediction of Eq. (33) to be of practical use, we still have to show that in an experiment it would be distinguishable from a Planck-distribution with an effective temperature $T_{\rm eff}$. In the following, we also use:

$$\beta_{\rm eff} = \frac{1}{k_B T_{\rm eff}} \, . \tag{34}$$

A suitable way to define such an effective temperature is to minimize the mean square deviations

$$\delta \equiv \int_0^\infty d\omega \left[I_{\rm NEST}(\omega, \{T_1, T_2\}) - I(\omega, T_{\rm eff}) \right]^2 \, . \tag{35}$$

In order to visualize the differences between the NEST-distribution and the equilibrium distribution with an effective temperature, we use:

$$\Delta \equiv \frac{I_{\rm NEST}(\omega, \{T_1, T_2\}) - I(\omega, T_{\rm eff})}{\max\{I(\omega, T_{\rm eff}), 0 \le \omega < \infty\}} \, . \tag{36}$$

According to Wien's displacement law [1], $I(\omega, T_{\rm eff})$ reaches its maximum at [1]

$$\hbar\omega_{\rm max} = \{3 + W_0[-3\exp(-3)]\} \, k_B T_{\rm eff} \simeq 2.82 k_B T_{\rm eff} \tag{37}$$

where W_0 is a Lambert W function. The value of the maximum is

$$\frac{\{W_0[-3\exp(-3)] + 3\}^3}{\{\exp(W_0[-3\exp(-3)] + 3) - 1\}} \frac{(k_B T_{\rm eff})^3}{\pi^2 c^3 \hbar^2} \simeq 1.42 \frac{(k_B T_{\rm eff})^3}{\pi^2 c^3 \hbar^2} \, . \tag{38}$$

Figure 1 displays the effective temperature obtained by minimizing Eq. (35) as a function of the ratio T_1/T_2. Without loss of generality we can assume

$$0 < T_2 \le T_1 < \infty \tag{39}$$

for our discussion. Figure 1 shows that the effective temperature approaches

$$T_{\rm eff} \simeq \begin{cases} \frac{T_1 + T_2}{2} & : \; T_2/T_1 \to 1 \\ 0.83 T_1 & : \; T_2/T_1 \to 0 \end{cases} \, . \tag{40}$$

However, the main point in this approach was not to calculate an effective temperature but rather to see if the NEST-prediction (33) can be distinguished from the effective temperature approach. Figure 2 shows the deviation of both approaches as defined in Eq. (36). As expected from closely inspecting Eq. (33), the differences are small if the temperature difference is small. However, they can be quite large for larger temperature differences: if the two temperatures differ by a factor of the order of two (e.g., room-temperature versus liquid nitrogen) the deviations are more than 10% (measured in units of the maximum of the curve) and should thus be easily detectable in an experiment.

[1] The analytic expression was obtained by using the computer algebra program Maple.

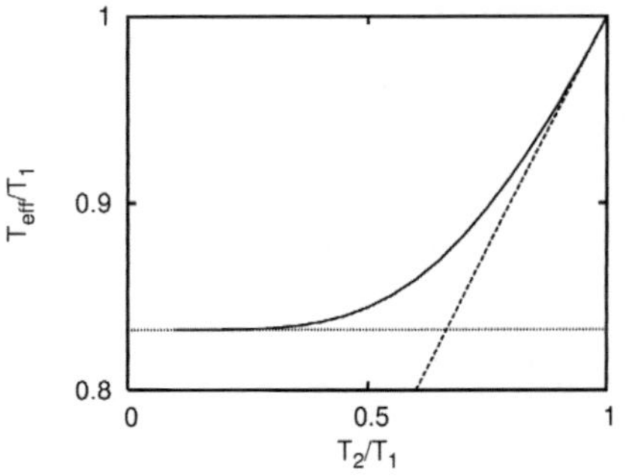

Figure 1. Effective temperature as a function of the ratio of the two temperatures. The effective temperature tries to describe the non-equilibrium steady state by an equilibrium distribution with an effective temperature. Solid line: the effective temperature was obtained by minimizing Eq. (35). Dotted line: the low temperature limit given in Eq. (40). Dashed line: the arithmetic mean of the two temperatures T_1 and T_2 [the high temperature limit of Eq. (40)]

4.1. Cavity

In order to obtain a NEST-equivalent of Planck's law we assumed in the previous section that the density of states between the two bodies is given by the free-space DOS (see. Eq. (31)). However in an experiment involving a cavity this cavity will strongly influence the density of states. As a generic example we assume, that the resonance has a Lorentzian shape [53]. Hence in following the density of states is given by

$$D(\omega) = D_0 \frac{\Delta\omega^2}{(\omega - \omega_0)^2 + \Delta\omega^2}, \tag{41}$$

with the resonance frequency ω_0 and the line width $\Delta\omega$ defined by $Q = \omega_0/\Delta\omega = 100$.

The effective temperature and the maximum derivation of the NEST distribution from an equilibrium one are shown in Fig. 3. As the density of the state has a sharp peak around ω_0 the dominant contribution to the integral in Eq. (35) steams from frequencies around ω_0. Hence

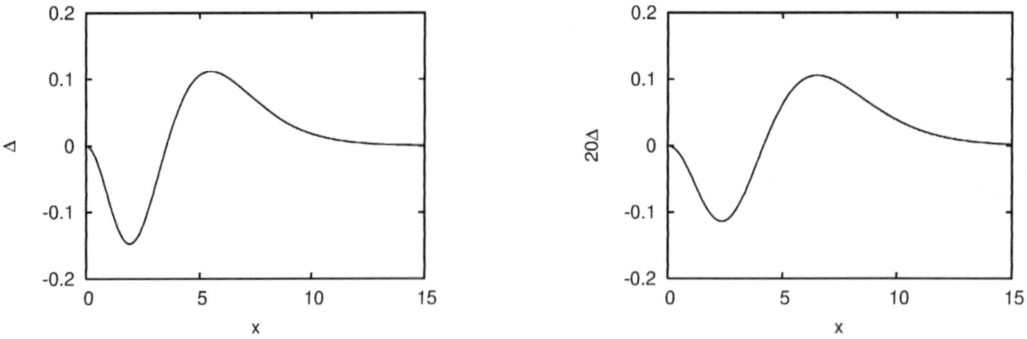

Figure 2. Trying to fit an equilibrium distribution to the non-equilibrium steady state with two temperatures (33) leads to experimentally detectable differences [cf. Eq. (36)]. The difference Δ is plotted as a function of dimensionless frequency $x \equiv \hbar\omega/(k_B T_{\rm eff})$ Left panel: $T_2 = 0.5T_1$, $T_{\rm eff} \simeq 0.8443T_1$. Right panel: $T_2 = 0.9T_1$, $T_{\rm eff} \simeq 0.9533T_1$. Note that the error is multiplied by 20 in the right panel. As expected, the deviation of the non-equilibrium steady state from thermal equilibrium thus is much larger if the ratio T_2/T_1 differs from one by an experimentally realistic [32] factor of two (left panel) than if the ratio is close to one (right panel).

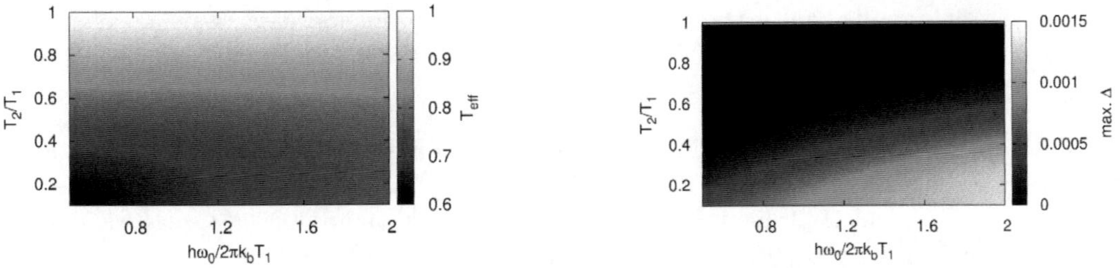

Figure 3. (Color online) Effective temperature (left panel) and maximum difference between the NEST- and equilibrium distribution (right panel) for a cavity with one resonance located at ω_0. Due to the single, sharp resonance the description via an effective temperature works quite well.

the description via an equilibrium distribution is quite good, as demonstrated by the very small errors in Fig. 3. The situation becomes more interesting by assuming that the cavity has two resonances, so that the DOS is given by

$$D(\omega) = D_0 \left(\frac{\Delta\omega_1^2}{(\omega - \omega_1)^2 + \Delta\omega_1^2} + \frac{\Delta\omega_2^2}{(\omega - \omega_2)^2 + \Delta\omega_2^2} \right). \tag{42}$$

In this case the differences between the NEST and the effective description are much more pronounced (see Fig. 4). And hence the results, presented in this section, provide a guideline, whether a description via an effective temperature is suitable or not. If the DOS is dominated by single resonance the system can be described by an effective temperature, while non-equilibrium effects have to be included when a broader frequency range comes into play.

5. Conclusion

We have investigated the non-equilibrium steady state (NEST) for photons in a cavity for which the boundaries are at two or more distinct temperatures. For this purpose we have derived an analytic expression for the NEST photon statistic and evaluated this expression for two different situations. Once assuming that the density of states can be described by the vacuum density and secondly for multi-mode cavities.

(i) If the density of states can be described by vacuum density, we predict that the NEST-equivalent of Planck's law can show deviations of more than 10% from the usual form of Planck's law if we take two distinct temperatures at the boundaries of the cavity that differ by a factor of two.

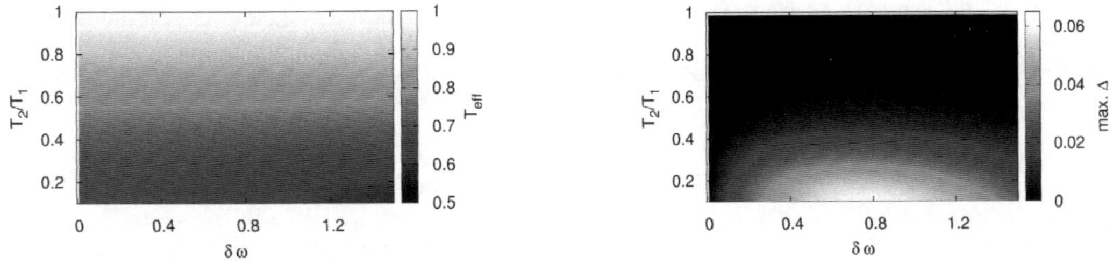

Figure 4. (Color online) Effective temperature (left panel) and maximum difference between the NEST- and equilibrium distribution (right panel) for a cavity with two resonances $\omega_1 = (0.8 - 0.5\delta\omega)\frac{k_b T_1}{\hbar}$ and $\omega_2 = (0.8 + 0.5\delta\omega)\frac{k_b T_1}{\hbar}$.

(ii) While for a single-mode cavity the description of the photonen statistics via an effective temperature captures the NEST quite well, for multi-mode cavities the photon statistics is dominated by the resonances and non-equilibrium effects have to be accounted for.

The approach presented here should have practical consequences for the energy and momentum transfer between bodies of comparable size in a NEST situation. Furthermore, black-body spectra influence the accuracy of state-of-the-art atomic clocks [39–41] for which temperature gradients near the atoms have been reported [39]. Thus, even in the absence of cavity resonances, calculating deviations from the black-body spectrum caused by more than one temperature can become important for high-precision atomic clocks.

For our approach to be valid, the time-scale on which the NEST is reached has to be shorter than the time-scales at which the temperatures at the boundaries of the cavity change.

Acknowledgments
We thank S. A. Biehs, E. M. Bridge, M. Holthaus, I. G. Hughes, C. Lienau, A. Kittel and V. Steenhoff for discussions.

Appendix A. Lindblad Master equation

A derivation of the Lindblad Master equation can be found, e.g., in Ref. [46]. If $a^{(\dagger)}$ is the annihilation (creation) operator of a photon in mode ω, the Lindblad equation for the cavity field reads:

$$\frac{d}{dt}\hat{\varrho} = \quad -i\omega[a^\dagger a, \hat{\varrho}] - \frac{\kappa_-}{2}\left(a^\dagger a\hat{\varrho} + \hat{\varrho}a^\dagger a - 2a\hat{\varrho}a^\dagger\right) - \frac{\kappa_+}{2}\left(aa^\dagger\hat{\varrho} + \hat{\varrho}aa^\dagger - 2a^\dagger\hat{\varrho}a\right) , \quad \text{(A.1)}$$

where $\hat{\varrho}$ is the density matrix. The rate equation (1) can be derived [45] by using the fact that the photon number distribution $p_n(t)$ is related to the density matrix via

$$p_n(t) = \langle n|\varrho|n\rangle \ . \quad \text{(A.2)}$$

In this paper, the Lindblad Master equation is used to describe emission and absorption of thermal photons, other possible applications include atom losses [54].

References

[1] Lifshitz E M and Pitaevskii L P 2002 *Landau and Lifshitz — Course of Theoretical Physics, Vol. 9: Statistical Physics, Part 1* (Oxford: Butterworth-Heinemann)
[2] Lin S Y, Fleming J G, Hetherington D L, Smith B K, Biswas R, Ho K M, Sigalas M M, Zubrzycki W, Kurtz S R and Bur J 1998 *Nature (London)* **394** 251
[3] Roll P G and Wilkinson D T 1966 *Phys. Rev. Lett.* **16** 405
[4] Peebles P J E and Ratra B 2003 *Rev. Mod. Phys.* **75**(2) 559
[5] Rauschenbeutel A, Nogues G, Osnaghi S, Bertet P, Brune M, Raimond J M and Haroche S 1999 *Phys. Rev. Lett.* **83**(24) 5166
[6] Lounis B and Moerner W E 2000 *Nature (London)* **407** 491
[7] Pinkse P W H, Fischer T, Maunz P and Rempe G 2000 *Nature (London)* **404**
[8] Englund D, Faraon A, Fushman I, Stoltz N, Petroff P and Vučković J 2007 *Nature (London)* **450** 857
[9] Lee H, Polynkin P, Scully M O and Zhu S Y 1997 *Phys. Rev. A* **55** 4454
[10] Cirac J I, Zoller P, Kimble H J and Mabuchi H 1997 *Phys. Rev. Lett.* **78**(16) 3221
[11] Chang D E, Sørensen A S, Hemmer P R and Lukin M D 2006 *Phys. Rev. Lett.* **97** 053002
[12] Greentree A D, Tahan C, Cole J H and Hollenberg L C L 2006 *Nat. Phys.* **2** 856–861
[13] Lifshitz E M 1956 *JETP* **2** 73
[14] Haakh H, Intravaia F, Henkel C, Spagnolo S, Passante R, Power B and Sols F 2009 *Phys. Rev. A* **80** 062905
[15] Haakh H, Intravaia F and Henkel C 2010 *Phys. Rev. A* **82** 012507
[16] Antezza M, Pitaevskii L P, Stringari S and Svetovoy V B 2006 *Phys. Rev. Lett.* **97** 223203
[17] Antezza M, Pitaevskii L P, Stringari S and Svetovoy V B 2008 *Phys. Rev. A* **77** 022901

[18] Mohideen U and Roy A 1998 *Phys. Rev. Lett.* **81** 4549
[19] Decca R S, Fischbach E, Klimchitskaya G L, Krause D E, López D and Mostepanenko V M 2003 *Phys. Rev. D* **68** 116003
[20] Decca R, López D, Fischbach E, Klimchitskaya G, Krause D and Mostepanenko V 2005 *Annals of Physics* **318** 37 – 80
[21] Obrecht J M, Wild R J, Antezza M, Pitaevskii L P, Stringari S and Cornell E A 2007 *Phys. Rev. Lett.* **98** 063201
[22] Pérez-Madrid A, Lapas L C and Rubí J M 2009 *Phys. Rev. Lett.* **103** 048301
[23] Chapuis P O, Laroche M, Volz S and Greffet J J 2008 *Phys. Rev. B* **77** 125402
[24] Dorofeyev I 2008 *Physics Letters A* **372** 1341 – 1347
[25] Pendry J B 1999 *Journal of Physics: Condensed Matter* **11** 6621
[26] Hu L, Narayanaswamy A, Chen X Y and Chen G 2008 *Appl. Phys. Lett.* **92** 133106
[27] Kittel A, Wischnath U F, Welker J, Huth O, Rüting F and Biehs S A 2008 *Appl. Phys. Lett.* **93** 193109
[28] Wischnath U F, Welker J, Munzel M and Kittel A 2008 *Rev. Sci. Instrum.* **79** 073708
[29] Narayanaswamy A, Shen S and Chen G 2008 *Phys. Rev. B* **78** 115303
[30] Rousseau E, Siria A, Jourdan G, Volz S, Comin F, Chevrier J and Greffet J J 2009 *Nature Photonics* **3** 514
[31] Shen S, Narayanaswamy A and Chen G 2009 *Nano Lett.* **9** 2909
[32] Altfeder I, Voevodin A A and Roy A K 2010 *Phys. Rev. Lett.* **105** 166101
[33] Ben-Abdallah P and Biehs S A 2014 *Phys. Rev. Lett.* **112** 044301
[34] Ottens R S, Quetschke V, Wise S, Alemi A A, Lundock R, Mueller G, Reitze D H, Tanner D B and Whiting B F 2011 *Phys. Rev. Lett.* **107**(1) 014301
[35] Dorofeyev I, Fuchs H and Jersch J 2002 *Phys. Rev. E* **65** 026610
[36] Diddams S A, Udem T, Bergquist J C, Curtis E A, Drullinger R E, Hollberg L, Itano W M, Lee W D, Oates C W, Vogel K R and Wineland D J 2001 *Science* **293** 825
[37] Takamoto M, Hong F L, Higashi R and Katori H 2005 *Nature (London)* **435**
[38] Rosenband T, Hume D B, Schmidt P O, Chou C W, Brusch A, Lorini L, Oskay W H, Drullinger R E, Fortier T M, Stalnaker J E, Diddams S A, Swann W C, Newbury N R, Itano W M, Wineland D J and Bergquist J C 2008 *Science* **319** 1808
[39] Bloom B J, Nicholson T L, Williams J R, Campbell S L, Bishof M, Zhang X, Zhang W, Bromley S L and Ye J 2014 *Nature (London)* **506**
[40] Middelmann T, Falke S, Lisdat C and Sterr U 2012 *Phys. Rev. Lett.* **109** 263004
[41] Hinkley N, Sherman J A, Phillips N B, Schioppo M, Lemke N D, Beloy K, Pizzocaro M, Oates C W and Ludlow A D 2013 *Science* **341** 1215
[42] Joulain K, Carminati R, Mulet J P and Greffet J J 2003 *Phys. Rev. B* **68** 245405
[43] Greffet J J, Carminati R, Joulain K, Mulet J P, Mainguy S and Chen Y 2002 *Nature (London)* **416** 61
[44] Joulain K, Mulet J P, Marquier F, Carminati R and Greffet J J 2005 *Surf. Sci. Rep.* **57** 59
[45] Haroche S and Raimond J M 2006 *Exploring the Quantum – Atoms, Cavities and Photons* (Oxford: Oxford University Press)
[46] Breuer H P and Petruccione F 2006 *The Theory of Open Quantum Systems* (Oxford: Clarendon Press)
[47] Biehs S A and Agarwal G S 2013 *J. Opt. Soc. Am. B* **30** 700
[48] Schmitt J, Damm T, Dung D, Vewinger F, Klaers J and Weitz M 2014 *Phys. Rev. Lett.* **112** 030401
[49] Marelic J and Nyman R A 2015 *Phys. Rev. A* **91** 033813
[50] Fani M and Naderi M H 2016 *J. Opt. Soc. Am. B* **33** 1242
[51] van der Wurff E C I, de Leeuw A W, Duine R A and Stoof H T C 2014 *Phys. Rev. Lett.* **113**(13) 135301
[52] Weiss C and Tempere J 2016 *Phys. Rev. E* **94** 042124
[53] Garraway B M and Knight P L 1996 *Phys. Rev. A* **54** 3592
[54] Stiebler K, Gertjerenken B, Teichmann N and Weiss C 2011 *J. Phys. B* **44** 055301

IOP Conf. Series: Journal of Physics: Conf. Series **826** (2017) 012023 doi:10.1088/1742-6596/826/1/012023

Peculiarities of excitation trapping in dense polyatomic ensemble in a Fabry-Perot cavity

A S Kuraptsev[1] and I M Sokolov[1,2]

[1]Peter the Great St. Petersburg Polytechnic University, 195251, St. Petersburg, Russia
[2]Institute for Analytical Instrumentation, Russian Academy of Sciences, Ivana Chernykh st. 31, St. Petersburg, 198095, Russia

E-mail: `ims@is12093.spb.edu`

Abstract. We report the investigation of the peculiarities of excitation trapping caused by the cooperative effects in a disordered ensemble of point-like impurity centers (atoms) embedded in a transparent dielectric and located in a Fabry-Perot cavity. On the basis of the general quantum microscopic theory we have calculated cooperative spontaneous decay of an excited atom located in the center of a sample. It is shown that in the case when the distance between the mirrors of a cavity d is less or comparable with resonant wavelength of the atomic transition λ_0, the character of cooperative effects in a cavity essentially differs from one without a cavity. For the opposite case $d \gg \lambda_0$ cooperative effects in a cavity manifest themselves almost like in the case of absence of a cavity.

1. Introduction

Since the seminal work of Purcell [1] light interface with atoms localized inside a cavity or waveguide as well as near its surface has attracted a considerable attention. Now it is understood that using of a cavity offers an exciting tool to control over the atomic radiative properties, in particular, the enhancement and inhibition of the spontaneous decay rate [2] – [5]. Furthermore, the spontaneous decay can be suppressed almost completely by choosing the appropriate geometry of a cavity. Light matter interaction in the presence of nanophotonic structures, such as nanofibers [6] – [8], photonic crystal cavities [9] and waveguides [10] – [11] propose future applications for quantum metrology, scalable quantum networks and quantum information science [12] – [13].

Besides the cavity, the interaction of active impurity atoms with the surrounding dielectric medium can influence on its radiative properties (see [14] – [24] and references therein). Even in the case of a transparent dielectric the internal fields of a medium cause spectral line shifts of the impurity atoms which can exceed the natural linewidth. These shifts depend mainly on the type of chemical bond of a dielectric, the symmetry of the internal fields and the temperature.

In many practical applications the efficiency of impurity centers can be improved by increasing of their density. For dense ensembles, in which the average distance between impurity centers is comparable with the wavelength of resonant radiation, the dipole-dipole interaction between themselves plays an important role [25] – [26]. Cooperative effects cause density-dependent shifts of atomic transition as well as distortion of spectral line shape [27]. This influences both on the properties of atoms [28] – [30] as well as on scattered radiation characteristics [31] – [36].

It is important that the radiative transfer between different atoms located inside the cavity is essentially modified compared with absence of a cavity [37]. Today the interatomic interaction in the atomic systems coupled to a nanofiber [38] as well as to phonic crystals [39] – [41] is intensively investigated. Nevertheless, multiatomic cooperative effects inside the cavity, including multiple and recurrent scattering have not yet been studied in detail. It is connected with both theoretical problems as well as some computational difficulties.

In this paper we study a disordered ensemble, which consists of $N \gg 1$ impurity centers in a Fabry-Perot cavity taking into account the transition frequency shifts of impurity centers caused by the internal fields of a dielectric medium. On the basis of the general theory we study the spontaneous decay dynamics of the local atomic excitation prepared inside a cavity depending on the distance between the mirrors. In particular, we analyze the case when the distance between two mirrors d is less than a half of the transition wavelength λ_0 as well as the case when $d \gg \lambda_0$. This first case attracts a special attention due to the fact that the spontaneous decay from some Zeeman sublevels of atomic exited state is suppressed. This indicates that the use of a cavity allows to achieve very large time of radiation trapping, much more than in the case of absence of a cavity [42].

2. Basic assumptions and approach

Let us consider a disordered ensemble, which consists of N motionless impurity atoms in a transparent dielectric closed in a Fabry-Perot cavity. The mirrors of a cavity are assumed to be perfectly conducting. The sizes of the mirrors are much larger than resonant transition wavelength λ_0, the distance between the mirrors d and the average distance between impurity atoms (see Fig. 1). It allows us to consider the sizes of the mirrors equal to infinity.

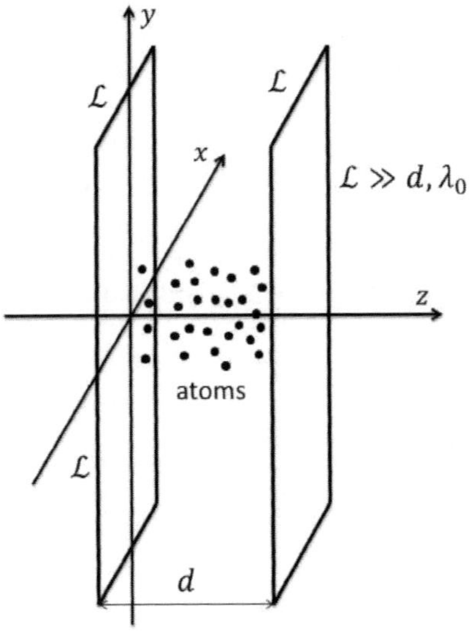

Figure 1. Sketch of the cavity and the atomic ensemble.

We assume that the temperature is low enough to neglect the interaction of impurity atoms with phonon reservoir. It allows us to assume the quantum state of the joint system, which consists of atoms and electromagnetic field to be pure, and describe it by the wave function.

Even for a transparent dielectric the transition frequency of impurity atoms in a dielectric ω_a can be different from the transition frequency of a free atom ω_0. It is connected with internal fields of a dielectric. Thus, $\omega_a = \omega_0 + \Delta_a$, where Δ_a is the frequency shift of the atom a which depends on its spatial position, $a = 1, ..., N$.

In this paper we use quantum microscopic approach described firstly in [43] and developed afterward in [31] for description of collective effects in dense and cold nondegenerate atomic gases. It was earlier successfully used for analysis of optical properties of dense atomic ensembles [28] – [30] as well as for studying of light scattering from such ensembles [34] – [42].

This approach is based on solution of the non-stationary Schrodinger equation for the wave function ψ of the joint system consisting of atoms and the electromagnetic field, including vacuum reservoir

$$i\hbar\frac{\partial\psi}{\partial t} = \widehat{H}\psi. \tag{1}$$

The Hamiltonian $\widehat{H}$ of the joint system can be presented as a sum of Hamiltonian $\widehat{H}_a$ of the atoms noninteracting with the field, the Hamiltonian $\widehat{H}_f$ of the free field in a Fabry-Perot cavity, and the operator $\widehat{V}$ of its interaction

$$\widehat{H} = \widehat{H}_0 + \widehat{V}, \tag{2}$$

$$\widehat{H}_0 = \widehat{H}_f + \sum_a \widehat{H}_a. \tag{3}$$

In the dipole approximation used here, we have

$$\widehat{V} = -\sum_a \widehat{\mathbf{d}}^{(a)}\widehat{\mathbf{E}}(\mathbf{r}_a). \tag{4}$$

In this equation $\widehat{\mathbf{d}}^{(a)}$ is the dipole momentum operator of the atom a, $\widehat{\mathbf{E}}(\mathbf{r}_a)$ is the electric field operator, and $\mathbf{r}_a$ is the position of the atom a.

The electric field operator $\widehat{\mathbf{E}}(\mathbf{r})$ in a cavity can be obtained in a standard way by quantization of the classical field $\mathbf{E}(\mathbf{r}, t)$. The latter is a solution of Maxwell equations with corresponding boundary conditions. Let us consider z axis perpendicular to the mirrors, x and y axis parallel to the mirrors (see Fig. 1). $z = 0$ corresponds to the position of the first mirror and $z = d$ – to the second mirror. The boundary conditions can be written as follows: $E_x|_{z=0} = E_x|_{z=d} = E_y|_{z=0} = E_y|_{z=d} = 0$. In the considered geometry we have

$$\widehat{\mathbf{E}}(\mathbf{r}) = \sum_{\mathbf{k},\alpha} \frac{i\omega_k}{c}\sqrt{\frac{\hbar}{2\omega_k}}\widehat{a}_{\mathbf{k},\alpha}B^0_{\mathbf{k},\alpha}\{\mathbf{e}_x u^x_{\mathbf{k},\alpha}\sin(k_n z) + \mathbf{e}_y u^y_{\mathbf{k},\alpha}\sin(k_n z) + \mathbf{e}_z u^z_{\mathbf{k},\alpha}\cos(k_n z)\}\exp(i\mathbf{k}_{||}\mathbf{r}_{||}) + \text{H.c.} \tag{5}$$

In this equation $B^0_{\mathbf{k},\alpha}$ is the normalization factor, given by

$$B^0_{\mathbf{k},\alpha} = \sqrt{\frac{8\pi c^2}{L^2 d}} \times \begin{cases} 1, & \text{if } n \in N \\ 1/\sqrt{2}, & \text{if } n = 0. \end{cases} \tag{6}$$

L is the longitudinal size of the quantization volume.

We will seek the wave function ψ as an expansion in a set of eigenfunctions ψ_l of the operator H_0:

$$\psi = \sum_l b_l(t)\psi_l. \tag{7}$$

Here, the subscript l defines the state of all atoms and the field. Using this representation of the wave function we convert the equation (1) to the system of linear differential equations for the quantum amplitudes of states

$$i\hbar \frac{\partial b_l(t)}{\partial t} - E_l b_l(t) = \sum_j V_{lj} b_j(t). \tag{8}$$

In this equation E_l is the energy of l state of the system, which consists of noninteracting atoms and electromagnetic field.

The total number of equations in the system (8) is equal to infinity. It is caused by the infinity number of the degrees of freedom of electromagnetic field.

The key simplification of the approach employed is in restriction of the total number of states $|l\rangle$ taken into account. We will calculate all radiative correction up to the second order of the fine structure constant. In this case we can consider only the following states (see [44]):

1. One-fold atomic excited states
$\psi_{e_a} = |g, ..., g, e, g, ..., g\rangle \otimes |vac\rangle$, $E_{e_a} = \hbar\omega_a$.
2. Resonant single-photon states
$\psi_g = |g, ..., g\rangle \otimes |\mathbf{k}, \alpha\rangle$, $E_g = \hbar\omega_k$.
3. Nonresonant states with two excited atoms and one photon
$\psi_{e_a e_b} = |g, ..., g, e, g, ..., g, e, g, ..., g\rangle \otimes |\mathbf{k}, \alpha\rangle$,
$E_{e_a e_b} = \hbar(\omega_a + \omega_b) + \hbar\omega_k$.

In the rotating wave approximation it is enough to take into account only the first and second group of states. Nonresonant states are necessary for a correct description of the dipole-dipole interaction at short interatomic distances, comparable with λ_0.

For a description of the coherent external light scattering, it is necessary to complete the set of quantum states by the vacuum state without excitation both in atomic and field subsystem
$\psi_{g'} = |g, ..., g\rangle \otimes |vac\rangle$, $E_{g'} = 0$.

In the framework of the assumptions considered here, the quantum amplitude of the state $\psi_{g'}$ does not change during the evolution of the considered system. It is explained by the fact that, any transitions between $\psi_{g'}$ and other quantum states taken into account are impossible. The Lamb shift is considered to be included in ω_0.

Despite the restriction of the total number of quantum states, the set of equations remains infinite because it contains the field states. We can, however, exclude amplitudes of states with one photon and obtain a finite closed system of equations for the atomic states b_e. For Fourier components $b_e(\omega)$ we have (at greater length; see [31])

$$\sum_{e'} \left[(\omega - \omega_a)\delta_{ee'} - \Sigma_{ee'}(\omega) \right] b_{e'}(\omega) = i\delta_{e'o}. \tag{9}$$

This set of equations was obtained under the assumption that at the initial time only one atom is excited. We denote it by the subscript o. All other atoms are in their ground state at $t = 0$ and electromagnetic field is in the vacuum state. The system (9) with the initial conditions considered here allows us to analyze both stationary light scattering as well as nonstationary problems. In particular, the time dependent excitation probability of initially excited atom $P_o(t) = |b_o(t)|^2$ can be calculated.

The size of the system (9) is determined by the number of atoms N and the structure of the energy levels. In this paper we consider that the ground state is characterized by the total angular momentum $J = 0$ and the excited state – $J = 1$. Note that, in the case considered, there are three excited sublevels for each atom $e = |J, m\rangle$, which differ by the value of angular momentum projection $m = -1, 0, 1$. Therefore, the total number of one-fold atomic excited states is $3N$. This scheme of levels corresponds to atoms with 2 valence electrons such as Sr, Yb, Ca.

The matrix $\Sigma_{ee'}(\omega)$ describes both spontaneous decay and photon exchange between the atoms. It plays a key role in the microscopic theory. The explicit expression for this matrix corresponding to the cavity was derived in Refs. [45] – [46]. It extremely differs from one corresponding to the free space. This is explained by the fact, that a cavity modifies the structure of modes of electromagnetic field.

In the next section, we will use the general approach to investigate the simultaneous influence of the cavity and interatomic dipole-dipole interaction on the spontaneous decay dynamics of an excited atom depending on the distance between two mirrors.

3. Results and discussion

The inverse Fourier transform of $b_e(\omega)$ allows us to obtain the time dependence of the quantum amplitudes of the one-fold atomic excited states $b_e(t)$. The spontaneous decay dynamics of initially excited atom is determined by $P_o(t) = |b_o(t)|^2$.

The influence of the dipole-dipole interaction on the properties of atomic ensemble is determined not only by the density of impurity atoms. The shifts of the energy levels caused by the internal fields of a dielectric are also very important. The value of these shifts depends on a number of factors, first of all, on the nature of the dielectric and its temperature. In the present paper we restrict ourselves by the assumption that the temperature is low enough to neglect the electron-phonon interaction. So the spectral lines of impurity centers are Zero-phonon.

The shift of the transition line Δ_a can be presented as a sum of its average value $\overline{\Delta}$ and some random contribution connected with the inhomogeneity of the internal fields of the dielectric. We consider this random contribution to be normally distributed with r.m.s. deviation δ. In this paper we will restrict ourselves by the case when $\overline{\Delta}$ is the same for all Zeeman sublevels of the excited state. This corresponds to the cubic symmetry of internal fields of a medium, for instance. Hereafter we will consider $\overline{\Delta}$ to be included in the resonant transition frequency ω_0. The ratio of δ to natural linewidth of the atoms γ_0 characterizes the degree of resonance between impurity atoms. This is one of a key parameters of the considered system in our theory. In particular, it can be considered the case when all atoms are resonant to each other $\delta \ll \gamma_0$ as well as opposite case when inhomogeneous broadening is much larger than natural linewidth $\delta \gg \gamma_0$. The criterion of the significance of cooperative effects for the last case was derived in Refs. [45] – [46]. It can be written as follows: $(n\gamma_0/\delta)^{-1/3} \leq \lambda_0$. This criterion can be generalized on the atomic media with inhomogeneous broadening of different nature. In particular, it can be used for the analysis of the gas cells with alkali atoms at room temperature that play a key role in frequency standards based on CPT [47] – [49].

In this paper we will focus our attention on the investigation of the cooperative spontaneous decay of an excited atom depending on the distance between the mirrors. Let us consider initially excited atom located in the middle between the mirrors $z_{exc} = d/2$ (reference point $z = 0$ corresponds to the left mirror) and in the center of atomic ensemble. All other atoms are considered to be in their ground states at initial time $t = 0$ and all field modes in the vacuum state.

Fig. 2 shows the spontaneous decay dynamics from Zeeman sublevel $m = -1$ or $m = 1$ of the initially excited central atom. Atomic density is chosen $n = 0.05$ (in this paper we consider $k_0^{-1} = c/\omega_0$ as a unit of length). Cooperative effects play an important role in the ensembles of such density even in the case of absence of a cavity [27]. Comparing the curves in Fig. 2 we conclude that for small distance between the mirrors the decay dynamics in a cavity essentially differs from the case of absence of a cavity. However, as d increases the cooperative decay in a cavity gradually tends to the case of absence of a cavity.

Fig. 3 shows the spontaneous decay dynamics from Zeeman sublevel $m = 0$. We can conclude

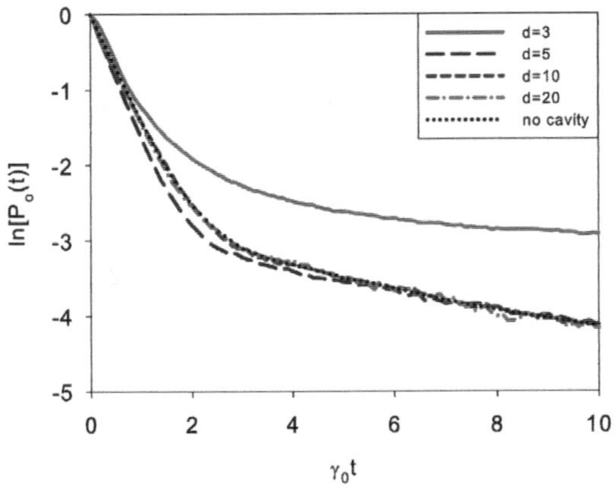

Figure 2. Spontaneous decay dynamics of an impurity atom excited at Zeeman sublevel $m = \pm 1$ and located in the center of a cylindrical atomic ensemble in a Fabry-Perot cavity. Radius of an ensemble is $R = 20$. The density of impurity atoms is $n = 0.05$. Inhomogeneous broadening $\delta \ll \gamma_0$.

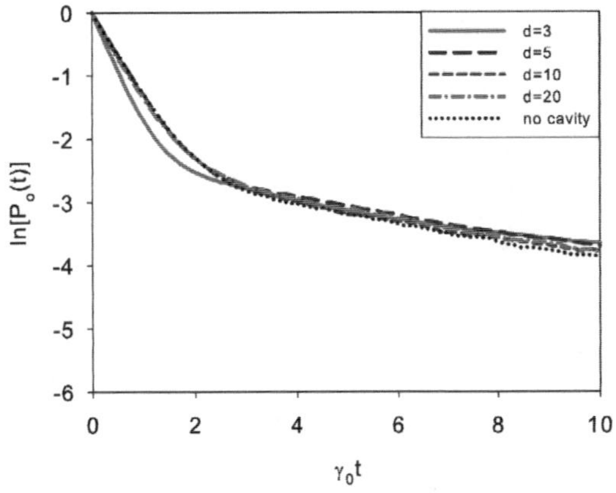

Figure 3. Spontaneous decay dynamics of an impurity atom excited at Zeeman sublevel $m = 0$. All other parameters are the same as for Fig. 2.

that for the parameters considered here the spontaneous decay in a cavity with $d > 10$ is almost the same as without a cavity. All results shown here are averaged over random spatial positions of impurity atoms. Very small fluctuations on the curves demonstrate the accuracy of calculations.

4. Conclusion

We have studied the cooperative spontaneous decay of an excited impurity atom embedded in a transparent dielectric and located in a Fabry-Perot cavity. It was performed on the basis of the quantum microscopic approach developed in Refs. [45] – [46]. In this paper the main attention was given to the investigation of the character of the cooperative effects depending on the distance d between the mirrors of a cavity. It was shown that in the case $d \leq \lambda_0$ cooperative spontaneous decay in a cavity essentially differs from one without a cavity and for $d \gg \lambda_0$ it is almost the same. In particular, it was demonstrated that for the atomic density $n = 0.05$ and inhomogeneous broadening $\delta \ll \gamma_0$ the discrepancy in the spontaneous decay in a cavity and without a cavity can be neglected if $d > 10$.

In our opinion, the approach used in the present paper allows us to investigate the influence of a cavity on the observability of Anderson localization of light in the ensembles of impurity centers. It can be done on the basis of the spectral analysis of the collective states in such ensembles [50] – [51].

The theory discussed here can be generalized to the case of the real refractive index of the dielectric. In addition, it can be further generalized to the atomic ensembles in the waveguide. The last attracts particular interest due to spontaneous decay suppression of all the Zeeman sublevels of the atomic excited state under the condition when the resonant frequency of atomic transition is less than the cut-off frequency of the waveguide.

Acknowledgments

We appreciate financial support by the Russian Foundation for Basic Research (Grants

No. RFBR-15-02-01013 and RFBR-15-02-08703), the Ministry of Education and Science of the Russian Federation (State Assignment 3.1446.2014K), Federal Special-Purpose Program "Research and development in priority areas of science and technology complex of Russia for 2014-2020 years" (contract No. 14.587.21.0211). A.S.K. also thanks State Assignment 2014/184, RFBR-16-32-00587 and the Council for Grants of the President of the Russian Federation.

References

[1] Purcell E M 1946 *Phys. Rev.* **69** 681
[2] Agarwal G S 1975 *Phys. Rev.* A **12** 1475
[3] Kleppner D 1981 *Phys. Rev. Lett.* **47** 233
[4] Barut A O and Dowling J P 1987 *Phys. Rev.* A **36** 649
[5] Dowling J P 1993 *Foundations of Physics* **23** 895
[6] Vetsch E, Reitz D, Sague G, Schmidt R, Dawkins S T and Rauschenbeute A 2010 *Phys. Rev. Lett.* **104** 203603
[7] Goban A, Choi K S, Alton D J, Ding D, Lacroute C, Pototschnig M, Thiele T, Stern N P and Kimble H J 2012 *Phys. Rev. Lett.* **109** 033603
[8] Kien F L, Gupta S D, Balykin V I and Hakuta K 2005 *Phys. Rev.* A **72** 032509
[9] Thompson J D, Tiecke T G, de Leon N P, Feist J, Akimov A V, Gullans M, Zibrov A S, Vuletic V and Lukin M D 2013 *Science* **340** 1202
[10] Yu S-P, Hood J D, Muniz J A, Martin M J, Norte R, Hung C-L, Meenehan S M, Cohen J D, Painter O and Kimble H J 2014 *Appl. Phys. Lett.* **104** 111103
[11] Goban A, Hung C-L, Yu S-P, Hood J D, Muniz J A, Lee J H, Martin M J, McClung A C, Choi K S, Chang D E, Painter O and Kimble H J 2014 *Nat. Commun.* **5** 3808
[12] Kalachev A 2008 *Phys. Rev.* A **78** 043812
[13] Kalachev A and Kocharovskaya O 2013 *Phys. Rev.* A **88** 033846
[14] Knoester J and Mukamel S 1989 *Phys. Rev.* A **40** 7065
[15] Glauber R J and Lewenstein M 1991 *Phys. Rev.* A **43** 467
[16] Milonni P W 1995 *J. Mod. Opt.* **42** 1991
[17] Barnett S M, Huttner B, Loudon R and Matloob R *J. Phys.* B **29** 3763
[18] Juzeliunas G 1997 *Phys. Rev.* A **55** R4015
[19] Crenshaw M E and Bowden C M 2000 *Phys. Rev. Lett.* **85** 1851
[20] Berman P R and Milonni P W 2004 *Phys. Rev. Lett.* **92** 053601
[21] Fu H and Berman P R 2005 *Phys. Rev.* A **72** 022104
[22] Pierrat R and Carminati R 2010 *Phys. Rev.* A **81** 063802
[23] Kuznetsov D V, Rerikh Vl K and Gladush M G 2011 *Journal of Experimental and Theoretical Physics* **113** 647
[24] Kuznetsov D V, Rerikh Vl K and Gladush M G 2011 *Theor. Math. Phys.* **168** 1078
[25] Anikushina T A, Gladush M G, Gorshelev A A and Naumov A V 2015 *EPJ Web of Conferences* **103** 05001
[26] Golovanova A V, Gubin M Yu, Gladush M G and Prokhorov A V 2016 *Bulletin of the Russian Academy of Sciences: Physics* **80** 808
[27] Kuraptsev A S and Sokolov I M 2014 *Phys. Rev.* A **90** 012511
[28] Fofanov Ya A, Kuraptsev A S, Sokolov I M and Havey M D 2011 *Phys. Rev.* A **84** 053811
[29] Fofanov Ya A, Kuraptsev A S and Sokolov I M 2012 *Opt. Spectrosc.* **112** 40
[30] Fofanov Ya A, Kuraptsev A S, Sokolov I M and Havey M D 2013 *Phys. Rev.* A **87** 063839
[31] Sokolov I M, Kupriyanov D V and Havey M D 2011 *Journal of Experimental and Theoretical Physics* **112** 246
[32] Bayram S B, Havey M D, Safronova M S and Sieradzan A 2006 *Journal of Physics* B **39** 2545
[33] Balik S, Win A L, Havey M D, Sokolov I M and Kupriyanov D V 2013 *Phys. Rev.* A **87** 053817
[34] Sokolov I M, Kuraptsev A S, Kupriyanov D V, Havey M D and Balik S 2013 *J. Mod. Opt.* **60** 50
[35] Pellegrino J, Bourgain R, Jennewein S, Sortais Y R P, Browaeys A, Jenkins S D and Ruostekoski J 2014 *Phys. Rev. Lett.* **113** 133602
[36] Kuraptsev A S and Sokolov I M 2015 *Phys. Rev.* A **91** 053822
[37] Agarwal G S and Gupta S D 1998 *Phys. Rev.* A **57** 667
[38] Kien Fam Le, Gupta S D, Nayak K P and Hakuta K 2005 *Phys. Rev.* A **72** 063815
[39] Douglas J S, Habibian H, Hung C-L, Gorshkov A V, Kimble H J and Chang D E 2015 *Nature Photonics* **9** 326
[40] Gonzalez-Tudela A, Hung C-L, Chang D E, Cirac J I and Kimble H J 2015 *Nature Photonics* **9** 320
[41] Goban A, Hung C-L, Hood J D, Yu S-P, Muniz J A, Painter O and Kimble H J 2015 *Phys. Rev. Lett.* **115** 063601

[42] Sokolov I M, Kupriyanov D V, Olave R G and Havey M D 2010 *J. Mod. Opt.* **57** 1833
[43] Heitler W 1954 *The Quantum Theory of Radiation* (London: Oxford University Press)
[44] Stephen M J 1964 *J. Chem. Phys.* **40** 669
[45] Kuraptsev A S and Sokolov I M 2016 *Journal of Experimental and Theoretical Physics* **123** 237
[46] Kuraptsev A S and Sokolov I M 2016 *Phys. Rev.* A **94** 022511
[47] Barantsev K A, Popov E N and Litvinov A N 2015 *Journal of Experimental and Theoretical Physics* **121** 758
[48] Barantsev K A and Litvinov A N 2014 *Journal of Experimental and Theoretical Physics* **118** 569
[49] Kazakov G A, Litvinov A N and Matisov B G 2012 *Quantum Electronics* **42** 185
[50] Skipetrov S E and Sokolov I M 2014 *Phys. Rev. Lett.* **112** 023905
[51] Skipetrov S E and Sokolov I M 2015 *Phys. Rev. Lett.* **114** 053902

IOP Conf. Series: Journal of Physics: Conf. Series **826** (2017) 012024 doi:10.1088/1742-6596/826/1/012024

Dynamic Entanglement Evolution of Multi-Qubits Systems

S nani ian a a yan M alban yan
l anian ational Science abo ato y l anian e evan menia

bst act e consi e fou atoms cou le to eac ot e by eisenbe g li e inte actions in an e te nal el of elect omagnetic a iation e e a e Sc o inge cat state using t e S an f-S tec ni ues

nt o uction

Quantum entanglement lays a ey ole in t e im o tant an a i ly evelo ing a eas of mo e n ysics li e uantum tele o tation uantum c y tog a y an uantum com uting

ne of t e most im o tant as ect in all t e above el s is an e ective e a ation of entangle uantum systems lot of o bot t eo etical an e e imental as evote to uantum entanglement e a ation an e ective cont ol of atomic co e ence n a ticula tec ni ues fo gene ating ola i ation-entangle oton ai s by means of a iative ecay of bie citons of uantum ots an a amet ic o n conve sion

n t e ot e an t e c eation of uantum-entangle systems in soli s is an essential tas an a la ge numbe of o s ave been evote to t e stu y of t e be aviou of uantum entanglement in soli s e en ing on t ei c a acte istics in a ticula in mo els fo systems of atoms nite o in nite connecte to eac ot e via eisenbe g inte actions

u o is base on ou evious esea c e e e consi e t e atoms cou le to eac ot e by eisenbe g li e inte actions in an e te nal el of elect omagnetic a iation Suc systems ave ysical eali ation an a ise in many systems suc as cou le semicon ucto uantum ots as ell as in t e bie citon system of a single semicon ucto uantum ot t at acts as a t o- ubit egiste su e con ucting ase an c a ge ubits an atoms ions t a e in a cavity ion t a it in t e is e sive limit ima ily e focus ou attention on t e c eation of uantum-entangle state of Sc o inge cat in t e system of fou linea ly cou le atoms

o t e Sc o inge cat state e a ation e use t e S an f-S tec ni ues Stimulate aman a iabatic assage S as been oven to be an e cient an obust tec ni ue fo t e com lete o ulation t ansfe f om one state to anot e state of atoms o molecules n t e ot e an f actional S a slig tly mo i e ve sion of S can be e loite to t ansfe a a t of o ulation f om one atomic state into anot e atomic state an simultaneously c eate an a bit a y co e ence bet een t em

n Sec e int o uce t e mo el of fou inte cou le ubits inte acting it inci ent lase el s n Sec e esent sc eme fo gene ating Sc o inge cat state by means of S an f-S tec ni ues e entanglement of t e system investigate in Sec e Mo el

o ou u ose e use fou t o-level atoms connecte by eisenbe g-li e cou lings ic a e esc ibe by

$$\lambda \quad S_i \; S \quad \frac{\omega}{\ } \quad S_i \, , \qquad i \qquad i$$

e e S_i i i i i S_i i i an S_i i i λ is t e st engt of inte ubit cou ling an ω is t e level s litting

e inte action it e te nal elect omagnetic el in i ole a o imation is esc ibe by

$$\vec{E}_i \quad \vec{d} \, , \qquad i$$

e e $\vec{E}$ $\vec{\varepsilon}$ t $e^{\ i\omega\ t}$ $\vec{\varepsilon}$ t $e^{i\omega\ t}$ a e t e fou incoming lase elect ic el intensities e also c oose all i ole moments e ual to e o

Sc o inge at State e a ation

n i ole a o imation t e allo e t ansitions fo m a linea connection bet een t e initial state an t e ig est state t oug t ee inte me iate states e co es on ing avefunctions an t e ene gies a e itten o n in t e table belo ic e enume ate as st t

able avefunctions an t e ene gies

ψ	E
, , ,	ω
, , , − , , , − , , , − , , ,	λ ω
, , , , , , , , , , , , , , , , , ,	λ
− , , , − , , , − , , , − , , ,	λ ω
, , ,	ω

e Sc o inge e uation in otating ave a o imation ta es t e follo ing fo m

$$i\frac{d}{dt} \quad a \quad e^{\ i\phi} \quad a \quad e^{i\phi} \quad e^{\ i\phi} \quad , \quad a \quad e^{i\phi} \quad a$$

e e

$$_i \; t \quad \varepsilon_i \; t \quad d \quad {}^i d$$

a e t e abi oscillations f e uencies a_i a e t e coe cients of t e avefunctions in t e t anslational inva iant basis ϕ_i a e t e ase s ifts of t e inci ent lase a iation an t e etuning coe cients a e c osen to be e ual to e o

e time se uence of t e t o initial aussian ulses is as follo ing t st e gene ate Stoc s ulse of - t ansition f e uency t an t en t e um ing one it - t ansition f e uency t suc t at t ey bot ecay at in nity it t e same ate counte intuitive

IOP Conf. Series: Journal of Physics: Conf. Series **826** (2017) 012024 doi:10.1088/1742-6596/826/1/012024

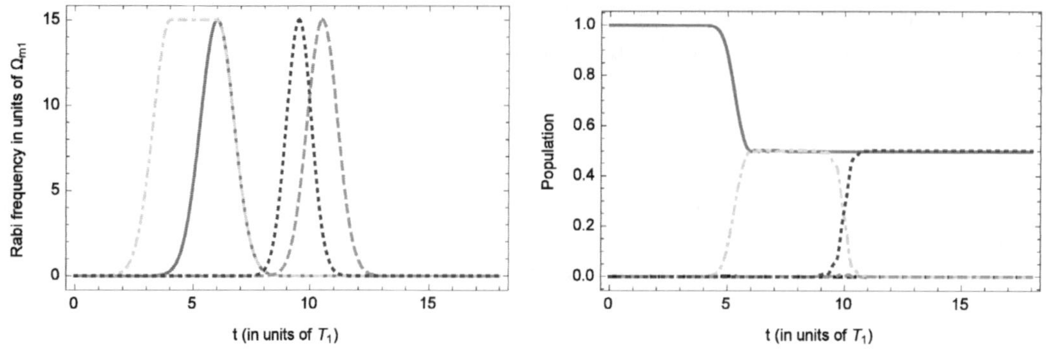

a b

igu e a ime se uence of t e ulses t soli e cu ve t ot- as e g een cu ve t as e o ange cu ve an t as e blue cu ve aime at c eating a fou - ubit state in a system of fou inte cou le ubits in t e st ong mutual cou ling egime e e T T T T T/T T/T . T/T . τ/T τ/T τ/T . an τ/T . b ime evolution of eigenstate o ulations soli e cu ve ψ as e blac cu ve ψ ot- as e g een cu ve ψ as e o ange cu ve ψ otte blue cu ve ψ

se uence f-S en e t ansfe o ulations f om to by means of o ina y se uence of aussian ulses t an t e e

$$_i\ t \quad {}_i e^{\ t\ \tau_i\ /T_i}, i \quad , \ ,$$

$$t \quad \begin{cases} e^{\ t\ \tau\ \ /T}, t < \tau \\ , \tau < t < \tau \\ e^{\ t\ \tau\ \ /T}, t > \tau \end{cases}$$

inally e obtain fou ubit Sc o inge cat state

$$\frac{\quad e^{\ i\phi}\quad}{-},$$

e e ϕ ϕ ... ϕ

e oga it mic egativity

o uantifying t e amount of entanglement u ing time evolution of t e system e ave use t e loga it mic negativity $Ne\ \rho$ ic is e ne as

$$Ne\ \rho \quad \log\ \ \rho^T \ ,$$

e e ρ^T is t e t ace no m of t e a tial t ans ose ρ^T of a bi a tite ensity mat i ρ

ote t at fo uantifying t e amount of entanglement u ing t is ocess by means of t e loga it mic negativity one as to e fo m a a tial t ace o e ation ove t e t o ubits o eve t e in e numbe of t e t ace out ubits can be c osen a bit a ily e e as t e system ossesses a t anslational symmet y

ig s o s t e time evolution of t e loga it mic negativity fo T T T T T/T T/T . T/T . τ/T τ/T τ/T . an τ/T .

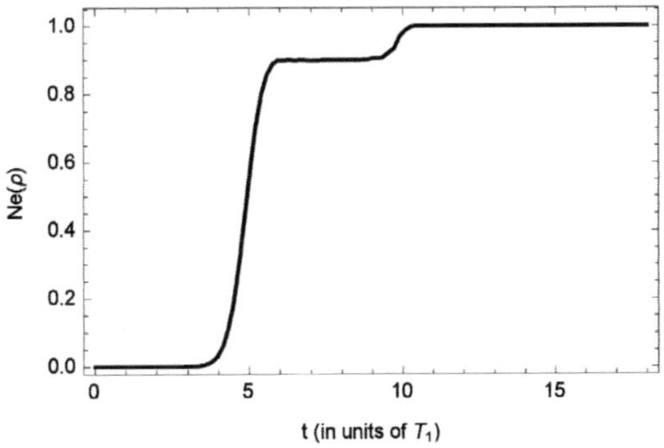

igu e ime evolution of t e loga it mic negativity Ne ρ

onclusion

e e a e Sc o inge cat state by means of t e S an f-S tec ni ues by using fou inte cou le atoms fo t is u ose an also foun time evolution of t e system loga it mic negativity u ing t is ocess

efe ences

ennett assa e eau o sa e es an ootte s ys ev ett
-S Ma e bst Sc ei l D ang S o atsc e aylo Mec ittmann o e E
nisimova Ma a ov enne ein sin an eilinge atu e on on
E ven outeau a amme an ei s t E ess
isin S i onio an Sangoua ys ev ett
o Mun o emoto al Do ling an Milbu n ev Mo ys
M Stevenson oung t inson oo e D itc ie an S iel s atu e on on
Salte M Stevenson a e icoll D itc ie an S iel s
ibi
M Mulle S ounoua D ons M l assl an Mic le at otonics
a ma c yan e oy nani ian an S ue in ys ev
iat E a s ite elbaum an Ebe a ys ev
ao - ang - en ang - Du ang an - an ys ev ett
u M Mi ailova ol ov an M e o ov ys ev
S ose ont ys
y at Meie D oss an D D sc alom ys ev
S oyle amsay ello iu M o inson M o an M S S olnic ys ev
enny esman o iano E oem D e s oni uong an M et o ys ev
aliaut inov ys ev
a is anting e ley o ansson M o nson uny E a i ins y a i ins y
an S an ys ev M o nson M S min an S il e t
atu e on on
S a aoanu ys ev
S - eng an - uo ys ev ys ev ett
Sac ett D iel ins i E ing et al atu e on on
a ma c yan ou nal of ontem o a y ysics
e gmann eue an S o e o e ent o ulation t ansfe among uantum states of atoms an molecules ev Mo ys -
Sauten ov - e ostovtsev elc an M Scully En ancement of el

gene ation via ma imal atomic co e ence e a e by fast a iabatic assage in b va o ys ev

Invariant Mass and Propagation Speed of Light Pulses in Vacuum

M V Fedorov[1], **S V Vintskevich**[2]

[1] A.M. Prokhorov General Physics Institute, Russian Academy of Sciences, Moscow, Russia
[2] Moscow Institute of Physics and Technology, Dolgoprudny, Moscow Region, Russia

E-mail: vintskevich@phystech.edu

E-mail: fedorovmv@gmail.com

Abstract. The concept of the Lorentz-invariant mass of groups of photons is shown to be applicable to classical light pulses with finite sizes and duration. Diffraction of light, providing non-collinearity of the photon motion in pulses, provides also nonzero values of their Lorentz-invariant masses. By considering light pulses as relativistic objects we find that owing to diffraction and finiteness of the Lorentz-invariant mass the mean propagation speed of pulses is always smaller than the light speed c.

1. Introduction

As known [1, 2, 3, 4], in the relativistic physics the mass m of any object is defined by the difference of its squared energy ε and momentum $\vec{p}$ (multiplied by the speed of light),

$$c^4 m^2 = \varepsilon^2 - c^2 \vec{p}^2. \tag{1}$$

Invariance of the mass with respect to Lorentz transformations is evident because the right-hand side of Eq. (1) can be presented as the scalar product of the object's four-momentum $p^{(4)} = \{\varepsilon/c, \vec{p}\}$ with itself, $c^2 m^2 = \left(p^{(4)} \cdot p^{(4)}\right)$, and scalar products are invariant with respect to rotations in the Minkowsky space. The same definition (1) is valid for combined systems of several non-interacting particles with the total energy and momentum given by sums of energies and momenta of particles, $\varepsilon = \sum_i \varepsilon_i$ and $\vec{p} = \sum_i \vec{p}_i$.

As discussed in details in Refs. [2, 3, 4], in the relativistic case the Lorentz-invariant is not related either to gravity or to inertia of objects. But the mass remains important for characterizing the object's propagation velocity

$$\vec{v} = c^2 \frac{\vec{p}}{\varepsilon}. \tag{2}$$

In the case of a nonzero mass, $m \neq 0$, Eq. (1) gives $c|\vec{p}| < \varepsilon$ and, hence, the object's velocity v is less than the light speed c, $v < c$. In this case one can identify the object's rest-frame in which the object does not move, and this is just the frame moving with the speed v (2) with respect to the laboratory frame. In the rest-frame the object's energy equals its invariant the mass times the squared speed of light, $\varepsilon_{r.f.} = mc^2$.

Oppositely, if $m = 0$, then in Eq.(1) gives $c|\vec{p}| = \varepsilon$ and, in accordance with Eq. (2), $v = c$.The massless objects move with the speed of light and they cannot be stopped, i.e. for such objects rest-frames do not exist.

Special and rather interesting cases are those of a single photon and of groups of photons. In the case of a single photon with a frequency ω and a wave vector $\vec{k}$ its energy and momentum are $\varepsilon = \hbar\omega$ and $\vec{p} = \hbar\vec{k}$. As in vacuum $k = \omega/c$, we get immediately $\varepsilon = c|\vec{p}|$ and $m = 0$, i.e., as well known, single photons are massless particles. The same is true for a group of collinearly moving photons with wave vectors $\vec{k}_i = \vec{n}\omega_i/c$, where $\vec{n}$ is the unit vector in the direction of propagation. In this case, again, $c|\vec{p}| = \hbar\sum_i \omega_i = \varepsilon$ and $m = 0$. This means that an arbitrary group of collinearly moving photons represents an object with a zero mass. But in the case of non-collinearly propagating photons the situation changes. In this case $\sum_i \omega_i$ can be different from c$|\sum_i \vec{k}_i|$ and, hence the invariant mass of a group of non-collinearly moving photons can have a non-zero invariant mass. Such system as whole propagates with a velocity v (2) smaller than the light speed c. The coordinate frame moving just with this velocity is the rest-frame for a group of non-collinearly moving photons. In the rest-frame the mean propagation velocity of the group of photon turns zero, $v_{r.f} = 0$, i.e., the translatory motion of a group of photons appears to be eliminated. The simplest group of photons with a nonzero invariant mass is that of two photons with equal frequencies ω and with the angle 2θ between directions of their wave vectors. In this case the invariant mass of the two-photon system is given by

$$mc^2 = 2\hbar\omega\,\sin\theta \approx 2\hbar\omega\,\theta. \tag{3}$$

All these results and related problems are discussed in details in the famous works by L.B. Okun [2, 3, 4].

In this work we consider classical light pulses of finite sizes. As such pulses consist of finite (though usually very large) numbers of photons, all the above-presented discussion must be applicable to these classical objects. Moreover non-collinearity of photon motion in light pulses is inherently present owing to diffraction. Thus, we expect, and this will be shown explicitly below, that classical light pulses can be characterized by the concept of a nonzero Lorentz-invariant mass and that their mean propagation speed is smaller than the light speed c.

2. Energy and momentum of a light pulse

The total energy and momentum of a light pulse can be defined as the well-known energy and momentum densities integrated over the whole space where the electric and magnetic components of the field are different from zero, $\vec{E}(\vec{r},t) \neq 0$ and $\vec{H}(\vec{r},t) \neq 0$,

$$\varepsilon = \int \frac{\vec{E}^{\,2}(\vec{r},t) + \vec{H}^{\,2}(\vec{r},t)}{8\pi} d\vec{r}, \qquad \vec{p} = \int \frac{[\vec{E}(\vec{r},t) \times \vec{H}(\vec{r},t)]}{4\pi c} d\vec{r}. \tag{4}$$

Note that the local $\vec{r}$-dependent energy and momentum densities of the field are components of the field tensor rather than components of any four-vector. But being integrated over $\vec{r}$ they turn into total energy and momentum of a pulse. As a light pulse as a whole can be considered as a kind of a relativistic object, its total energy and momentum form a four-momentum vector of a pulse, and they can be used for finding the Lorentz-invariant mass via Eq. (1).

It's convenient to present electric and magnetic field strengths in the form of expansions in plane waves obeying Maxwell equations [1]

$$\vec{E}(\vec{r},t) = \frac{1}{(2\pi)^{3/2}} \int d\vec{k} \left[\vec{E}_{\vec{k}}\, e^{i(\vec{k}\vec{r}-\omega_k t)} + \vec{E}^{\,*}_{-\vec{k}}\, e^{-i(\vec{k}\vec{r}-\omega_k t)}\right], \tag{5}$$

$$\vec{H}(\vec{r},t) = \frac{1}{(2\pi)^{3/2}} \int d\vec{k} \left[\vec{H}_{\vec{k}}\, e^{i(\vec{k}\vec{r}-\omega_k t)} + \vec{H}^{\,*}_{-\vec{k}}\, e^{-i(\vec{k}\vec{r}-\omega_k t)}\right], \tag{6}$$

where $\omega_k = c|\vec{k}|$, $\vec{H}_{\vec{k}} \perp \vec{E}_{\vec{k}}$, $\left|\vec{H}_{\vec{k}}\right| = \left|\vec{E}_{\vec{k}}\right| \equiv E_{\vec{k}}$ and $\vec{H}_{\vec{k}} \times \vec{E}_{\vec{k}} \| \vec{k}$. The expansion coefficients, for example $\vec{E}_{\vec{k}}$, are determined by the inverse Fourier transformations such as

$$\vec{E}_{\vec{k}}\, e^{-i\omega_k t} + \vec{E}^{\,*}_{-\vec{k}}\, e^{i\omega_k t} = \frac{1}{(2\pi)^{3/2}} \int d\vec{r}\, \vec{E}(\vec{r}, t) e^{-i\vec{k}\vec{r}}. \tag{7}$$

Substitution of expansions (5) and (6) into Eqs. (4), reduces the light-pulse energy and momentum to the form of integrals from the squared absolute values of the Fourier transforms of the electric field strength

$$\varepsilon = \frac{1}{2\pi} \int d\vec{k}\, E^2_{\vec{k}}, \qquad \vec{p} = \frac{1}{2\pi c} \int d\vec{k}\, E^2_{\vec{k}} \frac{\vec{k}}{|\vec{k}|}. \tag{8}$$

With these expressions, the general definition of the invariant mass (1) takes the form

$$m^2 = \frac{1}{(2\pi c^2)^2} \left[\left(\int d\vec{k} E^2_{\vec{k}} \right)^2 - \left(\int d\vec{k} E^2_{\vec{k}} \frac{\vec{k}}{|\vec{k}|} \right)^2 \right] =$$

$$= \frac{1}{(2\pi c^2)^2} \left[\int d\vec{k} E^2_{\vec{k}} + \left| \int d\vec{k} E^2_{\vec{k}} \frac{\vec{k}}{|\vec{k}|} \right| \right] \left[\int d\vec{k} E^2_{\vec{k}} - \left| \int d\vec{k} E^2_{\vec{k}} \frac{\vec{k}}{|\vec{k}|} \right| \right]. \tag{9}$$

3. Solution of the boundary problem with Gaussian boundary conditions

Equations of the previous section are general, and the results (8) and (9) are valid for any field distributions in a pulse. But for expressing the mass in terms of the pulse parameters, let us specify the field distribution by assuming that the pulse has a given Gaussian expression at the boundary $z = 0$ and, then, it propagates along the z-axis (in its positive direction) in accordance with the Maxwell equations. Thus, let the boundary condition for this problem be given by

$$E(\vec{r}, t)|_{z=0} = E_0 \exp\left\{ -\frac{\vec{r}^{\,2}_{\perp}}{2w^2} - \frac{t^2}{2\tau^2} \right\} \sin(\omega_0 t), \tag{10}$$

where $\vec{r}_{\perp} = \{x, y\}$ is the transverse part of radius-vector (in the (x, y)-plane, perpendicular to the z-axis), w is the pulse waist in the plane $z = 0$, and ω_0 and τ are the central frequency and duration of the pulse. In Eq. (10) and below we do not show vectorial structure of the field because, as it follows from the general expressions of Eqs. (8) and (9), the final results do not depend of the field polarization. The right-hand side of Eq. (10) can be presented in the form of Fourier expansions, both in transverse coordinates and time:

$$E(\vec{r}, t)|_{z=0} = -iE_0 \frac{w^2 \tau}{2(2\pi)^{3/2}} \int d\vec{k}_{\perp} d\omega \exp\left\{ -\frac{\vec{k}^{\,2}_{\perp} w^2}{2} - \frac{(\omega - \omega_0)^2 \tau^2}{2} \right\} e^{i\vec{k}_{\perp}\vec{r}_{\perp}} \left(e^{i\omega t} - e^{-i\omega t} \right). \tag{11}$$

For getting from this boundary condition the solution of the Maxwell equations at $z > 0$, it's sufficient to substitute the products ωt by $\omega t - k_z z$ with $k_z = \sqrt{\frac{\omega^2}{c^2} - \vec{k}^2_{\perp}}$ in the last bracket on the right-hand side of Eq. (11). The last step of these transformations consists in the replacement of the integration over ω by integration over k_z with ω substituted by $\omega_{\vec{k}} = c\sqrt{k_z^2 + \vec{k}^2_{\perp}} \equiv c|\vec{k}|$. The final result is given by

$$E(\vec{r}, t) = -iE_0 \frac{w^2 \tau}{2(2\pi)^{3/2}} \int d\vec{k} \frac{c^2 k_z}{\omega_{\vec{k}}} \exp\left\{ -\frac{\vec{k}^{\,2}_{\perp} w^2}{2} - \frac{(\omega_{\vec{k}} - \omega_0)^2 \tau^2}{2} \right\} \left(e^{i(\omega t - \vec{k}\vec{r})} - e^{-i(\omega t - \vec{k}\vec{r})} \right). \tag{12}$$

By comparing this expansion with the general one (5), we find the absolute value of the expansion coefficient $E_{\vec{k}}$ determining the invariant mass (9) for the light-pulse configuration which we consider here

$$E_{\vec{k}} = \frac{E_0 w^2 c\tau}{2} \frac{ck_z}{\omega_{\vec{k}}} \exp\left\{-\frac{\vec{k}_\perp^{\,2} w^2}{2} - \frac{(\omega_{\vec{k}} - \omega_0)^2 \tau^2}{2}\right\}. \tag{13}$$

4. Lorentz-invariant mass

Let us assume now that the length and transverse size of the pulse exceed significantly its wavelength and $k_z \gg k_\perp$ which corresponds to the paraxial approximation

$$w, c\tau \gg \lambda = \frac{2\pi c}{\omega_0},\; e^{-(\omega-\omega_o)^2\tau^2} \approx e^{-(k_z-\omega_0/c)^2(c\tau)^2},\; 1 - \frac{ck_z}{\omega_{\vec{k}}} \approx \frac{\vec{k}_\perp^2}{2(\omega_0/c)^2} \ll 1$$

Under these conditions the expression for the mass squared (9) takes the form:

$$m^2 \approx \frac{E_0^4 w^8 \tau^4}{(8\pi)^2}\, 2\int d\vec{k} \exp\left\{-\vec{k}_\perp^{\,2} w^2 - \left(k_z - \frac{\omega_0}{c}\right)^2 (c\tau)^2\right\} \times$$
$$\times \int d\vec{k} \frac{\vec{k}_\perp^{\,2}}{2(\omega_0/c)^2} \exp\left\{-\vec{k}_\perp^{\,2} w^2 - \left(k_z - \frac{\omega_0}{c}\right)^2 (c\tau)^2\right\}. \tag{14}$$

Gaussian integrals in this equation are easily calculated to give

$$m = \frac{\sqrt{\pi} E_0^2 w \tau}{8\omega_0} = \frac{E_0^2 w \tau \lambda}{16\sqrt{\pi} c}. \tag{15}$$

Calculated in a similar way, the total energy in the pulse (8) is given by

$$\varepsilon = \int d\vec{k} \frac{E_{\vec{k}}^2}{2\pi} = \frac{E_0^2 w^4 c^2 \tau^2}{8\pi} \int d\vec{k} \exp\left\{-\vec{k}_\perp^{\,2} w^2 - \left(k_z - \frac{\omega_0}{c}\right)^2 (c\tau)^2\right\} = \frac{\sqrt{\pi} E_0^2 w^2 c\tau}{8}. \tag{16}$$

By comparing Eqs. (15) and (16) we can establish relation between the invariant mass and energy of the pulse in the laboratory frame:

$$mc^2 = \frac{\lambda}{2\pi w} \varepsilon \ll \varepsilon. \tag{17}$$

This result reminds very closely that occurring in the case of two non-collinearly propagating photons [Eq. (3)]. The coincidence is almost complete if we take into account that $\lambda/2\pi w$ in Eq. (17) is the diffraction-divergence angle of a light beam. This comparison shows clearly that the physical reason for a nonzero invariant mass characterizing light pulses is in their diffraction divergence, or spreading.

5. The mean propagation speed of diverging laser pulses

As mentioned in the Introduction, in relativistic physics, the velocity v of any object is determined be the well-known equation (2), which can be rewritten as

$$v = c\sqrt{1 - \frac{m^2 c^4}{\varepsilon^2}}. \tag{18}$$

IOP Conf. Series: Journal of Physics: Conf. Series **826** (2017) 012025 doi:10.1088/1742-6596/826/1/012025

For a light pulse with the invariant mass of Eqs. (15), (17), its mean propagation velocity appears to be given by

$$v \approx c\left(1 - \frac{m^2c^4}{2\varepsilon^2}\right) = c\left(1 - \frac{\lambda^2}{8\pi^2 w^2}\right) < c. \tag{19}$$

It should be emphasized once again that v (19) is the mean velocity of the pulse, averaged over its field distribution or over distribution of velocities of all photons in the pulse. Each photon itself has a velocity exactly equal c. But directions of velocities of different photons can be different from each other. For this reason the averaged velocity is directed along the symmetry axis $0z$ and its absolute value is less than c. Thus, all finite-size diverging (spreading) light pulses propagate, as a whole, slower than with the light speed c.

The obtained conclusion means that, as well as for any objects moving with velocities smaller than c, for a light pulse one can identify its rest fame. The pulse's rest frame is the coordinate frame moving just with velocity v of Eq. (19) with respect to the laboratory frame. In the rest frame the translatory motion of a pulse becomes eliminated, i.e. its total momentum turns zero, $\vec{p}_{r.f.} = 0$, and the only remaining motion is spreading. In the rest frame the total energy of the pulse coincides with its invariant mass multiplied by c^2, $\varepsilon_{r.f.} = mc^2$. The distribution diagram of photon's wave vectors in the pulse in its rest frame is shown in Fig. 1. This diagrams is rather peculiar and it shows, in particular, that because of Lorentz transformations, in the rest frame the field of the pulse becomes strongly non-monochromatic.

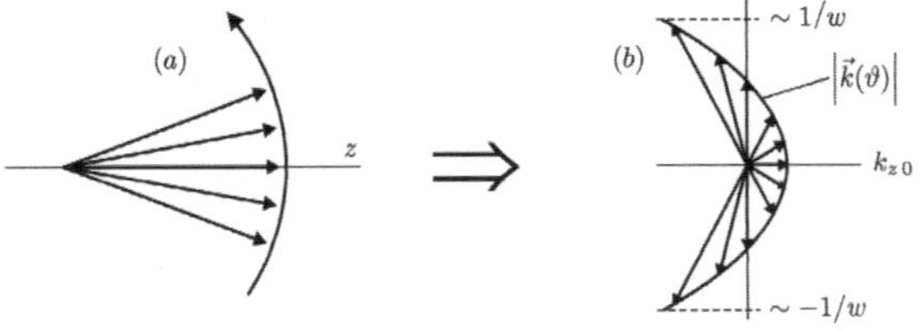

Figure 1. Distribution of photon wave vectors of a pulse (a) in laboratory and (b) rest frames.

6. Experiment

The difference between the pulse-propagation velocity v and the light speed c is relatively small. But its direct experimental measurement is not impossible. Some schemes adequate for such measurements will be discussed elsewhere. Here, instead, we note that the slowing down effect can be increased with the help of artificial structuring of the light pulse with the help of lenses. A scheme of such experiment is shown in Fig. 2. The laser beam is supposed to be split by a beam splitter (BS) for two identical parts propagating in the upper and lower channels. The pulse in the upper channel experiences no additional changes whereas the the pulse in the lower channel is assumed to be focused and, then, defocused again in a system of two confocal lenses. This system of lenses provides lengthening of the pulse pathway compared to propagation in a free space. In the case of narrow lenses the pathway lengthening is known to be identical for all rays forming a beam: if $s(x)$ is a pathway of a ray falling on the first lens in Fig. 2 at a distance x from the horizontal axis, then $s(x)$ is independent of x, $s(x) \equiv s$ [5]. It's known also that equality of pathways for all rays coming to a lens is related to existence of two reasons for pathway lengthening. The first of them is propagation of light in the body of a lens with the refractive index $n > 1$ and the related reduction of the propagation speed $c \to c/n$. The second reason of the pathway lengthening consists in changing the propagation direction of rays after

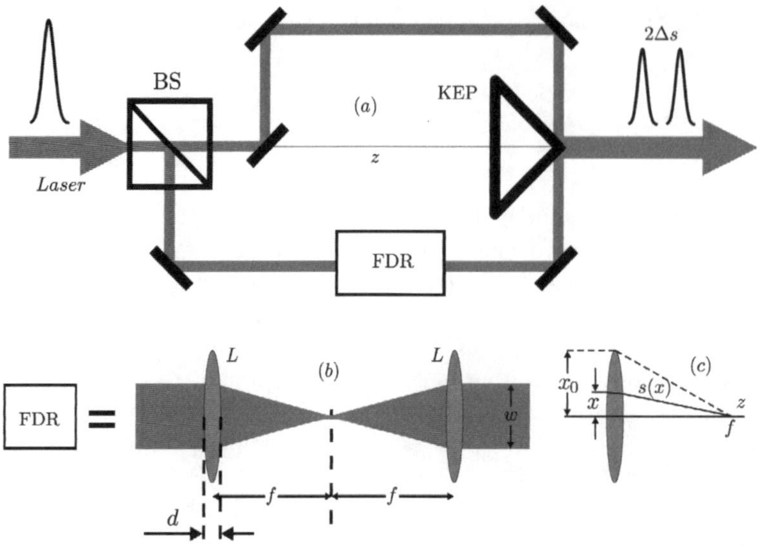

Figure 2. A suggested scheme for observation of the light-pulse slowing down owing to its focusing-defocusing. (a) BS and KEP are the beam splitter and the knife edge prism, δs is the distance between the slowed down and non-slowed pulses, FDR is the focusing-defocusing region, z is the central propagation axis;(b)light pulse in the focusing-defocusing region, L and L are confocal lenses, f is the focus distance of each of them, w is the pulse waist outside of the focusing-defocusing region; (c) pathways $s(x)$ of a ray falling on the lens at a distance x from the horizontal axis, the same for all x, $s(x) \equiv s = const.$

the lens. The sum of these two contributions makes the pathway $s(x)$ independent of x and gives the following expressions for the lengthening of the effective pathways at the focus compared to the lens-free pathway f

$$\Delta s = s(x) - f = (n-1)d(x) + \frac{x^2}{2f} \equiv (n-1)d(0) \equiv \frac{x_0^2}{2f}, \tag{20}$$

where $d(x)$ is the lens thickness at a hight x from the propagation axis, $d(0) \equiv d$ is its on-axis thickness, x_0 is the lens height (radius), and f is focal distance. In the parabolic approximation for the lens surfaces all these parameters are related to each other by equations

$$d(x) = d\left(1 - \frac{x^2}{x_0^2}\right), \quad f = \frac{x_0^2}{2d(n-1)}, \tag{21}$$

with the last formula following directly from Eqs. (27.9) and (27.11) of Ref. [5]. Substituted into Eq. (20) these formulas confirm that the pathway lengthening Δs is identically the same for all x, $\Delta s(x) \equiv const. \equiv \Delta s$. In other words, pathways to the focus for all rays of the beam, propagating at any distances x from the propagation axis, are identical. As said above, the term $(n-1)d(x)$ in the first line of Eq. (20) results from propagation of light in the lens body and the term $x^2/2f$ can be interpreted as arising owing to restructuring of the beam. Both of these two contributions to the pathway lengthening are equally important. The term related to propagation in the lens body dominates for parts of the beam close to the propagation axis, whereas the term resulting from the beam restructuring prevails for its peripheral parts. Summed together these two mechanisms provide equal lengthening of pathways for all parts of the beam, central and peripheral, and thus provide undistorted propagation of pulses through the focus, with the pulse shape unchanged.

If the pulse waist w is close to the hight of the lens, $w \approx x_0$ the final expression in the second line of Eq. (20) takes the form $\triangle s \approx w^2/2f$. At the double focal distances of two lenses in Fig. 2 the degree of lengthening of the effective pathways is doubled. This distance $2 \triangle s$ is just the distance, which will be seen at the exit of the setup of Fig. 2 between the pulse coming from the upper channel and the delayed pulse from the lower channel. As the pulse-restructuring effect in the focusing-defocusing region is one of two intrinsic parts of the described delay, observation of two well separated pulses in the suggested scheme can be considered as a direct demonstration that restructuring can give rise to at least 50% of the slowing down effect arising owing to lenses.

The expressions for the pathway lengthening (20) give immediately the following expression for the mean propagation speed of a pulse in the focusing-defocusing region:

$$v = c\left(1 - \frac{w^2}{2f^2}\right). \tag{22}$$

This result is similar to that of Eq. (19) with the replacement $\lambda/2\pi w \to w/f$, which indicates a deep analogy between diffraction and the focusing/defocusing restructuring. The slowing down effect occurring in the case of focusing-defocusing is much stronger than in the case of diffraction if $w^2 \gg f\lambda/2\pi$, or $f \ll L_D$, i.e., if the focal length f is much shorter than the diffraction length $L_D = 2\pi w^2/\lambda$. This condition is fulfilled by wide margins, for example, at $w \sim 1cm$ and $\lambda \sim 1\mu m$. As the propagation velocity and invariant mass of pulses are related to each other as shown in Eq. (19), the described decrease of the propagation velocity in the focusing-defocusing region is accompanied by increase of the invariant mass m which becomes equal to

$$m_{fdr} = \frac{\varepsilon}{c^2}\frac{w}{f}. \tag{23}$$

7. Conclusion

Thus, our analysis shows that any finite-size diverging or converging light pulses propagate in vacuum with a mean velocity smaller than the light speed c. This is true both for natural divergence (spreading) occurring owing to diffraction and for convergence and divergence imposed by lenses. The well-known relations between velocity, energy and momentum of relativistic objects are used for finding fundamental connection between Lorentz-invariant mass of light pulses and their mean velocity of propagation in vacuum. For pulses moving on average with a speed v smaller than c one can identify their rest-frame (moving just with the same velocity v with respect to the laboratory frame). In the rest-frame of a pulse its translatory motion vanishes and the only remaining motion is spreading. Energy of spreading in the rest-frame equals the pulse's Lorentz invariant mass multiplied by c^2, which elucidates the role of the Lorentz-invariant mass of light pulses as one of their important characteristics. Explicit expressions are found for both invariant the mass and mean propagation velocity for Gaussian pulses. It's shown also that the effects of slowing down light pulses and increasing their invariant mass can be strengthened significantly by focusing or/and defocusing by lenses. Based on this, a scheme of experimental set-up with two confocal lenses is proposed for measuring directly the pulse-propagation mean velocity and its deviation from the speed of light.

References

[1] Landau L D and Lifshitz E M 1980 *The Classical Theory of Fields* 4th edn (Oxford:ButterworthHeinemann)
[2] Okun L B 1989 *Sov. Phys.Usp.* 32 629-638
[3] Okun L B 2000 *Sov. Phys.Usp.* 43 1270
[4] Okun L B *Energy and Mass in Relativity Theory* (Singapore: World Scientific)
[5] Feynman R P, Leighton R B, Sands M 1964 *The Feynman Lectures on Physics* (Addison Weslei) Volume I, Chapter 27

Calculation of mutual information for nonlinear optical fiber communication channel at large SNR within path-integral formalism

A V Reznichenko[1,2], I S Terekhov[1,2], S K Turitsyn[2,3]

[1] Theory department, Budker Institute of Nuclear Physics of Siberian Branch Russian Academy of Sciences, Novosibirsk 630090, Russia
[2] Novosibirsk State University, Novosibirsk 630090, Russia
[3] Aston Institute of Photonics Technologies, Aston University, Aston Triangle, Birmingham B4 7ET, UK

E-mail: a.v.reznichenko@inp.nsk.su

Abstract. Using the path-integral technique we calculate the mutual information for the fiber optical channel modelled by the nonlinear Schrödinger equation with additive Gaussian noise. At large signal-to-noise ratio (SNR) we present the mutual information through the path-integral which is convenient for the perturbative expansion both in nonlinearity and dispersion. In the leading order in 1/SNR we demonstrate that the mutual information is determined through the averaged logarithm of the normalization factor Λ of the conditional probability density function $P[Y|X]$. In the limit of small noise and small nonlinearity we derive analytically the first nonzero nonlinear correction to the mutual information for the channel. For the arbitrary nonlinearity we restrict the mutual information by the low bound obtained from the Jensen's inequality and analyze the bound for the case of large dispersion.

1. Introduction

We consider the fiber optical channel described by the nonlinear Schrödinger equation (NLSE) with additive white Gaussian noise when the signal-to-noise ratio (SNR) is large:

$$\partial_z\psi + i\beta\partial_t^2\psi - i\gamma|\psi|^2\psi = \eta(z,t)\,, \tag{1}$$

where $\psi(z,t)$ is the outcoming signal, γ is the Kerr nonlinerity, $\beta = \beta_2/2$ is the dispersion parameter, $\eta(z,t)$ is the white Gaussian noise: $\langle\eta(z,t)\bar{\eta}(z',t')\rangle_\eta = Q\delta(z-z')\delta(t-t')$, where Q is a power of the white Gaussian noise (per unit length and frequency). The bar here and hereafter means complex conjugation.

Our goal is to calculate the mutual information [1] of the channel for given $P[X]$ — probability density function (PDF) of the input signal $X(t)$: $\psi(z=0,t) = X(t)$. In our approach we use the representation of the mutual information in the form of path-integral

$$I_{P[X]} = \int \mathcal{D}X\mathcal{D}Y\, P[X]P[Y|X]\log\left[\frac{P[Y|X]}{P_{out}[Y]}\right] = H[Y] - H[Y|X], \tag{2}$$

where $P[X]$ is the PDF of the input signal X with the fixed finite average power P_{ave}. The function $P[Y|X]$ here is the conditional probability density function, that is the probability density of receiving output signal $\psi(z = L, t) = Y(t)$ when the input signal is $\psi(z = 0, t) = X(t)$. The output signal PDF reads $P_{out}[Y] = \int \mathcal{D}X P[X]P[Y|X]$. The entropy $H[Y] = -\int \mathcal{D}Y P_{out}[Y] \log\left[P_{out}[Y]\right]$ is the output signal entropy (responsible for the signal transmission), and the entropy $H[Y|X] = -\int \mathcal{D}X\mathcal{D}Y P[X]P[Y|X] \log\left[P[Y|X]\right]$ is the conditional entropy responsible for the noise impact.

In our approach we use the formulation for the conditional PDF $P[Y|X]$ through the path-integral (Martin-Siggia-Rose formalism [2]). In Ref.[3] the following ("quasiclassical") representation for the conditional probability density function was obtained:

$$P[Y|X] = \Lambda \exp\left\{-\frac{S[\Psi]}{Q}\right\}, \qquad \Lambda = \int\limits_{\phi(z=0)=0}^{\phi(z=L)=0} \mathcal{D}\phi \exp\left\{-\frac{1}{Q}\left(S[\Psi+\phi] - S[\Psi]\right)\right\}, \tag{3}$$

where the action $S[\psi]$ in (3) reads in the time domain

$$S[\psi] = \int\limits_0^L dz \int\limits_T dt\, |\mathcal{L}[\psi(z,t)]|^2\,, \quad \mathcal{L}[\psi] = \partial_z\psi(z,t) + i\beta\partial_t^2\psi(z,t) - i\gamma\psi(z,t)|\psi(z,t)|^2. \tag{4}$$

Here T is a time interval containing both signals $X(t)$ and $Y(t)$. In what follows we use the discretization scheme in the time and in the frequency domain and the discrete Fourier transform. There are some relations between T and discretization intervals in the time domain (δ_t for the dense time grid with M' intervals and $\tilde{\delta}_t$ for the coarse time sub-grid with M intervals) and in the frequency domain (δ_ω): $T = 1/\delta_\omega = M'\delta_t = M\tilde{\delta}_t = 2\pi M/W = 2\pi M'/W'$. Here W is the bandwidth of the input signal, and W' is auxilary bandwidth: $W' \supset W$. In the frequency discretization scheme we should perform the following substitutions in our equations: $\partial_t^2\Phi(z,t) \to -\Omega_k^2\Phi_{\omega_k}(z)$, $\delta\left(\omega_{i_1} + \omega_{i_2} + \ldots\right) \to \frac{\Delta_{(M')}(i_1+i_2+\ldots)}{2\pi\delta_\omega}$, $\int\frac{d\omega}{2\pi}\ldots \to \delta_\omega\sum_{j'=0}^{M'-1}$, where $\omega_{n'} = -W'/2 + 2\pi\delta_\omega n'$, $\Omega_{n'} = 2\sin[\pi n'/M']M'/T = 2M'\delta_\omega\sin[\pi n'/M']$, $n' = 0, 1, \ldots, M'-1$, and we use the discrete analog of the Dirac delta-function $\Delta_{(M)}(k) = \frac{1}{M}\sum_{n=0}^{M-1}\exp\left\{-2\pi i\frac{n}{M}k\right\} = \sum_{m=-\infty}^{\infty}\delta_{k,mM}$.

The function Ψ in Eq. (3) is the solution (referred to as "the classical solution") of the Euler-Lagrange equation $\delta S[\Psi] = 0$ with the boundary conditions: $\Psi(0) = X$, $\Psi(L) = Y$. In the time domain this equation for $\Psi(z,t)$ has a notedly simple form [4]

$$\begin{aligned}&\left(\partial_z + i\beta\partial_t^2 - 2i\gamma|\Psi(z,t)|^2\right)\mathcal{L}[\Psi(z,t)] + i\gamma\Psi^2(z,t)\overline{\mathcal{L}[\Psi(z,t)]} = 0,\\ &\mathcal{L}[\Psi(z,t)] = \left(\partial_z + i\beta\partial_t^2 - i\gamma|\Psi(z,t)|^2\right)\Psi(z,t),\end{aligned} \tag{5}$$

and the function $\Psi(z,t)$ obeys the boundary conditions: $\Psi(0,t) = X(t)$, $\Psi(L,t) = Y(t)$.

It is convenient to introduce the function $\Phi(z,t)$ which is the solution of the nonlinear Schrödinger equation (NLSE) with zero noise, i.e., $\mathcal{L}[\Phi(z,t)] = 0$, and with the boundary condition $\Phi(0,t) = X(t)$. It is obvious that the function $\Phi(z,t)$ obeys the equation Eq. (5) and the boundary condition at $z = 0$, but it does not obey the boundary condition at $z = L$. It globally minimizes the action as well: $S[\Phi(z,t)] = 0$. Since we imply that the noise power is much less than the signal power we can present the solution of Eq. (5) in the form

$$\Psi(z,t) = \Phi(z,t) + \varkappa(z,t)\,, \tag{6}$$

where the function $\varkappa(z,t)$ is of order of $\sqrt{Q}$ for unsuppressed configurations $\Psi(z,t)$, since in the leading order in $1/\mathrm{SNR}$ the action is quadratic functional in $\varkappa(z,t)$, see details in [4]. Therefore we substitute the function Ψ in the form (6) to the Eq. (5), then linearizing Eq. (5) in $\varkappa(z,t)$ we obtain the following linear problem on $\varkappa(z,t)$:

$$\begin{aligned}&\left(\partial_z + i\beta\partial_t^2 - 2i\gamma|\Phi(z,t)|^2\right) l[\varkappa(z,t)] + i\gamma\Phi^2(z,t)\overline{l[\varkappa(z,t)]} = 0,\\ &l[\varkappa(z,t)] = \left(\partial_z + i\beta\partial_t^2\right)\varkappa(z,t) - i\gamma\left(2\varkappa(z,t)|\Phi(z,t)|^2 + \overline{\varkappa(z,t)}\Phi^2(z,t)\right),\end{aligned} \tag{7}$$

with the boundary conditions $\varkappa(z=0,t)=0$, $\varkappa(z=L,t)=Y(t)-\Phi(L,t)\equiv\delta Y(t)$.

In some sense, the introduction of the function $\varkappa(z,t)\sim\sqrt{Q}$ is the crucial idea of our consideration in the case of large SNR, since it allows us to reduce the difficult nonlinear problem of $P[Y|X]$ calculation to the linear one.

2. Factorization of pre-exponential factor Λ

Let us consider the separation of different scales in the conditional PDF $P[Y|X]$ when PDF $P[Y|X]$ is considered under an integral over X together with the input signal PDF $P[X]$ which has the following form in the frequency domain: $P[\vec{X}] = P_X^{(M)}[\vec{X}_1]\delta\left(\vec{X}_2\right)$, where $\delta\left(\vec{X}_2\right)$ means $2(M'-M)$-dimensional delta-function corresponding $M'-M$ complex remnant channels. The vector notations ($\vec{X}=\vec{X}_1\oplus\vec{X}_2$) relate to the frequency domain: $\vec{X}_1$ is $2M$-dimensional vector corresponding to M meaning complex channels in the frequency domain W, whereas $\vec{X}_2$ is $2(M'-M)$-dimensional vector corresponding to remnant $M'-M$ complex channels in the domain $W'\setminus W$. In the leading order in 1/SNR the conditional PDF $P[Y|X]$ reads

$$P[Y|X] \approx \Lambda[X]\, e^{-S_2[\varkappa]/Q}\,, \tag{8}$$

where the action $S_2[\varkappa] = \delta_t\sum_{k=0}^{M'-1}\Delta\sum_{n=1}^{N-1}\mathcal{L}_{eff}[\varkappa(z_n,t_k)]$ is quadratic functional in ϕ, where $\delta_t = T/M'$ is the discretization parameter in the time domain: $t_k = k\delta_t$, $k=0,1,\ldots,M'-1$. Here $\Delta = L/N$ is the distance discretization parameter: $z_n = n\Delta$, $n=1,2,\ldots,N-1$. We have introduced the "Lagrangian":

$$\mathcal{L}_{eff}[\varkappa] = \left|\partial_z\varkappa(z_n,t_k) + i\beta\partial_t^2\varkappa(z_n,t_k) - i\gamma\left(2\varkappa(z_n,t_k)|\Phi(z_n,t_k)|^2 + \bar{\varkappa}(z_n,t_k)\Phi^2(z_n,t_k)\right)\right|^2 . \tag{9}$$

Here derivatives should be regarded as difference derivatives in our discretization scheme. The function $\varkappa(z,t)$ in the exponent (8) is the solution of the Euler-Lagrange equation $-\delta\mathcal{L}_{eff}[\varkappa]/\delta\bar{\varkappa}=0$ with the boundary conditions $\varkappa(z=0,t)=0$, $\varkappa(z=L,t)=Y(t)-\Phi(L,t)$. The normalization factor $\Lambda[X]$ has the form

$$\Lambda[X] = \int\limits_{\phi(0,t)=0}^{\phi(L,t)=0}\mathcal{D}\phi\,\exp\left\{-\frac{\delta_t}{Q}\sum_{k=0}^{M'-1}\Delta\sum_{n=1}^{N-1}\mathcal{L}_{eff}[\phi(z_n,t_k)]\right\}. \tag{10}$$

Note that the sum in expression for $S_2[\varkappa]$ is performed over the dense time grid. To demonstrate the factorization we have to separate the scales in the action into the coarse and dense parts. In other words, we have to separate the summation over M meaning channels and $M'-M$ remnant channels. The scale separation procedure in some sense is similar to Wilson's renormalization procedure for the Lagrangian $\mathcal{L}_{eff}[\varkappa]$, see [5]. But in our approximation the Lagrangian (9) is quadratic functional in $\varkappa$ that is why there are no corrections to the effective action when we perform integration over remnant $2(M'-M)$ degrees of freedom $\varkappa(z,t_k)$ where t_k runs through values only on the dense grid without the coarse sub-grid. Let us demonstrate this fact.

First we perform the separation of variables:

$$\varkappa(z,t_k) = \varkappa^{(c)}(z,t_k) + \varkappa^{(d)}(z,t_k), \\ Y(t_k) = Y^{(c)}(t_k) + Y^{(d)}(t_k), \tag{11}$$

where $\varkappa^{(c)}(z,t_k)$, or $Y^{(c)}(t_k)$, is completely defined only by the values $\varkappa^{(c)}(z,T_{\tilde{i}})$, or $Y^{(c)}(T_{\tilde{i}})$, on the coarse time grid $T_{\tilde{i}} = \tilde{i}\tilde{\delta}_t$, $\tilde{i} = 0,1,\ldots,M-1$. Here and below superscript "(c)" means coarse variable. In other words, the function $\varkappa^{(c)}(z,t_k)$ evaluated at all grid points is the interpolation of some order (i.e., the interpolating polynomial degree) $N_0 > 2$ calculated on the base of values $\varkappa^{(c)}(z,T_{\tilde{i}})$ of the coarse time grid. The function $\varkappa^{(c)}(z,t_k)$ coincides with $\varkappa(z,T_{\tilde{i}})$ when t_k falls on the coarse time grid $T_{\tilde{i}}$ (i.e., $k = [\tilde{i}M'/M]$, $\tilde{i} = 0,1,\ldots,M-1$), i.e., $\varkappa^{(d)}(z,t_k) = 0$ on the coarse grid. In other grid points of the dense grid the function $\varkappa^{(c)}(z,t_k)$ smoothly interpolates the values of $\varkappa(z,t_k)$ with interpolation order $N_0 > 2$: $\varkappa^{(d)} = \mathcal{O}(\tilde{\delta}_t^{N_0})$ and $\partial_t^2\varkappa^{(d)}(z,t_k) = \mathcal{O}(\tilde{\delta}_t^{N_0-2})$, where we have used that $\partial_t^2\varkappa(z,t_k) = \partial_t^2\varkappa^{(c)}(z,t_k) + \partial_t^2\varkappa^{(d)}(z,t_k)$, here the derivatives are assumed as the difference derivatives on the dense grid. The boundary conditions are as follows:

$$\varkappa^{(c)}(0,t_k) = 0, \qquad \varkappa^{(c)}(L,t_k) = Y^{(c)}(t_k) - \Phi^{(c)}(L,t_k), \\ \varkappa^{(d)}(0,t_k) = 0, \qquad \varkappa^{(d)}(L,t_k) = Y^{(d)}(t_k) - \Phi^{(d)}(L,t_k), \tag{12}$$

where we have used that $\Phi(L,t_k)$ has the coarse and dense parts as well:

$$\Phi(z,t_k) = \Phi^{(c)}(z,t_k) + \Phi^{(d)}(z,t_k), \quad \Phi^{(d)}(z,t_k) = \mathcal{O}(\tilde{\delta}_t^{N_0}). \tag{13}$$

Note that if we consider (8) under the integral over $\mathcal{D}X$ with the input signal PDF, then the function $\Phi(z,t_k)$ is the (nonlinear) function of the input signal only on the coarse time grid $X(T_{\tilde{i}})$. This means that the dense part $\Phi^{(d)}(z,t_k) = \mathcal{O}(\tilde{\delta}_t^{N_0})$ is always small for sufficiently large M. Now we insert the representation (11) in the action S_2. The action fractionizes into three parts

$$S_2[\varkappa] = \delta_t \sum_{k=0}^{M'-1} \Delta \sum_{n=1}^{N-1} \mathcal{L}_{eff}[\varkappa(z_n,t_k)] = \delta_t \sum_{k=0}^{M'-1} \Delta \sum_{n=1}^{N-1} \mathcal{L}_{eff}[\varkappa^{(c)}(z_n,t_k)] + \\ \delta_t \sum_{k=0}^{M'-1} \Delta \sum_{n=1}^{N-1} \mathcal{L}_{eff}[\varkappa^{(d)}(z_n,t_k)] + \delta_t \sum_{k=0}^{M'-1} \Delta \sum_{n=1}^{N-1} \mathcal{L}_{int}[\varkappa^{(c)}(z_n,t_k), \varkappa^{(d)}(z_n,t_k)], \tag{14}$$

where the third part with interaction of coarse ($\varkappa^{(c)}$) and dense ($\varkappa^{(d)}$) degrees of freedom contains Lagrangian

$$\mathcal{L}_{int}[\varkappa^{(c)}(z,t_k), \varkappa^{(d)}(z,t_k)] = \\ \left(\partial_z\varkappa^{(c)}(z,t_k) + i\beta\partial_t^2\varkappa^{(c)}(z,t_k) - i\gamma\left(2\varkappa^{(c)}(z,t_k)|\Phi(z,t_k)|^2 + \bar{\varkappa}^{(c)}(z,t_k)\Phi^2(z,t_k)\right)\right) \times \\ \left(\partial_z\bar{\varkappa}^{(d)}(z,t_k) - i\beta\partial_t^2\bar{\varkappa}^{(d)}(z,t_k) + i\gamma\left(2\bar{\varkappa}^{(d)}(z,t_k)|\Phi(z,t_k)|^2 + \varkappa^{(d)}(z,t_k)\bar{\Phi}^2(z,t_k)\right)\right) + c.c. \tag{15}$$

Here "$c.c.$" means the same complex conjugated term.

The first part in the r.h.s. of Eq. (14) can be simplified as follows:

$$\delta_t \sum_{k=0}^{M'-1} \Delta \sum_{n=1}^{N-1} \mathcal{L}_{eff}[\varkappa^{(c)}(z_n,t_k)] = \tilde{\delta}_t \sum_{\tilde{i}=0}^{M-1} \Delta \sum_{n=1}^{N-1} \mathcal{L}_{eff}[\phi(z_n,T_{\tilde{i}})]\left(1 + \mathcal{O}(\tilde{\delta}_t^{N_0-2})\right), \tag{16}$$

where $\tilde{\delta}_t = \frac{2\pi}{W} = \delta_t \left\lceil M'/M \right\rceil$ is the grid spacing of the coarse time grid. Here we have replaced every term under the sum over the dense grid with its average value on the coarse grid. The accuracy in Eq. (16) is governed by the interpolation order of the second derivative in time.

The third part in the r.h.s. of Eq. (14) can be integrated (summed) over z by part resulting in the following expression:

$$\delta_t \sum_{k=0}^{M'-1} \Delta \sum_{n=1}^{N-1} \mathcal{L}_{int}[\varkappa^{(c)}(z_n,t_k), \varkappa^{(d)}(z_n,t_k)] = \delta_t \sum_{k=0}^{M'-1} \Delta \sum_{n=1}^{N-1} \left(\bar{\varkappa}^{(d)}(z_n,t_k) \frac{\delta \mathcal{L}_{eff}[\varkappa^{(c)}]}{\delta \varkappa} + c.c. \right) + S_{surf}, \tag{17}$$

where the variation $-\delta\mathcal{L}_{eff}[\varkappa^{(c)}]/\delta\varkappa$ is linear in $\varkappa^{(c)}$, and it represents the l.h.s. of the Euler-Lagrange equation, i.e., for the function $\varkappa^{(c)}$ we obtain $\delta\mathcal{L}_{eff}[\varkappa^{(c)}]/\delta\varkappa = \mathcal{O}(\tilde{\delta}_t^{N_0-2})$, i.e., it is always small. The term S_{surf} results from the surface term in integration by part over z in Eq. (17), and taking into account the boundary conditions (12) it reads

$$S_{surf} = \delta_t \sum_{k=0}^{M'-1} \left[\bar{Y}^{(d)}(t_k) - \bar{\Phi}^{(d)}(L,t_k) \right] \left(\partial_z \varkappa^{(c)}(L,t_k) + i\beta\partial_t^2 \varkappa^{(c)}(L,t_k) - i\gamma \left(2\varkappa^{(c)}(L,t_k)|\Phi(L,t_k)|^2 + \bar{\varkappa}^{(c)}(L,t_k)\Phi^2(L,t_k) \right) \right) + c.c. \tag{18}$$

We can omit the surface term (18), since it is linear both in the coarse and dense variables, but they are orthogonal when integrating over t (they have not intersecting supports in the frequency domain). It is obvious for the first two terms in the parentheses in Eq. (18). The last terms containing $\Phi(L,t_k)$ are coarse variables as well: we can replace $\Phi(L,t_k)$ with $\Phi^{(c)}(L,t_k)$ with the interpolation accuracy $\mathcal{O}(\tilde{\delta}_t^{N_0})$ and then replace $\Phi^{(c)}(L,t_k)$ with $Y^{(c)}(t_k)$ with the accuracy $\mathcal{O}(\sqrt{Q})$ (we remind that $\varkappa$ in Eq. (12) is of order of $\sqrt{Q}$). Then we can replace $Y^{(c)}(t_k)$ with the constants inside the whole interval of an coarse space with the interpolation accuracy $\mathcal{O}(\tilde{\delta}_t)$ and now use the orthogonality of the coarse and dense variables.

To summarize, with the accuracy of our interpolation $\mathcal{O}(\tilde{\delta}_t) = \mathcal{O}(1/M)$ we can omit the interaction term (17), and our action fractionizes into coarse and dense parts: $S_2[\varkappa] \approx S_2[\varkappa^{(c)}] + S_2[\varkappa^{(d)}]$. Both actions are expressed through the same Lagrangian (9) and are quadratic forms. The coefficients of these quadratic forms depend on input signal X only. The factorization of Λ can be shown using the normalization condition: $1 = \int \mathcal{D}Y P[Y|X] = \Lambda \int \mathcal{D}Y \exp\{-S_2[\varkappa]/Q\}$, where we have used that Λ does not depend on Y in the leading order in 1/SNR. Finally, one has

$$\Lambda^{-1} = \int \mathcal{D}Y e^{-\frac{S_2[\varkappa^{(c)}]}{Q} - \frac{S_2[\varkappa^{(d)}]}{Q}} = \int \mathcal{D}Y^{(c)} e^{-\frac{S_2[\varkappa^{(c)}]}{Q}} \mathcal{D}Y^{(d)} e^{-\frac{S_2[\varkappa^{(d)}]}{Q}} = \Lambda_1^{-1}\Lambda_2^{-1}, \tag{19}$$

or $\Lambda = \Lambda_1 \times \Lambda_2$. Here the normalization factor Λ_1 depends on the input signal X on the coarse grid only and it reads

$$\Lambda_1 = \int_{\phi(0,t)=0}^{\phi(L,t)=0} \left[\mathcal{D}\phi(z,t) \right]_M \exp\left\{ -\frac{\tilde{\delta}_t}{Q} \sum_{\tilde{i}=0}^{M-1} \Delta \sum_{n=1}^{N-1} \mathcal{L}_{eff}[\phi(z_n, T_{\tilde{i}})] \right\} \tag{20}$$

with the Lagrangian $\mathcal{L}_{eff}[\phi(z_n, T_{\tilde{i}})]$, see (9), considered on the coarse time grid $T_{\tilde{i}}$ only.

The measure $\left[\mathcal{D}\phi(z,t) \right]_M$ on the coarse time grid is defined as

$$\left[\mathcal{D}\phi(z,t) \right]_M = \lim_{\tilde{\delta}_t \to 0} \lim_{\Delta \to 0} \left(\frac{\tilde{\delta}_t}{\Delta\pi Q} \right)^M \prod_{\tilde{j}=0}^{M-1} \prod_{i=1}^{N-1} \left\{ \frac{\tilde{\delta}_t}{\Delta\pi Q} dRe\phi(z_i, T_{\tilde{j}})\, dIm\phi(z_i, T_{\tilde{j}}) \right\}. \tag{21}$$

The normalization factor Λ_1 corresponds to M meaning complex channels, and Λ_2 corresponds to $M' - M$ complex remnant channels. In this demonstration we have used that the quantity $P[Y|X]$ is considered under the integral over $\mathcal{D}X$ with the input signal PDF $P[X]$. The accuracy of our factorization is at least $\mathcal{O}(\tilde{\delta}_t) = \mathcal{O}(1/M)$.

3. Perturbative expansion vs Jensen's inequality for mutual information

The factorization of Λ guarantees, see details in [4], that the mutual information has the following form in the leading order in 1/SNR and for arbitrary nonlinearity:

$$I_{P[X]} = H[X] - M + \int d\vec{X}_1 P_X^{(M)}[\vec{X}_1] \log \Lambda_1[\vec{X}_1], \tag{22}$$

where $\vec{X}_1$ is $2M$ dimensional vector corresponding M complex meaning channels. We can calculate $\Lambda_1[\vec{X}_1]$ within the perturbation theory in dimensionless nonlinearity parameter $\tilde{\gamma} = \gamma LPW/(2\pi)$ and perform the averaging over Gaussian input signal PDF $P_X^{(M)}$: $P_X^{(M)}[\vec{X}_1] = P_G[\vec{X}_1] = \Lambda_P \, e^{-|\vec{X}_1|^2\delta_\omega/P}$, where $\Lambda_P = (\delta_\omega/(\pi P))^M$. We will denote the last term in (22) as $\int d\vec{X}_1 P_X^{(M)}[\vec{X}_1] \log \Lambda_1[\vec{X}_1] \equiv \langle \log \Lambda_1 \rangle_X$. For the averaging over Gaussian input signal PDF we use the Wick's theorem [5] and the following correlator

$$\langle X(\omega_k)\overline{X}(\omega_{k'})\rangle_X = P\delta_{k,k'}/\delta_\omega, \qquad k,\, k' = 0, 1, \ldots, M-1. \tag{23}$$

From the representation (20) we obtain the expression for $\Lambda_1[\vec{X}_1]$ in the frequency domain

$$\Lambda_1[\vec{X}_1] = \int\limits_{\phi(0,\omega)=0}^{\phi(L,\omega)=0} \Big[\mathcal{D}\phi(z,\omega)\Big]_M \exp\left\{ -\frac{\delta_\omega}{Q} \sum_{k=0}^{M-1} \Delta \sum_{n=1}^{N-1} \mathcal{L}_{eff}[\phi(z_n,\omega_k)] \right\}, \tag{24}$$

where the measure reads

$$\Big[\mathcal{D}\phi(z,\omega)\Big]_M = \lim_{\delta_\omega \to 0} \lim_{\Delta \to 0} \left(\frac{\delta_\omega}{\Delta \pi Q}\right)^M \prod_{n=1}^{N-1} \prod_{k=0}^{M-1} \left\{ \frac{\delta_\omega}{\Delta \pi Q} \, dRe\phi(z_n,\omega_k) \, dIm\phi(z_n,\omega_k) \right\}. \tag{25}$$

Now we present the "Lagrangian" $\mathcal{L}_{eff}$ as a sum: $\mathcal{L}_{eff} = \mathcal{L}_{eff}^{(0)} + \mathcal{L}_{eff}^{(1)} + \mathcal{L}_{eff}^{(2)}$, where the first term reads $\mathcal{L}_{eff}^{(0)}[\phi(z_n,\omega_k)] = \left|(\partial_z - i\beta\bar{\Omega}_k^2)\phi(z_n,\omega_k)\right|^2$. Here we have introduced $\bar{\Omega}_k = 2\sin[\pi k/M] M/T' = 2M\delta_\omega \sin[\pi k/M] = W \sin[\pi k/M]/\pi$. In the continuous limit $M \to \infty$ we can assume that $\bar{\Omega}_k \approx 2\pi\delta_\omega k = \omega_k - \omega_0$. The second and the third terms are the terms of $\mathcal{L}_{eff}$, see Eq. (9) in the frequency representation, explicitly proportional to $\tilde{\gamma}$ and $\tilde{\gamma}^2$, correspondingly.

Perturbative calculation. We present the perturbation expansion in $\tilde{\gamma}$ of the normalization factor Λ_1 in the form: $\Lambda_1 = \gamma^0\Lambda_1^{(0)} + \gamma\Lambda_1^{(1)} + \gamma^2\Lambda_1^{(2)} + \mathcal{O}(\tilde{\gamma}^3)$. Thus the last term in the expression (22) for the mutual information has the following expansion in γ: $\langle \log \Lambda_1 \rangle_X = \langle \log \Lambda_1^{(0)} \rangle_X + \gamma \left\langle \frac{\Lambda_1^{(1)}}{\Lambda_1^{(0)}} \right\rangle_X + \gamma^2 \left(\left\langle \frac{\Lambda_1^{(2)}}{\Lambda_1^{(0)}} \right\rangle_X - \frac{1}{2} \left\langle \left[\frac{\Lambda_1^{(1)}}{\Lambda_1^{(0)}} \right]^2 \right\rangle_X \right) + \mathcal{O}(\tilde{\gamma}^3)$. Retaining only the first term $\mathcal{L}_{eff}^{(0)}$ in Eq. (24) we have:

$$\Lambda_1^{(0)} = \int\limits_{\phi(0,\omega)=0}^{\phi(L,\omega)=0} \Big[\mathcal{D}\phi(z,\omega)\Big]_M \exp\left\{ -\frac{\delta_\omega}{Q} \sum_{k=0}^{M-1} \Delta \sum_{n=1}^{N-1} \left|(\partial_z - i\beta\bar{\Omega}_k^2)\phi(z_n,\omega_k)\right|^2 \right\} = \left(\frac{\delta_\omega}{\pi Q L}\right)^M. \tag{26}$$

This expression for $\log \Lambda_1^{(0)}$ results in the main (Shannon's) contribution $M \log [P/(QL)]$ to the mutual information (22) by taking into account that $H[X] = M - M \log [\delta_\omega/(\pi P)]$.

We introduce the averaging $\langle \dots \rangle_\phi$ over fields $\phi(z,\omega)$ defined as

$$\langle \dots \rangle_\phi = \frac{1}{\Lambda_1^{(0)}} \int_{\phi(0,\omega)=0}^{\phi(L,\omega)=0} \left[\mathcal{D}\phi(z,\omega)\right]_M \dots \exp\left\{ -\frac{\delta_\omega}{Q} \sum_{k=0}^{M-1} \Delta \sum_{n=1}^{N-1} \left|(\partial_z - i\beta\bar{\Omega}_k^2)\phi(z_n,\omega_k)\right|^2 \right\}. \quad (27)$$

The paired correlator has the form:

$$\langle \phi(z,\omega_k)\bar{\phi}(z',\omega_{k'})\rangle_\phi = -\frac{Q}{\delta_\omega}\delta_{k,k'} G(z,z') \exp\left[i\beta\bar{\Omega}_k^2(z-z')\right], \quad (28)$$

where the Green function reads $G(z,z') = z\frac{z'-L}{L}\theta(z'-z) + z'\frac{z-L}{L}\theta(z-z')$. In what follows for brevity sake we will write the sum over z as the integral. Let us stress that in Eq. (27) and hereinafter the derivative with respect to z is assumed in the "causative" manner, $\partial_z\phi(z_n,\omega_k) = (\phi(z_{n+1},\omega_k) - \phi(z_n,\omega_k))/\Delta$, as provided by our approach [3].

In the notations (27) we can present the mutual information (22) in the form

$$I_{P[X]} = M \log\left[\frac{P}{QL}\right] + \left\langle \log\langle e^{-S_{nl}[\phi(z,\omega)]/Q}\rangle_\phi \right\rangle_X, \; S_{nl} = \int_0^L dz\delta_\omega \sum_{k=0}^{M-1} \left(\mathcal{L}_{eff}^{(1)} + \mathcal{L}_{eff}^{(2)}\right). \quad (29)$$

It is easy to see by the direct calculation that there are no corrections to the mutual information (22) of order of $\tilde{\gamma}$: $\gamma\left\langle \frac{\Lambda_1^{(1)}}{\Lambda_1^{(0)}} \right\rangle_X = 0$. The second order calculation is cumbersome but straightforward. One can find the details of this perturbative calculations in Ref. [4]. Finally we obtain the following expression for the mutual information

$$I_{P_G[X]} = M \log \mathrm{SNR} - M\frac{\tilde{\gamma}^2}{3} g(\tilde{\beta}) + \mathcal{O}(\tilde{\gamma}^3), \quad (30)$$

where M is the number of complex meaning channels. This number is implied to be large $M \gg 1$. The parameter $\tilde{\beta} = \beta L W^2$ is the dimensionless dispersion parameter. The function $g(\tilde{\beta})$ in discretization scheme can be presented as the triple sum:

$$g(\tilde{\beta}) = \frac{1}{M^3} \sum_{k_1,k_2,k_3=0}^{M-1} F\left(\frac{\tilde{\beta}}{2}\left[\bar{\Omega}_{k_1}^2 + \bar{\Omega}_{k_2}^2 - \bar{\Omega}_{k_3}^2 - \bar{\Omega}_{k_1+k_2-k_3}^2\right]\right). \quad (31)$$

The elementary function $F(\mu)$ in Eq. (31) is the result of integration of the derivatives of the dimensionless Green function, $G_0(\zeta_1,\zeta_2) = \zeta_1(\zeta_2-1)\theta(\zeta_2-\zeta_1) + \zeta_2(\zeta_1-1)\theta(\zeta_1-\zeta_2)$:

$$F(\mu) = -12 \int_0^1 d\zeta_1 \int_0^1 d\zeta_2 \frac{\partial G_0(\zeta_1,\zeta_2)}{\partial \zeta_1} \frac{\partial G_0(\zeta_1,\zeta_2)}{\partial \zeta_2} e^{-2i\mu(\zeta_1-\zeta_2)} = 3\frac{\mu^2 - \sin^2(\mu)}{\mu^4}, \quad (32)$$

and for convenience it is normalized as $F(0) = 1$. In the continuous limit of sufficiently large M

we can present $g(\widetilde{\beta})$ as the integral, as the series, and as the explicit hypergeometric expression:

$$g(\widetilde{\beta}) = \int_0^1 dx_1 \int_0^1 dx_2 \int_0^1 dx_3 F\left(\frac{\widetilde{\beta}}{4}(x_1 - x_3)(x_2 - x_3)\right) =$$
$$4! \sum_{n=0}^{\infty} \frac{(-1)^n \widetilde{\beta}^{2n} \left[(4n+2)! + (1+2n)!^2\right]}{2^{2n-1}(2n+4)!(4n+3)!(1+2n)^2} =$$
$$\frac{1}{33075\widetilde{\beta}^4}\Bigg\{ -3528000\widetilde{\beta}^2\, {}_2F_3\left(1,1;\frac{1}{4},\frac{3}{4},2;-\frac{\widetilde{\beta}^2}{256}\right) + 140\widetilde{\beta}^6\, {}_2F_3\left(\frac{3}{2},2;\frac{9}{4},\frac{11}{4},\frac{7}{2};-\frac{\widetilde{\beta}^2}{256}\right) -$$
$$576\widetilde{\beta}^6\, {}_2F_3\left(\frac{7}{4},2;\frac{9}{4},\frac{11}{4},\frac{11}{4};-\frac{\widetilde{\beta}^2}{256}\right) + 210\widetilde{\beta}^6\, {}_2F_3\left(2,\frac{5}{2};\frac{9}{4},\frac{11}{4},\frac{7}{2};-\frac{\widetilde{\beta}^2}{256}\right) -$$
$$14\widetilde{\beta}^6\, {}_3F_4\left(\frac{3}{2},\frac{3}{2},2;\frac{9}{4},\frac{5}{2},\frac{11}{4},\frac{7}{2};-\frac{\widetilde{\beta}^2}{256}\right) - 588000\widetilde{\beta}^4\, {}_1F_2\left(\frac{1}{2};\frac{3}{2},\frac{5}{2};-\frac{\widetilde{\beta}^2}{16}\right) +$$
$$564480\widetilde{\beta}^4\, {}_1F_2\left(\frac{3}{4};\frac{7}{4},\frac{5}{2};-\frac{\widetilde{\beta}^2}{16}\right) - 270480\widetilde{\beta}^4\, {}_1F_2\left(1;\frac{5}{4},\frac{7}{4};-\frac{\widetilde{\beta}^2}{256}\right) +$$
$$176400\widetilde{\beta}^4\, {}_2F_3\left(\frac{1}{2},\frac{1}{2};\frac{3}{2},\frac{3}{2},\frac{5}{2};-\frac{\widetilde{\beta}^2}{16}\right) - 294000\widetilde{\beta}^4\, {}_2F_3\left(\frac{1}{2},1;\frac{5}{4},\frac{7}{4},\frac{5}{2};-\frac{\widetilde{\beta}^2}{256}\right) +$$
$$564480\widetilde{\beta}^4\, {}_2F_3\left(\frac{3}{4},1;\frac{5}{4},\frac{7}{4},\frac{7}{4};-\frac{\widetilde{\beta}^2}{256}\right) + 33075\widetilde{\beta}^4\, {}_2F_3\left(1,1;\frac{5}{4},\frac{7}{4},3;-\frac{\widetilde{\beta}^2}{256}\right) -$$
$$147000\widetilde{\beta}^4\, {}_2F_3\left(1,\frac{3}{2};\frac{5}{4},\frac{7}{4},\frac{5}{2};-\frac{\widetilde{\beta}^2}{256}\right) + 88200\widetilde{\beta}^4\, {}_3F_4\left(\frac{1}{2},\frac{1}{2},1;\frac{5}{4},\frac{3}{2},\frac{7}{4},\frac{5}{2};-\frac{\widetilde{\beta}^2}{256}\right) +$$
$$352800\widetilde{\beta}^2 + 6209280\widetilde{\beta}\sin\left(\frac{\widetilde{\beta}}{2}\right) - 564480\cos\left(\frac{\widetilde{\beta}}{2}\right) + 564480\Bigg\}, \tag{33}$$

where ${}_pF_q$ is generalized hypergeometric function.

For the large $\widetilde{\beta}$ we can deal with the asymptotics of the function $g(\widetilde{\beta})$ at large $\widetilde{\beta}$:

$$g(\widetilde{\beta}) = \frac{16\pi}{\widetilde{\beta}}\left(\log[\widetilde{\beta}] - \log[2] + \gamma_E - \frac{23}{6}\right) + \mathcal{O}\left(\frac{1}{\widetilde{\beta}^{3/2}}\right), \qquad \widetilde{\beta} \to \infty. \tag{34}$$

The function $g(\widetilde{\beta})$, see Eq. (33), and the asymptotics (34) for large $\widetilde{\beta}$ are presented in Fig. 1.

For $\widetilde{\beta} = 0$ we have $g(\widetilde{\beta} = 0) = 1$ and we arrive at the result [6] for the nondispersive channel in expansion in $\widetilde{\gamma}$:

$$I_{P[X]}^{(\beta=0)} = M\log\mathrm{SNR} - \frac{M}{2}\int_0^{\infty} d\tau e^{-\tau}\log\left(1 + \frac{\tau^2\widetilde{\gamma}^2}{3}\right) = M\log\left[\mathrm{SNR}\right] - M\frac{\widetilde{\gamma}^2}{3} + \mathcal{O}(\widetilde{\gamma}^4). \tag{35}$$

Let us estimate the mutual information $I_{P[X]}$ for typical fiber optical links [7]: $\beta = 20\,\mathrm{ps}^2/\mathrm{km}$, $L = 1000\,\mathrm{km}$, $\gamma = 1.31(\mathrm{Wkm})^{-1}$, $W = 100\,\mathrm{GHz}$, $P_{noise} = QLW/(2\pi) = 5.3\times 10^{-4}\mathrm{mW}$. For these parameters one has $\widetilde{\beta} = \beta LW^2 \approx 200$, and $g(\widetilde{\beta}) \approx 0.42$. From Eq. (30) we obtain

$$I_{P[X]} \approx M\left\{\log\left[\mathrm{SNR}\right] - 7\times 10^{-8}\times \mathrm{SNR}^2\right\}. \tag{36}$$

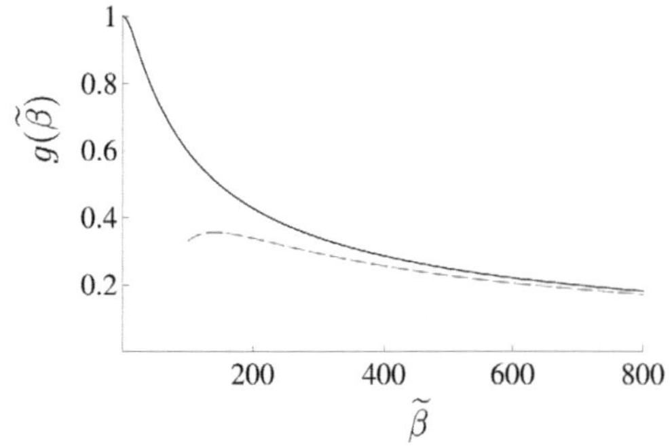

Figure 1. The function $g(\widetilde{\beta})$ (the solid black line), see Eq. (33), and the asymptotics of $g(\widetilde{\beta})$ for large $\widetilde{\beta}$ (red dashed line), see Eq. (34).

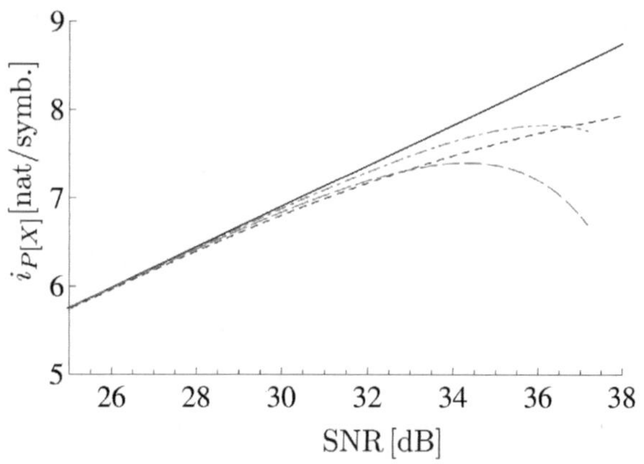

Figure 2. The spectral efficiency $i_{P[X]} = I_{P[X]}/M$ corresponding to Eq. (30) for different $\widetilde{\beta}$: Shannon's result (the solid black line), $\widetilde{\beta} = 200$ (red long-dashed line), $\widetilde{\beta} = 0$ (blue dashed line), $\widetilde{\beta} = 800$ (red chain line).

For these parameters we present the spectral efficiency $i_{P[X]} = I_{P[X]}/M$ for different dispersion parameters $\widetilde{\beta}$ in Fig. 2. Increasing parameter $\widetilde{\beta}$ the first nonlinear correction, see Eq. (30), and the asymptotics, see Eq. (34), goes to zero as $\tilde{\gamma}^2 \log(\widetilde{\beta})/\widetilde{\beta}$. Therefore for larger $\widetilde{\beta}$ the result (30) is closer to Shannon's result than the result (35) in wider region in SNR.

Low bound from the Jensen's inequality. From the direct calculation using Eq. (28) and Wick's theorem [5] we have the following average for the action part with nonlinearity terms

$$-\left\langle \frac{\langle S_{nl}[\phi(z,\omega)]\rangle_\phi}{Q} \right\rangle_X = 5\gamma^2 \int_T dt \int_0^L dz G(z,z) \left\langle |\Phi(z,t)|^4 \right\rangle_X . \tag{37}$$

In the first nonvanishing order for small $\widetilde{\gamma}$ we have $-\left\langle \langle S_{nl}[\phi(z,\omega)]\rangle_\phi / Q \right\rangle_X = -\frac{5}{3} M \widetilde{\gamma}^2 + \mathcal{O}(\widetilde{\gamma}^4)$.

The relation (37) plays significant role since it delivers the low bound for the capacity. Indeed, from the Jensen's inequality we have

$$\langle \exp\left[-\frac{S_{nl}[\phi(z,\omega)]}{Q}\right]\rangle_\phi \geq \exp\left[-\frac{\langle S_{nl}[\phi(z,\omega)]\rangle_\phi}{Q}\right], \tag{38}$$

and it means that we can find the low bound for the mutual information:

$$I_{P[X]} \geq M \log[\mathrm{SNR}] - \left\langle \frac{\langle S_{nl}[\phi(z,\omega)]\rangle_\phi}{Q} \right\rangle_X . \tag{39}$$

The representation (39) is convenient for the nonperturbative analysis. For instance, it is useful for the arbitrary nonlinearity at large or at small $\widetilde{\beta}$, i.e., when the explicit expression for the NLSE solution $\Phi(z,t)$ is known. Let us consider large dispersion parameter $\widetilde{\beta}$. From the asymptotic representation of the NLSE solution [8] we obtained that for large z, $z \gg z_0 = L/\widetilde{\beta}$,

$$\Phi(z,t) \sim \frac{1}{\sqrt{4\pi\beta z}} X(\omega = \frac{t}{2\beta z}) \exp\left\{ -i\frac{t^2}{4\beta z} - i\frac{\pi}{4} + i\frac{\gamma}{4\pi\beta} \left| X(\omega = \frac{t}{2\beta z}) \right|^2 \log\left[\frac{z}{z_0}\right] \right\}. \tag{40}$$

This equation demonstrates that for the case of large $\widetilde{\beta} \gg 1$ the low bound, i.e., r.h.s. of Eq. (39), tends to the Shannon limit. The same result is obvious in the perturbative expansion (30) from the asymptotics (34): see Fig. 2. Physically it means that in the case of large $\widetilde{\beta}$ the dispersion leads to signal spreading in time domain. It results in the amplitude decreasing and thereby effectively decreasing of the nonlinear term in the equation (1).

4. Concluding remarks

We h derived the analytical expression for the mutual information $I_{P[X]}$ of the channel modelled by the nonlinear Schrödinger equation with the additive Gaussian noise at large SNR. We used the path-integral approach to the calculation of the conditional probability density function based on Martin-Siggia-Rose formalism. In the leading order in 1/SNR we obtained the general representation (in the case of arbitrary nonlinearity and dispersion) for the mutual information for the Gaussian input signal PDF. The correction to Shannon's result has the form of the averaged logarithm of the normalization factor Λ of the conditional PDF $P[Y|X]$ expressed in terms of the path-integral. In the case of small nonlinearity (in expansion in $\tilde{\gamma} = \gamma L P_{ave} \lesssim 1$) we obtained the first nonvanishing correction to Shannon's result. It is of order of $\tilde{\gamma}^2$ and decreases for increasing dispersion parameter. Using Jensen's inequality we obtained the low bound for the mutual information in the case of arbitrary nonlinearity and dispersion.

Acknowledgments

The general expressions for the entropies and the mutual information have been obtained with the support of the Russian Science Foundation (RSF) (grant No. 16-11-10133). Part of the work (perturbative calculations, mutual information estimations) was supported by the Russian Foundation for Basic Research (RFBR), Grant No. 16-31-60031/15. A. V. Reznichenko thanks the President program (SP-2415.2015.2) for support.

References

[1] Shannon C E 1949 *Communication in the presence of noise. Proc. Institute of Radio Engineers* vol 37 p 1
[2] Martin P C, Siggia E D, and Rose H A 1973 *Phys. Rev.* A **8** 423
[3] Terekhov I S, Vergeles S S, and Turitsyn S K 2014 *Phys. Rev. Lett.* **113** 230602
[4] Terekhov I S, Reznichenko A V, and Turitsyn S K 2016 *Phys. Rev.* E **94** 042203
[5] Itzykson C, Zuber J B 1980 *Quantum Field Theory* (New York: McGraw-Hill Book Company)
[6] Terekhov I S, Reznichenko A V, Kharkov Ya A, and Turitsyn S K, *Preprint* arXiv: 1508.05774v2
[7] Essiambre R -J *et al.* 2010 *J. of Lightwave Technol.* **28** 0733
[8] Novikov S, Manakov S V, Pitaevskii L P, and Zakharov V E 1984 *Theory of Solitons: The Inverse Scattering Method (Monographs in Contemporary Mathematics)* (Rumford: Springer)

Conjecture of perturbative QED breakdown at $\alpha\chi^{2/3} \gtrsim 1$

Alexander Fedotov

National Research Nuclear University "MEPhI" (Moscow Engineering Physics Institute), Kashirskoe sh. 31, 115409 Moscow, Russian Federation

E-mail: am_fedotov@mail.ru

Abstract. The old Ritus-Narozhny conjecture of possible breakdown of Intense Field QED perturbation theory for ultrarelativistic particles passing transversely through a strong electromagnetic field is reviewed with a special emphasis on its possible significance for near-future experiments.

Dedicated to memory of Nikolay Borisovich Narozhny (1940-2016)

1. Introduction

Professor Nikolay Borisovich Narozhny, the eminent Russian theoretician and the organizer of the Symposium "Extreme Light Technologies, Science, and Applications", passed away on 15 February 2016 [1,2]. I was lucky to work under his supervision on PhD thesises in 1997–2000 and to remain one of his main coauthors since then. Most of his pioneer works are well recognized by the community, e.g. calculation of probabilities for photon emission and pair photoproduction in circularly polarized electromagnetic wave [3], first calculation of polarization operator in a constant crossed field [4], first direct calculation of spontaneous pair production in electric field [5], or the effect of collapses and revivals in cavity QED [6,7].

However, his probably most deep and significant contribution (or at least claimed as such in his Dr.Sc. dissertation back in 1982), the α^3-order IFQED calculations [8–10] proving the original Ritus conjecture [11] of possible breakdown of perturbative QED at $\alpha\chi^{2/3} \gtrsim 1$, still remains rather unknown. In this talk I am going to give the review of that old idea. My particular aims are: to explain some known arguments in favor of the conjecture, to give some insights into its meaning, and to stress its significance for the near future progress of laser-matter interaction studies at extreme intensities.

2. Radiation corrections in Classical Electrodynamics and ordinary QED

Let me start with brief reminder of the basics of radiation corrections issues in Classical Electrodynamics and in ordinary QED. In Classical Electrodynamics the self-energy of a particle at rest reads

$$\mathscr{E}_{\rm em} = \frac{1}{2}\int d^3r \int d^3r' \, \frac{\rho(\vec{r})\rho(\vec{r}')}{|\vec{r}-\vec{r}'|} \simeq \frac{e^2}{r_0},$$

IOP Conf. Series: Journal of Physics: Conf. Series **826** (2017) 012027 doi:10.1088/1742-6596/826/1/012027

and becomes $\gtrsim mc^2$ at $r_0 \lesssim r_e \equiv \frac{e^2}{mc^2}$. The length r_e is traditionally called the 'classical electron radius'. The radiation reaction force (in proper reference frame 'p')

$$F_{\rm rad} = \frac{2}{3}\frac{e^4}{m^2c^4}E_{\rm p}^2,$$

produces over the distance r_e the work

$$A = F_{\rm rad} r_e \simeq \frac{e^6 E_{\rm p}^2}{m^3c^6},$$

which also $\gtrsim mc^2$ at $E_{\rm p} \gtrsim E_{\rm cr} \equiv \frac{m^2c^4}{e^3}$. The distance r_e and the field strength $E_{\rm cr}$ are considered as limits of applicability of Classical Electrodynamics.

Things are changed drastically in QED, though. To simplify formulas, let us consider scalar QED (sQED) as an example. The free quantized scalar charged field reads

$$\hat{\Psi}(x) = \sum_{\vec{p}} \frac{1}{\sqrt{2V\varepsilon_{\vec{p}}}}\left(e^{-ipx}\hat{a}_{\vec{p}} + e^{ipx}\hat{b}^{\dagger}_{\vec{p}}\right),$$

where $\hat{a}_{\vec{p}}$ and $\hat{b}_{\vec{p}}$ are charged particle and antiparticle annihilation operators, respectively. The sQED 4-current operator

$$\hat{j}_\mu(x) = ie\ :\hat{\Psi}^{\dagger}(x)\ \overleftrightarrow{\partial_\mu}\ \hat{\Psi}(x): = \sum_{\vec{p},\vec{p}'} \frac{e}{2V\sqrt{\varepsilon_{\vec{p}}\varepsilon_{\vec{p}'}}}\left\{(p_\mu + p'_\mu)\left[\underbrace{e^{i(p'-p)x}\hat{a}^{\dagger}_{\vec{p}'}\hat{a}_{\vec{p}}}_{\text{particle current}} - \underbrace{e^{-i(p'-p)x}\hat{b}^{\dagger}_{\vec{p}}\hat{b}_{\vec{p}'}}_{\text{antiparticle current}}\right] + \right.$$
$$\left. + (p_\mu - p'_\mu)\left[\underbrace{e^{-i(p'+p)x}\hat{b}_{\vec{p}'}\hat{a}_{\vec{p}} - e^{i(p'+p)x}\hat{a}^{\dagger}_{\vec{p}'}\hat{b}^{\dagger}_{\vec{p}}}_{\text{non-diagonal terms}}\right],\right\}$$

contains a non-diagonal part, which corresponds to annihilation and creation of a virtual pair.

By passing to collective coordinate $\vec{R} = \frac{\vec{r}+\vec{r}'}{2}$ and separation $\vec{\xi} = \vec{r} - \vec{r}'$, the self-energy can be cast into form

$$\mathscr{E}_{\rm em} = \frac{1}{2}\int d^3\xi\,\frac{C(\vec{\xi})}{|\vec{\xi}|},$$

where $C(\vec{\xi})$ is difference between the expectation values of the charge density correlator in 1-particle state and in vacuum:

$$C(\vec{\xi}) = \int d^3R\,\langle 1_{\rm rest}|\hat{j}^0\left(\vec{R}+\frac{\vec{\xi}}{2}\right)\hat{j}^0\left(\vec{R}-\frac{\vec{\xi}}{2}\right) - :\ :|1_{\rm rest}\rangle =$$
$$= \frac{e^2}{2}\int\frac{d^3p}{(2\pi)^3}\underbrace{\left(1+\frac{m}{\varepsilon_{\vec{p}}}\right)e^{i\vec{p}\vec{\xi}}}_{\text{from } \hat{a}^{\dagger}\hat{a}\hat{a}^{\dagger}\hat{a}} - \frac{e^2}{2}\int\frac{d^3p}{(2\pi)^3}\underbrace{\left(1-\frac{m}{\varepsilon_{\vec{p}}}\right)e^{i\vec{p}\vec{\xi}}}_{\text{from } \hat{b}\hat{a}^{\dagger}\hat{a}\hat{b}^{\dagger}}.$$

In classical limit ($m \to \infty$) the first (particle) contribution in braces reduces (as expected) to $e^2\delta^{(3)}(\vec{\xi})$, while the second term vanishes. This results in linear divergency as above: $\mathscr{E}_{\rm em} \propto \frac{e^2}{2}\int d^3\xi\frac{\delta^{(3)}(\vec{\xi})}{|\vec{\xi}|} \simeq \frac{e^2}{r_0}$, $r_0 \to 0$.

However, in general setting the leading divergency is canceled by the virtual pairs contribution [12]:

$$C(\vec{\xi}) = \frac{e^2}{2}\left(\cancel{\delta^{(3)}(\vec{\xi})} - \frac{m^2}{2\pi^2|\vec{\xi}|}K_1(m|\vec{\xi}|)\right) - \frac{e^2}{2}\left(\cancel{\delta^{(3)}(\vec{\xi})} + \frac{m^2}{2\pi^2|\vec{\xi}|}K_1(m|\vec{\xi}|)\right) =$$
$$= -\frac{e^2m^2}{2\pi^2|\vec{\xi}|}K_1(m|\vec{\xi}|) \simeq \frac{e^2 m}{\pi^2|\vec{\xi}|^2}, \quad \vec{\xi}\to 0,$$

so that

$$\mathscr{E}_{\rm em} \simeq \frac{e^2 m}{\pi^2}\int\frac{d^3\xi}{|\vec{\xi}|^3} \propto e^2 m \log\left(\frac{1}{mr_0}\right), \quad r_0 \to 0.$$

Thus, a pointlike charge is replaced by a cloud of virtual pairs of Compton size $\simeq l_C = \frac{1}{m} \simeq 137 r_e$. Divergency is still present but now gets much weaker (logarithmic vs linear) than in Classical Electrodynamics. After renormalization (which is all the same required for physical reasons, albeit $\mathscr{E}_{\rm em} \simeq \alpha m \log\left(\frac{1}{mr_0}\right) \ll m$ for any reasonable value of r_0), the coupling constant becomes effectively 'running', and its energy dependence essentially mimics the character of divergency: $\alpha(\varepsilon) \simeq \alpha\log\left(\frac{\varepsilon}{m}\right)$, $\varepsilon \gg m$ (high energy 'stripping'). Note that $\alpha(\varepsilon)$ nevertheless remains small for all reasonable values of energy. Review and classification of the variety of high-energy QED processes [13,14] demonstrates that all the cross sections remain also small $\sigma(\varepsilon) \lesssim \alpha^n r_e^2 \log^k\left(\frac{\varepsilon}{m}\right)$ within all the reasonable energy range. Thus perturbation theory in ordinary QED works extremely well within all the reasonable range of parameters.

3. Radiation corrections in Intense Field QED

However, in external field with $a_0 \equiv \frac{e\sqrt{-A_\mu^2}}{m} \gg 1$ (for optical lasers at intensity $I_L \gtrsim 10^{18}\mathrm{W/cm^2}$)perturbation theory *with respect to interaction with that field* breaks down and all-order summation is needed, which reduces only to replacement of free external lines and propagators by the exact ('dressed') ones in external field:

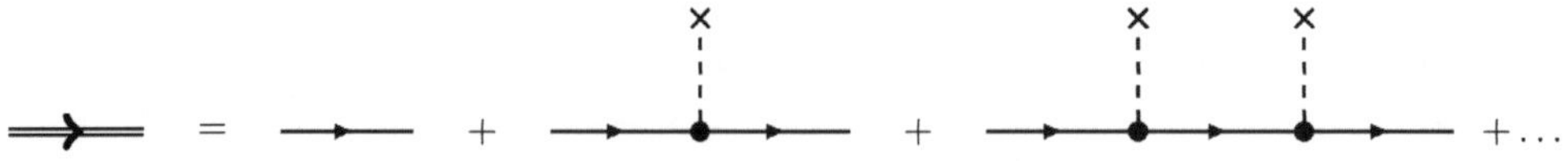

For several cases (including the most important paradigmatic case of constant crossed field, which corresponds to $a_0 \gg 1$ and relativistic motion across the field), the equation

can be solved in closed form. Note that in CCF electrons / photons are characterized by a single Lorentz- and gauge-invariant parameter $\chi = \frac{e}{m^3}\sqrt{-(F_{\mu\nu}p^\nu)^2}$ / $\varkappa = \frac{e}{m^3}\sqrt{-(F_{\mu\nu}k^\nu)^2}$ - for electron this is just its proper acceleration in Compton units.

It was noticed already at its birth [4,11,15] that in IFQED radiation corrections are growing unusually fast with χ or $\varkappa$ (i.e. with both energy and field strength):

$M^{(2)}(\chi) =$ $\simeq \alpha m\chi^{2/3}, \quad W_{e^\pm\to e^\pm\gamma}(\chi) = \frac{2m}{p_0}\,\mathrm{Im}\,M^{(2)} \simeq \frac{\alpha m^2}{p_0}\chi^{2/3}, \quad \chi \gg 1;$

IOP Conf. Series: Journal of Physics: Conf. Series **826** (2017) 012027 doi:10.1088/1742-6596/826/1/012027

$$\mathscr{P}^{(2)}(\varkappa) =$$  $$\simeq \alpha m^2 \varkappa^{2/3}, \quad W_{\gamma\to e^+e^-}(\varkappa) = \frac{2}{k_0}\,\mathrm{Im}\,\mathscr{P}^{(2)} \simeq \frac{\alpha m^2}{k_0}\varkappa^{2/3}, \quad \varkappa \gg 1;$$

This implies that for $\chi,\ \varkappa \gtrsim \alpha^{-3/2} \simeq 1.6\times 10^3$ (or, equivalently, $E_{\mathrm{p}} \gtrsim 12E_{\mathrm{cr}} \simeq 1.6\times 10^3 E_S$)

$$M^{(2)} \simeq m, \quad \mathscr{P}^{(2)} \simeq m^2$$

and that in proper reference frame

$$t_e \sim W^{-1}_{e^\pm\to e^\pm\gamma} \simeq t_C, \quad t_\gamma \sim W^{-1}_{\gamma\to e^+e^-} \simeq t_C$$

These means that radiation corrections become not small and radiation-free motion could show up only at Compton scale (where localization is all the same impossible).

For high-energy electrons counterpropagating a laser pulse we have $\chi \sim \frac{E\gamma_{\mathrm{in}}}{E_S}$. Some typical values of parameters required to fulfill the condition $\alpha\chi^{2/3} \gtrsim 1$ are listed in Table 1. Observe that this threshold could be almost overcome experimentally by combining state-of-the-art laser systems with the future ILC–class TeV lepton colliders. It is also worth noting that the table assumes transverse propagation across the field. For self-sustained (A-type) cascades [16, 17] $E \gtrsim \alpha E_S$ and

$$\measuredangle(\vec{p},\vec{E}) \sim \left(\frac{\alpha E_S}{E}\right)^{1/4} \lesssim 1, \quad \chi \sim \left(\frac{E}{\alpha E_S}\right)^{3/2} \gtrsim 1, \quad \text{but} \quad \alpha\chi^{2/3} \sim \frac{E}{E_S} \ll 1$$

During 1972–1981 some higher-order radiation corrections in IFQED were either obtained or estimated by Ritus, Narozhny and Morozov [8, 9, 18–20]. Namely, for mass corrections it was obtained:

$$\frac{M}{m} =$$

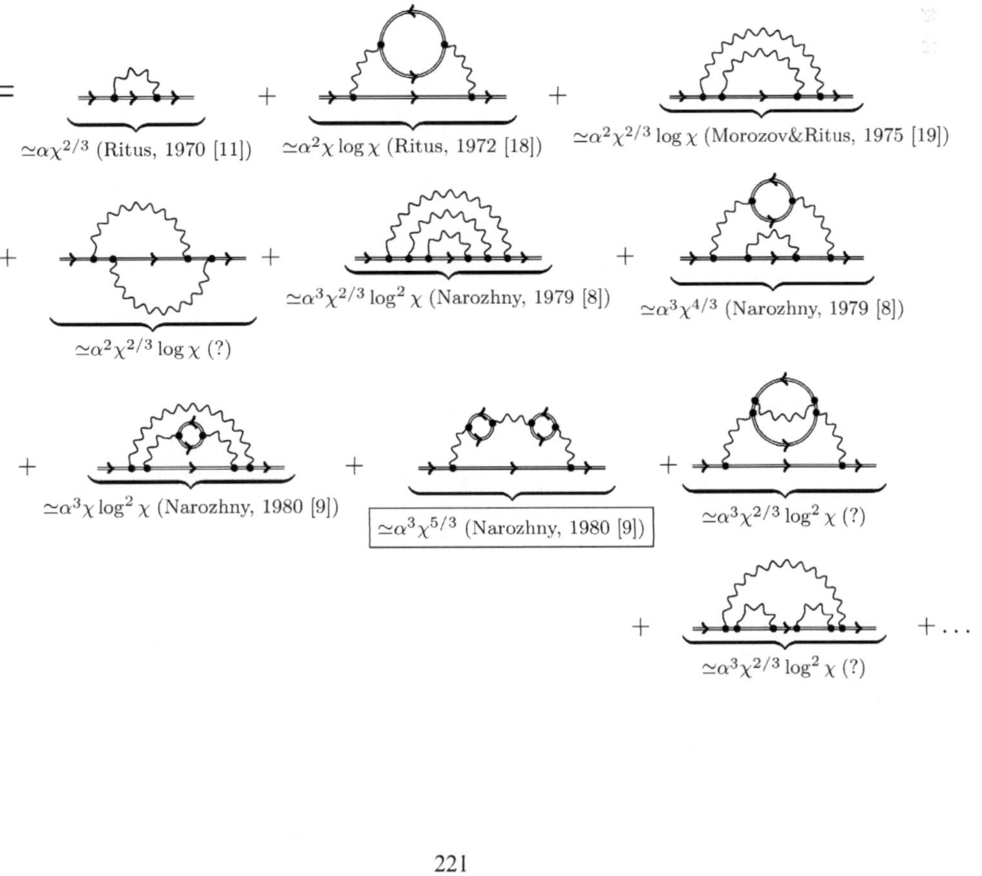

IOP Conf. Series: Journal of Physics: Conf. Series **826** (2017) 012027 doi:10.1088/1742-6596/826/1/012027

(here and below the contributions of the diagrams shown with question mark were either calculated in draft but unpublished, or guessed basing on their infrared behavior). For polarization correction it was obtained:

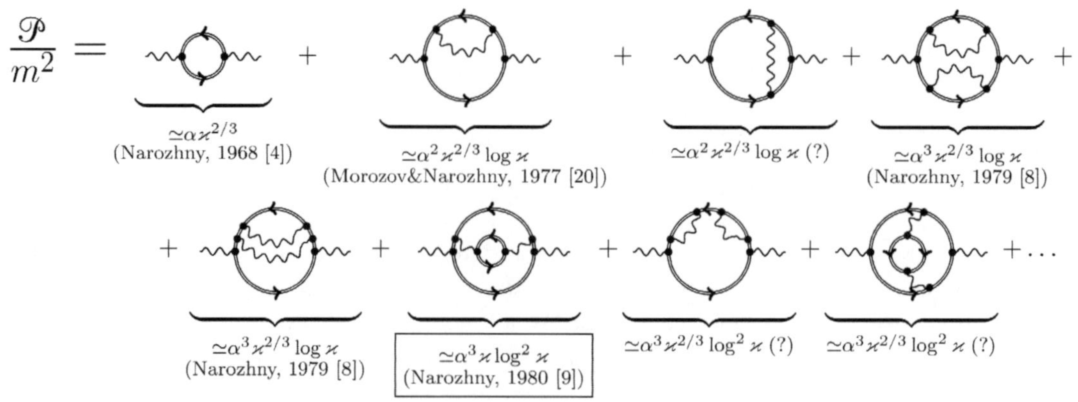

One can observe that within each order different types of diagrams depend on χ or $\varkappa$ rather differently: some higher-order diagrams acquire just additional logarithmic factors as in ordinary QED, but some others reveal power dependence, with power growing in higher orders. In fact, when Ritus first obtained his e^4-order polarization correction to mass operator [18], he initially suggested $\alpha\chi^{1/3}$ as roughly an expansion parameter in IFQED perturbation theory (which should be also puzzling, in fact). The main goal of Narozhny [9] was demonstrating that the e^6-order two-loop polarization diagram grows even faster and that the true expansion parameter should be $\alpha\chi^{2/3}$ instead. Moreover, he argued that the 4-th order contribution to the mass operator and 2-nd and 4-th order contributions to polarization operator are just occasionally suppressed and that corrections to mass operator outstrip those to polarization operator by one order, so that the same parameter $\alpha\chi^{2/3}$ should show up in polarization operator in 8-th order as well.

Basing on these results, one could get an impression that the leading contributions probably originate in diagrams containing maximal number of successive uncorrected polarization loops (if true, it could drastically simplify further considerations). However, the only analyzed vertex correction [10]

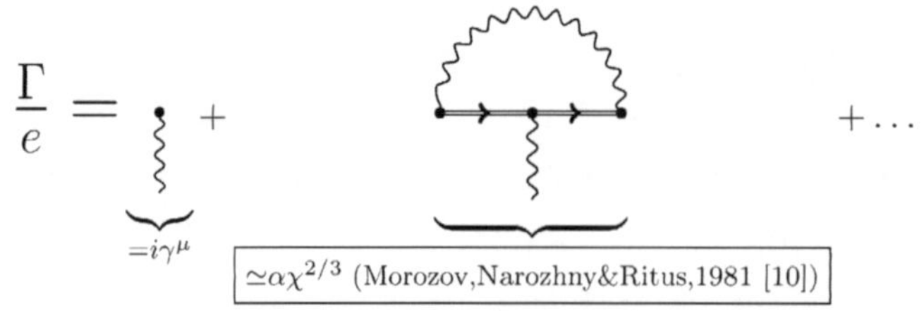

also grows as $\alpha\chi^{2/3}$, so that some further calculations are definitely needed to support this idea. According to my knowledge, no further progress was made on these calculations since 1981.

4. Qualitative estimates and physical meaning

On defense, Nikolay Borisovich was asked for a 'simple words' physical reasoning (interpretation) for appearance of the parameter $\alpha\varkappa^{2/3}$. And his answer was about that in ultrarelativistic case $\mathcal{P} = \alpha m^2 F(\varkappa)$ should not depend on m. Then, since $\varkappa \propto m^{-3}$, it should be $F(\varkappa) \propto \varkappa^{2/3}$

IOP Conf. Series: Journal of Physics: Conf. Series **826** (2017) 012027 doi:10.1088/1742-6596/826/1/012027

Table 1. Typical values of electron energy ($\varepsilon_{\rm in}$) and laser intensity (I_L) required for attaining the regime $\alpha\chi^{2/3} \simeq 1$ in counterpropagating setup.

$\varepsilon_{\rm in} = m\gamma_{\rm in}$, GeV	800	80	8	0.8
E/E_S	10^{-3}	10^{-2}	0.1	1
I_L,W/cm^2	5×10^{23}	5×10^{25}	5×10^{27}	5×10^{29}

unambiguously for $\varkappa \gg 1$. Unfortunately, this argument doesn't work for M, Γ and higher orders.

However, recently a more visual and direct explanation was seemingly found [21]. Consider for definiteness[1] formation time t and length l for the polarization operator $\mathscr{P}^{(2)}(\varkappa \gg 1)$. Assuming k, p and $p-k$ all large and also that initially $\vec{k} \perp \vec{E}$, the energy uncertainty of the virtual process $\gamma \to e^-e^+ \to \gamma$

$$\Delta\varepsilon(t) \;=\; \sqrt{p^2+e^2E^2t^2+m^2} \;+\; \sqrt{(k-p)^2+e^2E^2t^2+m^2} \;-\; k \;\geq\; \frac{2\left(e^2E^2t^2+m^2\right)}{k}$$

Assume additionally (to be confirmed by the result) that $eEt \gg m$. Then $\Delta\varepsilon(t) \simeq \frac{e^2E^2t^2}{k}$ and from the uncertainty principle $\Delta\varepsilon \cdot t \sim 1$ we obtain

$$t, l_\parallel \simeq \left(\frac{k}{e^2E^2}\right)^{1/3} \equiv \frac{k}{m^2\varkappa^{2/3}} \equiv \frac{m}{eE}\varkappa^{1/3},$$

where $\varkappa = eEk/m^3$. Note that $eEt \simeq m\varkappa^{1/3} \gg m$, exactly as assumed. It turns out that this key simple estimate is confirmed by direct derivation of effective formation region for the integrals that define the quantum amplitude [19].

Transverse separation $l_\perp \simeq \frac{eEt^2}{k} \simeq \frac{1}{m\varkappa^{1/3}} \equiv \frac{1}{(eEk)^{1/3}}$ (note it is m-independent). Maybe a bit counterintuitively, charge separation reduces (rather than increases) with the field – this quantum effect arises because t reduces too fast. Moreover, strong ($\varkappa \gg 1$) field is capable for confining virtual pairs to distances smaller than $l_C = \frac{1}{m}$ (this is very reminiscent to the Ritus's observation [22] of strong field–small distance correspondence). In 'proper' reference frame[2] $l'_\parallel \sim \frac{m}{k}l_\parallel \sim \frac{1}{m\varkappa^{2/3}} \ll l_\perp$, thus $l'_\parallel$ is the smallest scale. Surprisingly, for $\alpha\chi^{2/3} \sim 1$ it coincides to the classical electron radius r_e. Now polarization operator should be defined by these scales: $\mathscr{P}(\varkappa) \simeq e^2/{l_\perp}^2(\varkappa)$. Similarly, $M \simeq e^2/l'_\parallel(\varkappa)$ and the parameter $\alpha\chi^{2/3} \equiv \frac{e^2/l'_\parallel}{m}$ may be viewed as just Coulomb energy to rest energy ratio.

5. Nonperturbative regime and cascades

QED cascades may be in direct relation with the nonperturbative regime under discussion. Indeed, at $\chi, \varkappa \gg 1$ the work produced by the field during the formation time t, $A \simeq eEl_\perp \simeq \Delta\varepsilon$. This means that the virtual intermediate states in such a regime are in fact rather close to mass shell. This corresponds to the fact that at $\chi, \varkappa \gg 1$ the real and imaginary parts of the mass and polarization operators become of the same order. Moreover, according to the optical theorem, these imaginary parts define the probabilities of the processes described by diagrams that can be obtained by making cuts over intermediate states. By restricting to diagrams containing polarization corrections to propagators, i.e. those of type:

[1] Ultrarelativistic kinematics is in fact similar for all the processes.

[2] I.e. where the photon is 'soft' ($k' \sim m$) and $E_P \sim \varkappa E_S$.

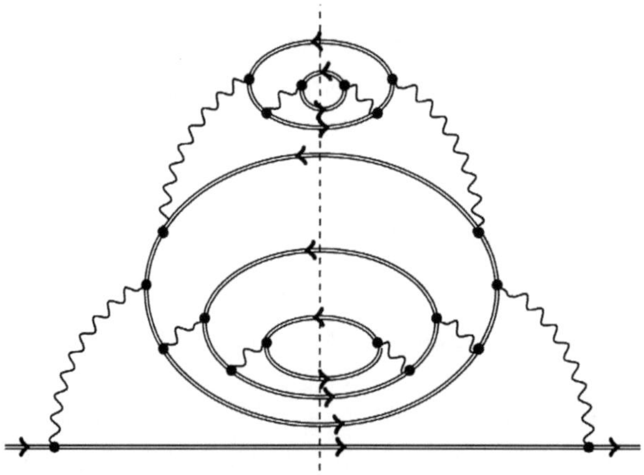

we obtain by such cuts exactly the diagrams for cascade processes (at that cutting corrections for mass operator corresponds to electron-seeded cascades, and for polarization operator to photon-seeded cascades). Intuitively, passing from perturbative to non-perturbative regime might correspond to transition from 'ordinary' (S-type) to self-sustained (A-type) cascades [23]. As already mentioned, for self-sustained cascades $\alpha\chi^{2/3} \simeq 1$ corresponds to $E \simeq E_S$. According to our recent results [16], attaining of such field strengths can be questionable (notably due to cascade generation), provided the field is capable for spontaneous pair creation. This conclusion, however, is not related to fields that are incapable for that (e.g. plane wave field or CCF in which, on other hand, self-sustained cascades can not originate at all). Besides, our cascade model was developed by neglecting radiation corrections. Since development of self-sustained cascades necessarily requires field variation, possible significance of self-sustained cascades for non-perturbative dynamics yet remains unclear.

6. Conclusion

The conjecture that radiation corrections in IFQED are growing as a power of energy and field strength is really puzzling and challenging for theoreticians. First, in such a regime QED may become a truly non-perturbative theory and a lot of numerous results published by now may become invalid. Second, at least for a mass operator, for $\alpha\chi^{2/3} \sim 1$ the known leading radiation corrections arise from diagrams containing the maximal number of successive uncorrected polarization loops. At the same time, the longitudinal formation length $l'_{\parallel}$ becomes comparable to the classical electron radius r_e (which may indicate that classical theory is somehow partially recovered). Finally, possible relations are expected via optical theorem between the IFQED radiation corrections and QED cascades at $E \simeq E_S$. These remarks may serve as possible hints for further studies.

Importantly, the regime $\alpha\chi^{2/3} \gtrsim 1$ may 'soon' appear observable for experimentalists. But unfortunately, potential significance of the conjecture has been still underestimated by the community.

Acknowledgements

The work was supported by the MEPhI Academic Excellence Project (contract No. 02.a03.21.0005, 27.08.2013) and RFBR grant 16-02-00963. I am grateful to A. Ilderton, T. Heinzl and A.I. Titov for inspiring questions and comments to be properly addressed in future studies.

References

[1] Popruzhenko S 2016 Tribute to Nikolay Narozhny, In: The International Committee on Ultra-high Intensity Lasers (ICUIL) newsletter URL `http://www.icuil.org/newsletter.html?download=255:icuil-news-volume-7-june-2016`

[2] Avrorin E, Andriyash A, Babichev N, Belinski V, Bisnovatyi-Kogan G, Voskresensky D, Nikishov A, Kagan Y, Lozovik Y, Mur V, Pevnitski A, Popov V, Popov V, Popruzhenko S, Ritus V, Rykovanov G, Sergeev A, Simonenko V, Solov'ev V, Strihanov M, Trutnev Y, Fedotov A, Chetverushkin B, Chechetkin V and Yakovlev V 2016 In memory of Nikolay Borisovich Narozhny [official obituary in Russian] URL `http://ufn.ru/dates/inmemoria/narozhny.pdf`

[3] Narozhnyi N, Nikishov A and Ritus V 1965 *Sov. Phys. JETP* **20** 622

[4] Narozhny N 1969 *Sov. Phys. JETP* **28** 371

[5] Narozhny N and Nikishov A 1970 *Soviet. J. Nucl. Phys* **11** 596

[6] Eberly J, Narozhny N and Sanchez-Mondragon J 1980 *Physical Review Letters* **44** 1323

[7] Narozhny N, Sanchez-Mondragon J and Eberly J 1981 *Physical Review A* **23** 236

[8] Narozhny N 1979 *Physical Review D* **20** 1313

[9] Narozhny N 1980 *Physical Review D* **21** 1176

[10] Morozov D, Narozhnyj N and Ritus V 1981 *Sov. Phys. JETP* **53** 1103

[11] Ritus V 1970 *Sov. Phys. JETP* **30** 1181

[12] Weisskopf V 1939 *Physical Review* **56** 72

[13] Gorshkov V 1973 *Physics-Uspekhi* **16** 322

[14] Baier V, Fadin V, Khoze V and Kuraev E 1981 *Physics Reports* **78** 293

[15] Nikishov A and Ritus V 1964 *Sov. Phys. JETP* **19** 529

[16] Fedotov A, Narozhny N, Mourou G and Korn G 2010 *Physical Review Letters* **105** 080402

[17] Elkina N, Fedotov A, Kostyukov I Y, Legkov M, Narozhny N, Nerush E and Ruhl H 2011 *Physical Review Special Topics-Accelerators and Beams* **14** 054401

[18] Ritus V 1972 *Annals of Physics* **69** 555

[19] Morozov D and Ritus V 1975 *Nuclear Physics B* **86** 309

[20] Morozov D and Narozhny N 1977 *Sov. Phys. JETP* **45** 23

[21] Fedotov A 2015 *arXiv:1507.08512*

[22] Ritus V 1975 *Sov. Phys. JETP* **42** 774

[23] Mironov A, Narozhny N and Fedotov A 2014 *Physics Letters A* **378** 3254

IOP Conf. Series: Journal of Physics: Conf. Series **826** (2017) 012028 doi:10.1088/1742-6596/826/1/012028

Weighted Particle Monte Carlo Method for Effective Simulation of Electromagnetic Cascades

Nina Elkina

Ludwig-Maximilians Universität München, 80539, Germany

E-mail: N.Elkina@gsi.de

Abstract. We present a new version of the event generator for simulation of quantum electromagnetic cascades generated in plasma irradiated by ultra-intense laser field. The new code is capable to account for very soft photons as well as for polarization effects in nonlinear vacuum. Numerical implementation differs significantly from the original version aiming to provide a guaranteed accuracy of computations. The code is benchmarked against the semi-analytical solution describing electromagnetic cascade in a rotating field.

1. Introduction

The Extreme Light Infrastructure (ELI) is anticipated to provide an experimental access to study intense field quantum electrodynamic (QED) effects in laser driven plasmas. Since analytical results for relativistic plasma with QED effects are rather limited, numerical simulation has become the only tool to explore various scenarios of such phenomena. Thus it is important to verify numerical accuracy of numerical plasma codes which are applied to study the QED regime. For this purpose one can think of a unified testbed case to benchmark different codes. The QED cascade can serve this purpose since in particular field configurations it can be treated analytically. The example is provided in [1], where analytical theory of electromagnetic cascade was developed for homogeneous rotating electric field. This analytical theory was later validated by the numerical simulation presented in [2]. Although good qualitative agreement has been observed, the difference between theoretical and numerical predictions for electron-positron yields remained unexplained. In this contribution we report on a benchmark study of the new version of our code and compare it with the original one [2].

2. Basic equations

The mathematical basis for study of QED cascades is provided by a set of semi-classical transport equations for $\gamma e^{\pm}$ plasma. For example, equation for electron distribution function $f(\boldsymbol{r}, \boldsymbol{p}, t)$ in the phase space reads

$$\frac{df(\boldsymbol{r},\boldsymbol{p},t)}{dt} \equiv \frac{\partial f(\boldsymbol{r},\boldsymbol{p},t)}{\partial t} + \boldsymbol{v}\cdot\nabla f(\boldsymbol{r},\boldsymbol{p},t) - e\left\{\boldsymbol{E}(\boldsymbol{r},t) + \boldsymbol{v}\times\boldsymbol{B}(\boldsymbol{r},t)\right\}\cdot\frac{\partial f(\boldsymbol{r},\boldsymbol{p},t)}{\partial \boldsymbol{p}} \tag{1}$$
$$= \int w_{rad}(\boldsymbol{r},t,\boldsymbol{p}+\boldsymbol{k}\to\boldsymbol{k}) f(\boldsymbol{r},\boldsymbol{p}+\boldsymbol{k},t)\, d^3k - f(\boldsymbol{r},\boldsymbol{p},t)\int w_{rad}(\boldsymbol{r},t,\boldsymbol{p}\to\boldsymbol{k})\, d^3k,$$

where $\boldsymbol{v} = \boldsymbol{p}/\varepsilon$ and $\varepsilon(\boldsymbol{p}) = \sqrt{p^2+m^2}$ are the velocity and energy for given momentum and $\hbar = c = 1$. By definition, $w_{rad}(\boldsymbol{p}\to\boldsymbol{k})$ is the differential probability rate for emission of a

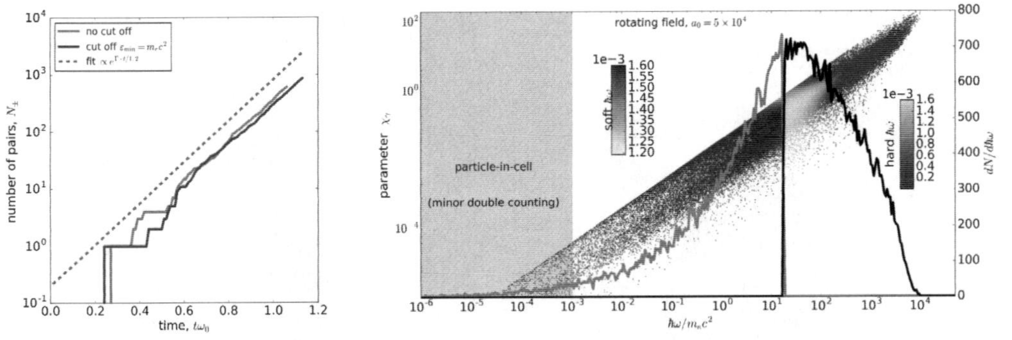

Figure 1. Left: growth of the number of pairs as a function of time with and without soft photons cutoff; right: hard and soft photon spectra, together with spread of emitted photons in χ_γ.

photon with momentum $\boldsymbol{k}$ by an electron with a given momentum $\boldsymbol{p}$. The common approach to solve these equations numerically combines particle-in-cell method [3] for plasma dynamics with the Monte Carlo (MC) method [2] for sampling quantum processes. In the simplest realization all individual particles are following their classical trajectories, while event generation algorithm is applied on each time step. In the new code all particle trajectories are integrated in covariant form using implicit Runge-Kutta methods, as suggested in [4]. Time symmetric form of these integrators allows to preserve the Minkowski norm $u_\mu u^\mu = 1$ up to round off precision.

3. Soft photons

While the probability rate w_e of electron-positron pair production

$$w_e = \frac{d\mathscr{W}_{e^\pm}}{d\varepsilon_e} = \frac{+\alpha m^2}{\varepsilon_\gamma^2}\left\{Ai_1(x) + \left[\frac{g(\phi)}{x} - \chi_\gamma\sqrt{x}\right]Ai'(x)\right\}, \quad x = \left(\frac{\chi_\gamma}{\chi_e(\chi_\gamma - \chi_e)}\right)^{2/3}, \qquad (2)$$

is relatively easy for computations, the photon emission one

$$w_{rad} = \frac{d\mathscr{W}_{rad}}{d\varepsilon_\gamma} = -\frac{\alpha m^2}{\varepsilon_e^2}\left[Ai_1(x) + \left[\frac{g(\phi)}{x} + \chi_\gamma\sqrt{x}\right]Ai'(x)\right], \quad x = \left(\frac{\chi_\gamma}{\chi_e(\chi_e - \chi_\gamma)}\right)^{2/3}, \qquad (3)$$

may be problematic at small energies, possessing a weak singularity $w_{rad}(\varepsilon_\gamma) = \mathcal{O}(\varepsilon_\gamma^{-2/3})$ at $\varepsilon_\gamma \to 0$. In the previous version [2] of our event generator the soft photon singularity was isolated by putting a cutoff energy below which photons were not sampled. In the new version the impact of soft photons can be estimated by splitting of the rate integrals into singular and nonsingular parts as follows

$$\mathscr{W}_{rad} = \int_0^{\varepsilon_e} \frac{d\mathscr{W}_{rad}}{d\varepsilon'_\gamma} d\varepsilon'_\gamma = \int_0^{\varepsilon_{\min}} + \int_{\varepsilon_{\min}}^{\varepsilon_e} .$$

The aforementioned singularity can then be integrated using an appropriate substitution. The results of simulation runs with and without soft photons sampling are presented in the ledt panel of Fig. 1, where the growth of the number of pairs is shown as a function of time. It is clear from the picture that soft photons do not play important role in strong cascading regime. However, the increment of growth turns out to be smaller than the one reported in [2]. This can be attributed

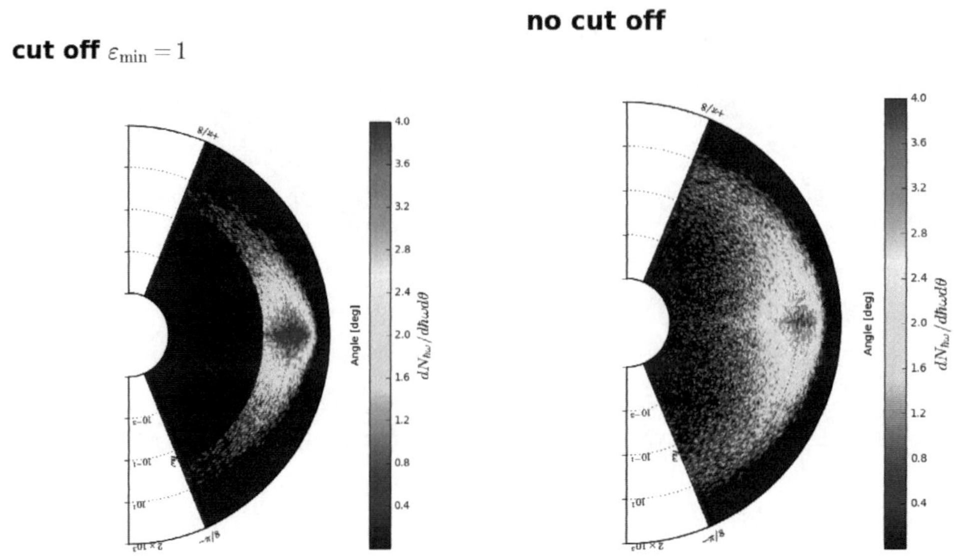

Figure 2. Angular distribution of photons produced via nonlinear Compton scattering of energetic electrons with $\gamma = 10^3$ on counter-propagating laser pulse with $a_0 = 10^2$, with and without cutoff.

to more accurate numerics used to calculate integrals in the sampling equation. The results are in good agreement with theoretical estimate for the growth rate $\Gamma_{th} = \alpha\mu^{1/4}\sqrt{m_e c^2\omega_0/\hbar}$, where $\mu = \frac{\hbar\omega_0}{\alpha m_e c^2}a_0$. Thus for high a_0 one can neglect soft photons by imposing an appropriate energy cut off. More insights on formation of the photon spectra is presented in the right panel of Fig. 1, where each dot of the scatter plot represents a photon emitted with energy $\hbar\omega$ and quantum parameter χ_γ. The spectrum curve $dN/d\hbar\omega$ splits into the soft photons (red) and hard photons one (black). One can see that the number of really soft photons below 0.1 MeV is not significant enough to interfere with low-frequency plasma or laser waves. Although impact of soft photons turns out to be unimportant at higher field intensity, their influence can be still significant at less severe parameters. For example, this is the case when photon beam is produced via nonlinear Compton scattering of electrons counter-propagating laser pulse with intensity insufficient for pair photoproduction. The difference in angular distributions of emitted photons computed with and without imposing a soft photons cutoff is presented in Fig. 2. This resulting photon beam has been used in [5] to probe the vacuum birefringence effect.

In order to describe briefly the basic functionality of the code, let us consider the method of sampling of emission of a polarized photon as an example. Over each time step we first calculate the total probability rate $\mathcal{W}^u_{tot}$ for electron to produce an unpolarized photon. The probability for a particle to generate photon during the time step Δt if given by $\Delta t\mathcal{W}^u_{tot}$. A new photon is produced if the condition $\Delta t\mathcal{W}^u_{tot} > r_1$ holds, where $0 < r_1 < 1$ is a random number. Using two other random numbers $0 < r_{2,3} < 1$ we sample the polarization state and energy of secondary photon as following. The discrete polarization state $p \in \{0, \pi/2\}$ is picked up according to the relative ratio between the corresponding total probabilities $\mathcal{W}^0_{tot}/(\mathcal{W}^{\pi/2}_{tot} + \mathcal{W}^0_{tot}) \gtrless r_2$. Finally, the energy ε_γ of polarized photon is selected by solving the sampling equation

$$\frac{1}{W^p_{tot}}\int_0^{\varepsilon_\gamma}\frac{d\mathcal{W}^p(\hbar\omega')}{d\hbar\omega'}d\hbar\omega' = r_3. \tag{4}$$

In order to speed up the nonlinear sampling equation solver we adopt a modified false position method instead of the bisection root finding algorithm which was previously applied in the original version of our code. However, in rare cases of slow convergence the code can still call back the bisection method. In this case the root is finally located in the interval returned by the false position method after it hits the maximal allowed number of iterations (usually $N \simeq 10$). The code also provides two options to compute integrals in the sampling equation by either by the Gauss-Kronrod [6] or Double Exponentiation rule [7]. Each method is supplemented with built in error control allowing to preserve numerical accuracy in the course of computation. The code is further optimized by gradual creation of a sparse 2d array containing the values of both total and differential probability rates for variables (ε, χ) on a grid, with further interpolation to the actually required values of that variables.

4. Summary and conclusions

In conclusion, we have carried out benchmark study of the new version of event generator. The main task of this work was to determine the impact of soft photons, which was neglected in the earlier version of the code. Our results suggest that soft photons do not affect significantly the plasma dynamics in a very strong laser field capable to support exponentially growing cascades. Moreover, the photon spectrum is very sparsely populated at low frequencies with almost no photon emission at plasma and laser wave scales. This suggests that semi-classical modeling of QED effects is justified at least for very high intensities ($I > 10^{24}$ W/cm^2). However, soft photons should be taken into account at moderate intensities, for example in various scenarios with electron beams scattering in laser field, where the photon flux is generated via nonlinear Compton scattering. If soft photons are included, the resulting photon flux exhibits noticeable angular spread compared to simulations with imposing a cutoff.

In summary, we have presented a list of new features of the event generator which is a part of the PIC-ANTARES code under development. The main difference with previous version is that the impact of soft photons below 1 MeV can be now taken into account if needed. Polarization effects are now also implemented in the code, thus allowing to study various nonlinear effects like vacuum birefringence. Finally, the built in automatic accuracy control is now implemented in each part of the code.

Acknowledgements

This research was supported by the Grant No. SFB TR18 project B13. The author thanks Dr. Sergey Rykovanov for warm hospitality at the Helmholtz Institute Jena (HI Jena), where this contribution was written, and Dr. Alexander Fedotov for numerous discussions and helpful suggestions.

References

[1] Fedotov A M, Narozhny N B, Mourou G and Korn G 2010 *Phys. Rev. Lett.* **105**(8) 080402
[2] Elkina N V, Fedotov A M, Kostyukov I Y, Legkov M V, Narozhny N B, Nerush E N and Ruhl H 2011 *Physical Review Special Topics Accelerators and Beams* **14** 054401
[3] Birdsall C and Langdon A 2004 *Plasma Physics via Computer Simulation* Series in Plasma Physics and Fluid Dynamics (Taylor & Francis)
[4] Elkina N V, Fedotov A M, Herzing C and Ruhl H 2014 *Physical Review E* **89** 053315
[5] King B and Elkina N 2016 *Physical Review A* **94** 062102
[6] Kronrod A 1965 *Nodes and weights of quadrature formulas: sixteen-place tables* (Consultants Bureau)
[7] Mori M and Sugihara M 2001 *Journal of Computational and Applied Mathematics* **127** 287 – 296

IOP Conf. Series: Journal of Physics: Conf. Series **826** (2017) 012029 doi:10.1088/1742-6596/826/1/012029

Observable Features of QED Cascades in Collisions of GeV Electrons with Intense Laser Pulses

Arseny Mironov, Alexander Fedotov and Nikolay Narozhny

National Research Nuclear University MEPhI (Moscow Engineering Physics Institute), Kashirskoe sh. 31, 115409 Moscow, Russian Federation

E-mail: mironov.hep@gmail.com

Abstract. We consider interaction of multi-GeV electron beam with electromagnetic field formed by intense focused laser pulses. At intensity $\sim 10^{24} \div 10^{25}$ W/cm^2 such a scheme can be used to produce both S(Shower)-type and A(Avalanche)-type (or "self-sustained") QED cascades. The proposed experimental design provides a solution to the problem of injection of seed particles into the strong field region. Here we are focusing on discussion of some peculiarities of cascade dynamics and evaluation of such quantities as final angular and spectral particle distributions, that could be observed in a realistic experiment. It turns out that one can unambiguously attribute some features of such distributions to either an S- or A-type cascade. We also discuss the possibility to initiate A-type cascades by collision of GeV electrons with a single tightly focused laser pulse. As a byproduct, it would be possible to produce in this way bright collimated γ-ray beams of GeV energies.

1. Introduction

A-type or self-sustained electron-positron-photon (QED) cascades have become a rather widely discussed phenomena since their first predictions [1–3]. Most authors assume that a slow electron is somehow placed in a strong field region formed by a few laser pulses [1–12]. Examples of discussion of more realistic scenarios, like irradiation of solid or gas targets by strong laser pulses, are also known [13–17], but such studies require rather involved models and simulations. We discuss another realistic option of injecting seed electrons into the strong field region, which is based on the effect of collapse and revival of cascades caused by collision of GeV electron beam with focused laser pulses [18, 19]. Though having access to the whole simulation data it is typicaly evident whether or not an A-type cascade occurs, we are interested in developing criterions that rely on such features as final distributions, which should be directly observable in a realistic experimental setup.

2. Two laser pulses setup

Assume that two circularly polarized laser pulses synchronously counter-propagate along the same axis as shown in Fig. 1, and that laser field satisfies the conditions of A-type cascade initiation by a slow e^- in the focus. This condition is satisfied e.g. if lasers are optical ($\hbar\omega \simeq 1$ eV), and the total intensity is $I \sim 10^{25}$ W/cm^2 (corresponding to electric field strength $E_0 \sim 10^{-3} E_{cr}$, where $E_{cr} = m^2c^3/e\hbar$). Now suppose that e^--beam with energy $\varepsilon_0 \sim 1 \div 10$ GeV is injected into the focus. We study cascade dynamics using 3D Monte Carlo code, which

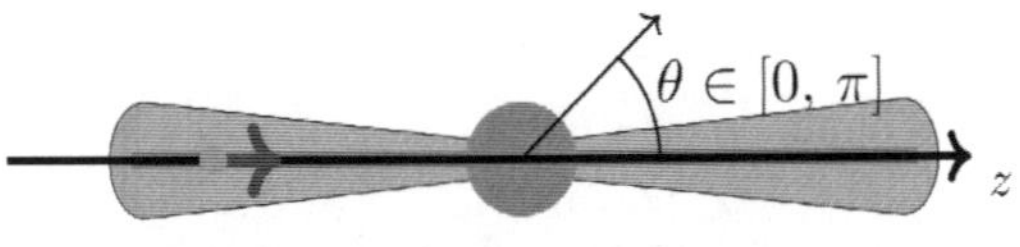

Figure 1. Scheme of experiment. Red shapes sketch the laser field and blue dot respresents the initial GeV e^-.

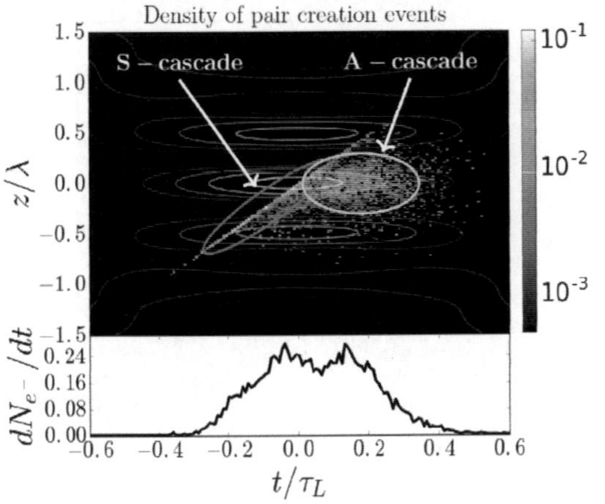

Figure 2. Evolution of the density $d^2N_{e^-e^+}/dtdz$ of e^-e^+ creation events along z-axis (in a.u., on top), and the electron creation rate dN_{e^-}/dt (in a.u., on bottom) for $\varepsilon_0 = 3$ GeV, $E_0 = 3.2 \times 10^{-3} E_S$ (total $I \approx 4.8 \times 10^{24}$ W/cm^2). Thin solid lines indicate the structure of electric field distribution.

was developed and tested in Refs. [18,19]. We neglect all plasma effects since we study cascades near their formation threshold when plasma density remains rather low, and assume that each laser pulse is tightly focused and of finite duration $\tau_L = 10$ fs.

Under the above conditions an S-type cascade is initiated first. Assume the cascade is induced by electrons moving along the focal z-axis towards the focal center, see Fig. 1. As can be seen from Fig. 2, as they approach the strong field region, the particle creation rate tends to grow. As S-type cascade develops, the emitted γ and secondary $e^\pm$ propagate along z-axis at almost speed of light. However the charged particles are losing energy and longitudinal momentum $p_\parallel$ randomly, hence their distribution after some time is broadened, particle creation rate is decreased and the S-cascade starts to collapse, as is seen for $t \gtrsim -0.1\tau_L$ in Fig. 2. Though according to Pomeranchuk's theorem initial electrons cannot reach the focal center, photons still propagate freely, and by reaching the focus have a chance for creating a pair of slower $e^\pm$, which further gain large transverse momentum $p_\perp \gg p_\parallel$ from the field. These secondary particles act as seeds for an A-type cascade which takes place at the central antinode of the field. Particle creation rate is growing again after $t > 0$, i.e. cascade revives and continues until laser field in the focal plane remains strong enough.

Though the assumption that direction of the initial e^--beam is aligned along the optical axis of laser pulses is not pleasurable from experimental point of view, the mechanism of collapse and revival does not change qualitatively even if electrons hit the focus at some nonzero tilt angle. Moreover, when the initial direction of e^--beam is set orthogonal to focal axis, almost all particles of the initial S-type cascade fall into the central antinode of the field. As the number of seeds increases, this results in enhancement of the subsequent A-type cascade, see Fig. 3 and further details in Ref. [19].

Evidence for subdivision of a cascade into S- and A- stages and occurrence of the latter is clear from evolution of spatial distributions presented in Fig. 2. However it is hard to imagine that such instant distributions could be ever resolved in a real experiment. Therefore one should rather study final distributions of γ and $e^\pm$ trying to identify signatures of both cascades. For an S-type cascade, multiplicity depends on both the initial electron energy and laser field intensity, and its dynamics strongly depends on initial direction of the particles, whereas properties of

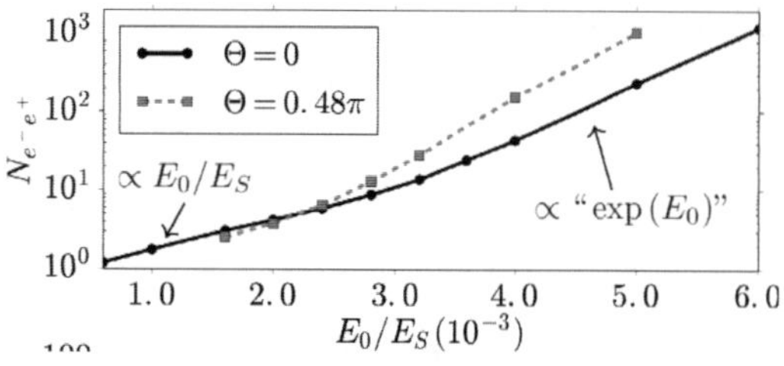

Figure 3. Number of e^-e^+ pairs produced in a cascade normalized to the initial number of electrons in the beam for $\varepsilon_0 = 3$ GeV, $E_0 = 3.2 \times 10^{-3} E_S$. Θ is the tilt angle of e^--beam with the focal axis.

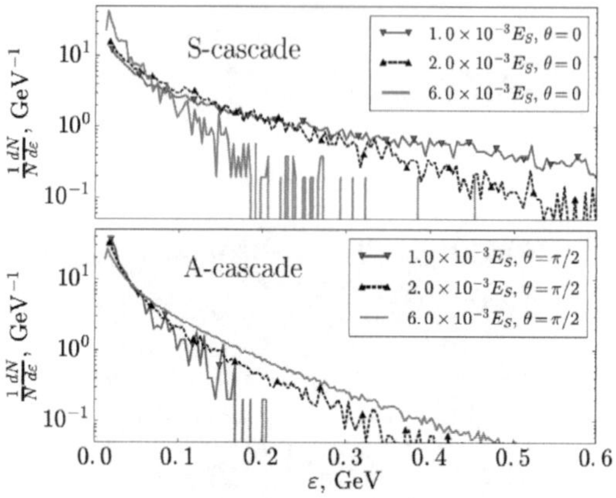

Figure 4. Normalized spectra of photons emitted at angles $\theta = 0$ (top) and $\theta = 90°$ (bottom) within spread $\Delta\theta \approx 10°$ for different intensities and $\varepsilon_0 = 3$ GeV.

an A-type cascade depend solely on the laser field structure, and in addition for them there exists some threshold in laser intensity. Bearing all that in mind, let us imagine that we can vary the intensity from low (when A-cascade does not occur) to higher (when it prevails) values, looking for those changes in final particle distributions that may allow to identify occurrence of an A-type cascade as well as to distinguish it from the S-type one.

The simplest quantity to measure is the total number N of $e^{\pm}$ created in a cascade. Multiplicity of S-type cascade is proportional to $\propto \varepsilon_0 E_0$ [18], while for A-type cascade it depends exponentially on E_0 [4]. This means that when intensity is gradually increased, the slope of the dependence $N(E_0)$ should change above the threshold E_{th}, which is exactly the case observed in our simulations for both longitudinal and transverse incidence of e^- beam at $E_{th} \approx 2.5 \div 3.0 \times 10^{-3} E_S$ (or $I \approx 4 \times 10^{24}$ W/cm^2), see Fig. 3.

More information can be extracted from the final photon spectra. For example, as discussed above, for the setup in Fig. 1 almost all the particles of an S-type cascade are propagating along z-axis, while A-type cascade develops approximately in an orthogonal central antinode plane $z = 0$. Hence photons of these cascades are emitted in different directions. Let θ be the angle between z-axis and the direction of propagation of γ. If the detectors measure photons emitted along directions $\theta = 0$ and $\theta = \pi/2$, they count photons from S- and A-type cascade, respectively. Spectra of photons emitted in these directions is shown in Fig. 4. When the intensity is not enough for initiation of an A-type cascade, almost all γ are emitted along $\theta = 0$, but once one increases E_0 openning the A-type "channel", more photons are detected at $\theta = \pi/2$. At the same time spectrum of photons from S-type cascade becomes narrower since less energy is needed for a photon to create a pair, and hence smaller amount of energetic photons survive up to the end. Spectrum of photons from A-type cascade behaves in the opposite manner: since $e^{\pm}$ are accelerated by the field, as E_0 increases, they tend to emit more energetic photons contributing to the final distribution. Such diverse changes in photon spectra can be considered as another evidence of collapse and revival of a cascade.

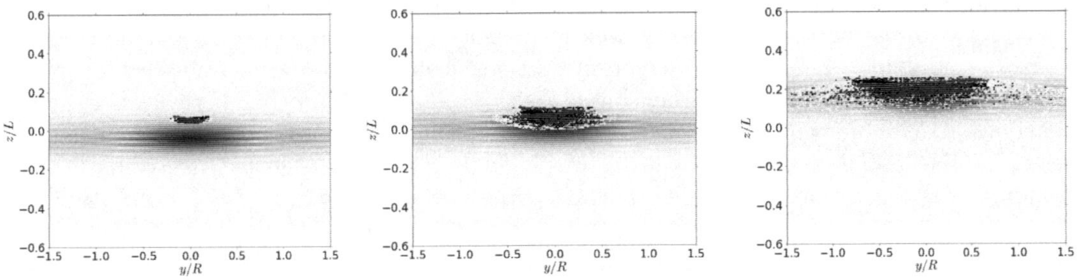

Figure 5. Distribution of particles and laser field at different moments of time, from left to right: $t = -0.16\tau_L$, $t = 0.20\tau_L$, $t = 1.40\tau_L$. Dark red and black dots correspond to e^- and e^+, yellow to γ. Blue distribution depicts laser field (in a.u.) of intensity $I \approx 1.2 \times 10^{26}$ W/cm^2. Initial e^- with energy $\varepsilon_0 = 2$ GeV counter propagate the laser pulse so that in absence of the field they would reach center of the focus at $t = 0$.

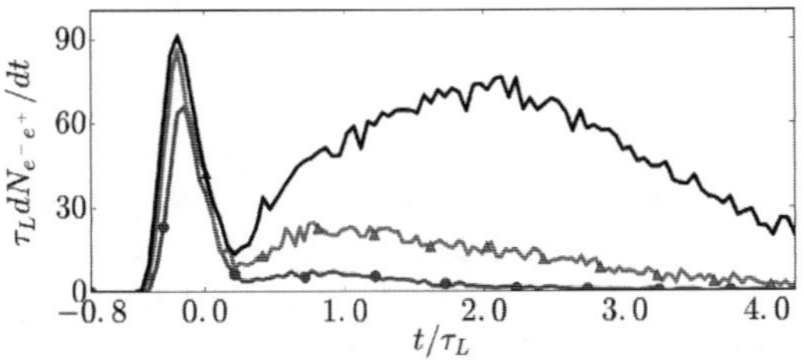

Figure 6. Pair creation rate in a cascade at intensities $I \approx 3 \times 10^{25}$ W/cm^2 (blue), $I \approx 7 \times 10^{26}$ W/cm^2 (green), and $I \approx 1.2 \times 10^{26}$ W/cm^2 (black) for $\varepsilon_0 = 2$ GeV.

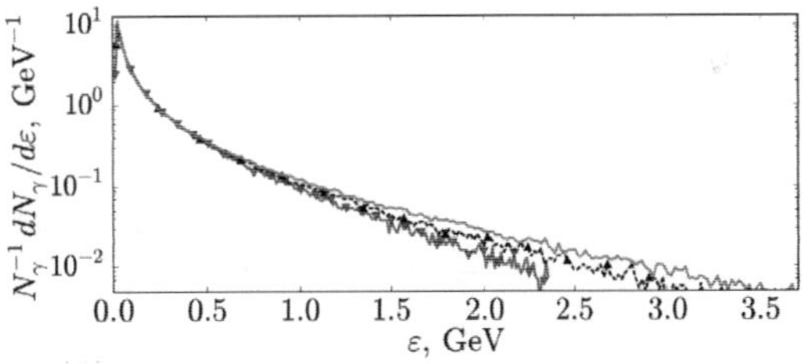

Figure 7. Normalized final spectra of photons at intensities $I \approx 3 \times 10^{25}$ W/cm^2 (magenta), $I \approx 7 \times 10^{26}$ W/cm^2 (black-dashed) and $I \approx 1.2 \times 10^{26}$ W/cm^2 (red) for $\varepsilon_0 = 2$ GeV.

3. Single pulse setup

Up to now A-type QED cascades have been almost always considered in standing waves formed by a couple or more of laser pulses, mainly in order to lower the required laser intensity per single pulse. Nevertheless, it is worth studying whether self-sustained cascades can occur in a single laser pulse. Obviously, for that laser field should be first of all tightly focused, since a weakly focused, i.e. plane wave-like pulse is not capable for particle acceleration in the sense of increasing the key parameter χ. Secondly, while in case of several colliding laser pulses there are some alternative scenarios of injecting seed electrons into the focus, for a single laser pulse there is no alternative other than head-on collision of GeV e^-- or photon beam with the laser pulse.

Particle distributions at different moments of time and the corresponding pair creation rate for such a scheme are shown in Fig. 5 and Fig. 6, respectively. We assume that laser pulse propagates along z-axis, and a bunch of e^- of the same initial position and momentum propagates in the opposite direction. S-type cascade starts immediately after the electron bunch reaches the strong field region, as indicated by a drastic increase of particle production rate. The high-energy γ are emitted towards the center of the focus, and create e^-e^+ pairs close to the focal region, which is non-accessible for initial electrons. As particles approach the center of the focus, they lose their energy. In addition, the field contorts their trajectories and accelerates them in opposite directions (at least on average). As a result, the initial S-type cascade starts collapsing and the particle creation rate reduces substantially. But after that the laser pulse reaccelerates them in the direction of its own propagation carrying away along z-axis. Since the pulse is tightly focused, in a proper frame of the focal center it looks as a standing spherical wave, hence the particles moving along with the pulse are under optimal conditions for seeding an A-type cascade, and the cascade starts over again. The second long hump on Fig. 6 corresponds to such a secondary A-type cascade. We note that its occurrence is suppressed for $I \lesssim 3 \times 10^{25}$ W/cm^2, while S-cascade has almost the same multiplicity for all the values of I in our simulations. Such behavior confirms that the mechanism of collapse and revival is the same as in case of two laser pulses considered above.

The main observed difference with the case of cascades in a two pulses setup is sharper separation of the S- and A-type cascade stages. This happens since S-type cascade develops as long as particles are capable for penetration inside the field, while A-type cascade cannot start until they turn around starting to propagate along with the pulse. Though such a scheme requires laser intensity $I \sim 10^{25} \div 10^{26}$ W/cm^2, one order of magnitude higher than the two pulses setup, sharper separation may be useful to study properties of each stage more thoroughly. As in two pulses setup, it makes sense to study e.g. distributions of photons emitted at different angles.

The final spectrum of photons in case of single laser pulse, which is presented in Fig. 7, shares some common features with the spectra of photons from the A-type cascade in Fig. 4. In particular, the spectrum becomes broader as intensity is growing. But unlike the two-pulse setup, where the photons are emitted uniformly in all directions orthogonal to the focal axis, here they form a narrow bunch of collimated γ-quanta, of few μm length, propagating along with the laser pulse. The energies of photons in such a bunch are in GeV range. This implies that QED cascades produced by a single laser pulse may serve as a source of short collimated hard γ-rays.

4. Conclusion

Collision of high-energy GeV electrons with laser pulses of intensity $10^{24} \div 10^{26}$ W/cm^2 (depending on laser field configuration) is a realistic experimental scenario for observation of A-type cascades. Though one could expect A-type cascade to be shaded by a preceding S-type cascade, according to our results final particle distributions provide enough information to separate them, in particular to verify occurrence of an A-type cascade. Besides, we demonstrate that the collapse and revival scenario reveals even with a single laser pulse. Even though such simplified scheme requires higher laser intensity, it is still favorable in certain aspects, for example provides sharper separation of the S- and A-type stages. It can also serve as sources of short collimated bunches of γ-rays or high-energy electrons and/or positrons.

Acknowledgments

The work was performed using resources of NRNU MEPhI high-performance computing center. The research was supported by the MEPhI Academic Excellence Project (Contract No. 02.a03.21.0005) and the Russian Fund for Basic Research (Grants No. 16-32-00863mol_a and No. 16-02-00963a).

References

[1] Bell A R and Kirk J G 2008 *Phys. Rev. Lett.* **101** 200403
[2] Kirk J G, Bell A R and Arka I 2009 *Plasma Phys. Controlled Fusion* **51** 085008
[3] Fedotov A M, Narozhny N B, Mourou G and Korn G 2010 *Phys. Rev. Lett.* **105** 080402
[4] Elkina N V, Fedotov A M, Kostyukov I Yu, Legkov M V, Narozhny N B, Nerush E N and Ruhl H 2011 *Phys. Rev.* STAB **14** 054401
[5] Duclous R, Kirk J G and Bell A R 2011 *Plasma Phys. Control. Fusion* **53** 015009

[6] Nerush E N, Bashmakov V F, Kostyukov I Yu 2011 *Physics of Plasmas* **18** 083107
[7] Nerush E N, Kostyukov I Yu, Fedotov A M, Narozhny N B, Elkina N V, and Ruhl H 2011 *Phys. Rev. Lett.* **106** 035001
[8] King B, Elkina N, and Ruhl H 2013 *Phys. Rev.* A **87** 042117
[9] Bashmakov V F, Nerush E N, Kostyukov I Yu, Fedotov A M, Narozhny N B 2014 *Phys. of Plasmas* **21** 013105
[10] Grismayer T, Vranic M, Martins J L, Fonseca R, Silva L O 2015 Seeded QED cascades in counter propagating laser pulses *Preprint* arXiv:1511.07503
[11] Jirka M, Klimo O, Bulanov S V, Esirkepov T Zh, Gelfer E, Bulanov S S, Weber S and Korn G 2016 *Phys. Rev.* E **93** 023207
[12] Gelfer E G, Mironov A A, Fedotov A M, Bashmakov V F, Nerush E N, Kostyukov I Yu, Narozhny N B 2015 *Phys. Rev.* A **92** 022113
[13] Ridgers C P, Brady C S, Duclous R, Kirk J G, Bennett K, Arber T D, Robinson A P L and Bell A R 2012 *Phys. Rev. Lett.* **108** 165006
[14] Kirk J, Bell A and Ridgers C 2013 *Plasma Phys. Control. Fusion* **55** 095016
[15] Ji L L, Pukhov A, Nerush E N, Kostyukov I Yu, Shen B F and Akli K U 2014 *Physics of Plasmas* **21** 023109
[16] Brady C S, Ridgers C P, Arber T D and Bell A R 2014 *Physics of Plasmas* **21** 033108
[17] Tamburini M, Di Piazza A and Keitel C H 2015 Laser-Pulse-Shape Control of Seeded QED Cascades *Preprint* arXiv:1511.03987
[18] Mironov A A, Narozhny N B and Fedotov A M 2014 *Phys. Lett.* A **378** 3254
[19] Mironov A A, Fedotov A M and Narozhny N B 2016 *Quantum Electronics* **46** 305

Conversion of Transverse to Longitudinal Plasma Waves Induced by Radiation Reaction

E G Gelfer[1], A M Fedotov[1], N V Elkina[2]

[1] National Research Nuclear University "MEPhI" (Moscow Engineering Physics Institute), 115409 Moscow, Russia

[2] Arnold Sommerfeld Center, LMU Munich, Theresienstrasse 37, 80333 Munich, Germany

E-mail: egelfer@gmail.com

Abstract. We investigate the evolution of a circularly polarized wave in a cold plasma in a regime when radiation reaction is essential. A new effect is that circularly polarized transverse waves become unstable with respect to partial conversion to the longitudinal ones. This effect appears due to radiation reaction entirely and in principle could be implemented for particle acceleration. We provide analytical expression for arising longitudinal field in the case, when it is much smaller, than transverse one.

1. Introduction

One of the promising possible applications of the new powerful laser facilities such as ELI Beamlines [1] in Czech Republic or VULCAN [2] in the UK is particle acceleration by longitudinal electric fields generated by nonlinear propagation of intense laser pulses through plasma. For example, in the most popular LWFA scheme [3–5] these fields are generated by pushing plasma electrons via ponderomotive force of a short focused laser pulse.

As is well known, both longitudinal and circularly polarized transverse modes are capable for propagation through a cold plasma [6] and are independent in linear regime. Since for modern facilities Lorentz factor of electron quiver oscillations becomes already of the order of hundreds, radiation reaction effects should also come into play [7]. In particular, we will show that in such highly ultrarelativistic regime circularly polarized transverse waves in plasma become unstable with respect to conversion to longitudinal modes, and that this effect is provided entirely by radiation reaction. As a result, we suggest a new radiation reaction based mechanism for generation of longitudinal field in plasma, which should compete with the usually employed ponderomotive mechanism and in principle could be implemented for particle acceleration.

Note that previously impact of radiation reaction was studied in a number of single particle problems, see e.g. [7–11], but here we examine how it affects collective motion in plasma. Here we investigate evolution of a circularly polarized one dimensional wave in a cold plasma in a regime when radiation reaction is essential, but can be still treated as perturbation with respect to the Lorenz force. It corresponds to dimensionless field amplitude a of the order of few hundreds, i.e. to laser intensity of around 10^{22}–10^{23} W/cm^2, i.e. the value that is expected to be attainable with the laser facilities of the near future. Besides, in this case we can neglect quantum corrections and use classical picture of particle motion based on Lorenz equation with

Landau-Lifshitz radiation friction force [12], leaving in addition just the dominant contribution to the latter.

In what follows, we are using a cold plasma model, which in hydrodynamic approximation is described by the equations

$$\nabla \times \boldsymbol{E} = -\frac{\partial \boldsymbol{B}}{\partial t}, \tag{1}$$

$$\nabla \cdot \boldsymbol{E} = -4\pi e(n - n_0), \tag{2}$$

$$\nabla \times \boldsymbol{B} = \frac{\partial \boldsymbol{E}}{\partial t} - 4\pi e n \boldsymbol{v}, \tag{3}$$

$$\nabla \cdot \boldsymbol{B} = 0, \tag{4}$$

$$\frac{\partial \boldsymbol{p}}{\partial t} + (\boldsymbol{v} \cdot \nabla)\boldsymbol{p} = -e(\boldsymbol{E} + \boldsymbol{v} \times \boldsymbol{B}) + \boldsymbol{F}_{RR}, \tag{5}$$

$$\boldsymbol{F}_{RR} = -\frac{2e^4}{3m^2}\gamma^2 \left\{(\boldsymbol{E} + \boldsymbol{v} \times \boldsymbol{B})^2 - (\boldsymbol{E} \cdot \boldsymbol{v})^2\right\}\boldsymbol{v}. \tag{6}$$

Here we assume such units that $c = 1$; $e > 0$ and m are the magnitude of electron charge and mass respectively; $\mathbf{E}$ and $\mathbf{B}$ are electric and magnetic fields; n_0 is the density of ions (which we assume immobile); n, $\boldsymbol{v}$, $\boldsymbol{p}$ and γ are local density, velocity, momentum and γ-factor of electron fluid.

2. One dimensional wave solution

We will seek for a solution of equations (1)–(6) in the form of functions, which depend only on the phase $\xi = \omega_p(t - z/\beta)$, where $\omega_p = \sqrt{4\pi e^2 n_0/m}$ is the plasma frequency, β is the phase velocity of a plasma wave propagating along z direction. Then it follows from Maxwell equations that $B_y = E_x/\beta$, $B_x = -E_y/\beta$, and $n = \frac{n_0}{1-v_z/\beta}$. Besides we introduce dimensionless field $\boldsymbol{a} = e\boldsymbol{E}/m\omega_p$ and four-velocity $\mathbf{u} = \mathbf{p}/m$ and take into account that particles motion is ultrarelativistic, $u \gg 1$, so that $\gamma \approx u$. Then we obtain equations for transverse and longitudinal motions in the form

$$\dot{\boldsymbol{u}}_\perp = -\boldsymbol{a}_\perp + \boldsymbol{f}_{RR\perp}, \quad \dot{\boldsymbol{a}}_\perp = \frac{\boldsymbol{u}_\perp}{(1 - 1/\beta^2)(u - u_z/\beta)}, \tag{7}$$

$$\dot{u}_z = -\frac{u a_z + \boldsymbol{u}_\perp \boldsymbol{a}_\perp/\beta}{u - u_z/\beta} + \boldsymbol{f}_{RRz}, \quad \dot{a}_z = \frac{u_z}{u - u_z/\beta}, \tag{8}$$

where

$$\boldsymbol{f}_{RR} = -K\boldsymbol{u}\left[(u - u_z/\beta)|\boldsymbol{a}_\perp|^2 + \frac{(u a_z + \boldsymbol{u}_\perp \boldsymbol{a}_\perp/\beta)^2 - (\boldsymbol{u}\boldsymbol{a})^2}{u - u_z/\beta}\right], \quad K = \frac{2}{3}e^2\omega_p/m. \tag{9}$$

Note that parameter K is actually very small. For ω_p in optical range K is of the order of 10^{-8}.

2.1. Radiation reaction turned off

First consider the case when radiation reaction can be completely neglected. If the wave was initially transverse, $u_z(0) = a_z(0) = 0$, and circularly polarized, then following [6] Eq. (7) has the solution

$$u_x = -u_0 \sin\omega_0\xi, \quad u_y = u_0\cos\omega_0\xi, \quad a_x = \omega_0 u_0 \cos\omega_0\xi, \quad a_y = u_0\omega_0 \sin\omega_0\xi, \tag{10}$$

which is a transverse circularly polarized wave with the frequency[1]

$$\omega_0 = \frac{1}{\sqrt{u_0(1-1/\beta^2)}}. \tag{11}$$

Since $\boldsymbol{u}_\perp \boldsymbol{a}_\perp = 0$, longitudinal component does not appear according to Eq. (8).

2.2. Radiation reaction turned on

The situation changes when we take radiation reaction into account. To solve equations (7), (8) we will use several assumptions:

- Radiation reaction force is much smaller than the Lorentz force

$$Ku^2a^2 \ll a; \tag{12}$$

- Generated longitudinal field is still much smaller than the transverse field $a_z \ll a_\perp$;
- Phase velocity is close to the speed of light, $\beta \approx 1$.

Then the radiation reaction term can be rewritten in the form

$$\boldsymbol{f}_{RR} = -r\boldsymbol{u}, \quad r = Ka_\perp^2 u_-, \tag{13}$$

where we introduced $u_- = u - u_z$.

First consider transverse motion, which is described by the equation

$$\ddot{\boldsymbol{u}}_\perp + r\dot{\boldsymbol{u}}_\perp + \omega_\perp^2 \boldsymbol{u}_\perp = 0, \quad \omega_\perp = \frac{1}{\sqrt{1-1/\beta^2 u_-}}, \tag{14}$$

where we neglected $\dot{r}$ against $\omega_\perp^2$ (it follows from the condition $a_z \ll a_\perp$, as we will show below). Eq. (14) has the WKB solution

$$u_x = -u_0^{3/4}u_-^{1/4}e^{-\frac{1}{2}\int r d\xi}\sin\int\omega_\perp d\xi, \quad u_y = u_0^{3/4}u_-^{1/4}e^{-\frac{1}{2}\int r d\xi}\cos\int\omega_\perp d\xi, \tag{15}$$

and for transverse field we obtain

$$a_x = \frac{u_0^{3/4}}{u_-^{1/4}\sqrt{1-1/\beta^2}}e^{-\frac{1}{2}\int r d\xi}\cos\int\omega_\perp d\xi, \quad a_y = \frac{u_0^{3/4}}{u_-^{1/4}\sqrt{1-1/\beta^2}}e^{-\frac{1}{2}\int r d\xi}\sin\int\omega_\perp d\xi. \tag{16}$$

It follows from (7), (8) that

$$\dot{u}_- = a_z - ru_-, \tag{17}$$

then from the conditions $a_z \ll a_\perp$ and (12) we obtain

$$\frac{\sqrt{1-1/\beta^2}|\dot{u}_-|}{\sqrt{u_-}} \ll 1, \tag{18}$$

or $|\dot{\omega}_\perp/\omega_\perp^2| \ll 1$. Hence our assumptions provide validity of WKB approximation as well as the condition $\dot{r} \ll \omega_\perp^2$ assumed above.

[1] Note, that if plasma wave is generated by external electromagnetic wave of frequency ω, then Eq. (11) actually defines phase velocity $\beta = (1-(\omega_p/\omega)^2/u_0)^{-1/2}$.

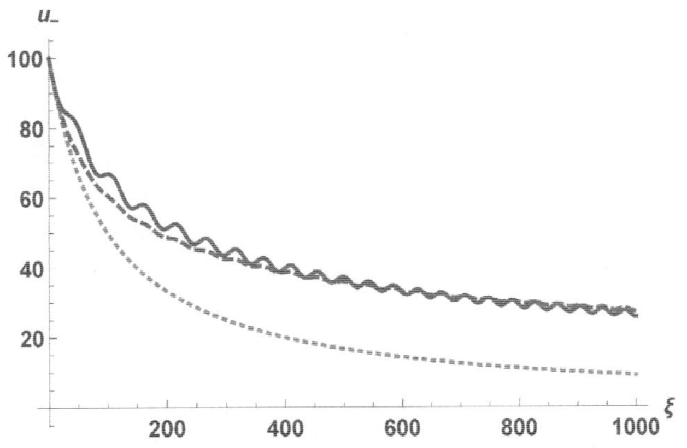

Figure 1. $u_-(\xi)$ for $u_0 = a_0 = 10^2$, $1 - 1/\beta^2 = 10^{-2}$. Solid blue line: numerical solution of Eqs. (19), (20); dotted orange line: solution (21) for single electron a circularly polarized plane wave [8]; dashed red line: self-consistent formula (22).

Consider now longitudinal component of a plasma wave. The equation for u_- can be represented in the form

$$\ddot{u}_- = \frac{1}{2}\left[\left(\frac{u_0}{u_-}\right)^{3/2} R - 1\right] - \frac{3Ku_0^{3/2}}{2(1-1/\beta^2)}\dot{u}_-\sqrt{u_-}R + \frac{K^2u_0^3}{(1-1/\beta^2)^2}u_-^2R^2, \tag{19}$$

$$\dot{R} = -KR^2\sqrt{u_-}\frac{u_0^{3/2}}{1-1/\beta^2}, \tag{20}$$

where we defined $R = e^{-\int r d\xi}$.

In [8] A. Di Piazza obtained an exact solution for electron motion in an external plane wave field with radiation reaction taken into account. In our notations

$$u_-^{DP} = \frac{u_0}{1 + Ku_0a_0^2\xi}, \tag{21}$$

where a_0 is the dimensionless external field amplitude. In our case this formula cannot be implemented straightforwardly, because damping of an external plane wave field was not taken into account in [8]. However, if we substitute into Eq. (22) $a(\xi)$ from the numerical solution of Eqs. (7), (8), then the self-consistent version of the expression (22)

$$u_-^{SC} = \frac{u_0}{1 + Ku_0a^2(\xi)\xi}, \tag{22}$$

perfectly fits the numerical solution, see Fig. 1.

In order to solve equations (19), (20), note that the derivative

$$\frac{d}{d\xi}\left(Ru_-^{-3/2}\right) = -KR^2\frac{u_0^{3/2}}{u_-(1-1/\beta^2)} - \frac{3}{2}\frac{\dot{u}_-}{u_-^{5/2}}R \ll 1 \tag{23}$$

for those values of u_- that radiation reaction force is important, therefore we can assume that in the zeroth order $Ru_-^{-3/2}$ remains constant. Taking into account that $u_-(0) = u_0$ and $R(0) = 1$,

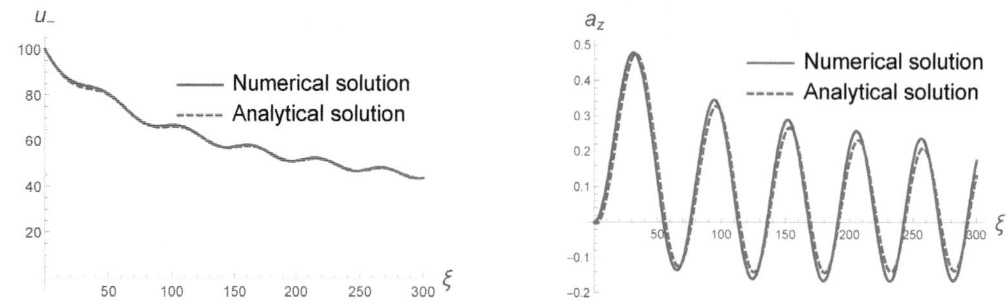

Figure 2. Left panel: numerical solution for $u_-(\xi)$ (solid blue) vs. analytical solution (26) (dashed red). Right panel: numerical solution for $a_z(\xi)$ (solid blue) vs. analytical solution (27) (dashed red). $u_0 = 10^2$ ($\alpha\sqrt{u_0} \approx 0.13$), $1 - 1/\beta^2 = 10^{-2}$.

as a zeroth approximation we get

$$R^{(0)} = \left(\frac{u_-^{(0)}}{u_0}\right)^{3/2} \tag{24}$$

Substituting this expression into Eq. (20), we obtain the solution

$$u_-^{(0)} = \frac{u_0}{\sqrt{1+\alpha\xi}}, \quad R^{(0)} = \frac{1}{(1+\alpha\xi)^{3/4}}, \tag{25}$$

where $\alpha = \frac{4}{3}\frac{K\rho_0^2}{1-1/\beta^2}$.

Solving Eqs. (19), (20) to the next order we obtain

$$u_- = \frac{u_0}{\sqrt{1+\alpha\xi}} - \frac{\alpha u_0^{3/2}}{2\sqrt{3}(1+\alpha\xi)^{11/16}} \sin\left[\frac{2\sqrt{3}}{5\alpha\sqrt{u_0}}((1+\alpha\xi)^{5/4} - 1)\right], \tag{26}$$

and from (17) derive the expression for the longitudinal field

$$a_z = \frac{\alpha u_0}{4(1+\alpha\xi)^{3/2}} \left(1 - (1+\alpha\xi)^{17/16} \cos\left[\frac{2\sqrt{3}}{5\alpha\sqrt{u_0}}((1+\alpha\xi)^{5/4} - 1)\right]\right). \tag{27}$$

These analytical formulas are compared with numerical solutions in Figs. 2 and 3.

Note that the expansion parameter in eq. (26) is $\alpha\sqrt{u_0}$, and that the condition $\alpha\sqrt{u_0} \ll 1$ is stronger than (18). However, analytical expressions (26) and (27) remain qualitatively valid and give correct estimations for longitudinal field amplitude even for $\alpha\sqrt{u_0} > 1$, see Fig. 3.

In order to illustrate the obtained results, consider the plasma wave generated by a circularly polarized electromagnetic pulse with a smooth front, so that one can neglect impact of the ponderomotive mechanism on longitudinal mode generation. We can estimate $\dot{u}_\perp \sim a_0\frac{\omega}{\omega_p}$, where a_0 is the dimensionless electric field of the external pulse (normalized as usually to the wave frequency). Hence $\sqrt{u_0/(1-1/\beta^2)} \sim a_0\omega/\omega_p$ and $u_0 \sim a_0$. Then from (27) we can estimate the amplitude of longitudinal mode as

$$a_z \sim \alpha u_0 \sim K a_0^4 \left(\frac{\omega}{\omega_p}\right)^2. \tag{28}$$

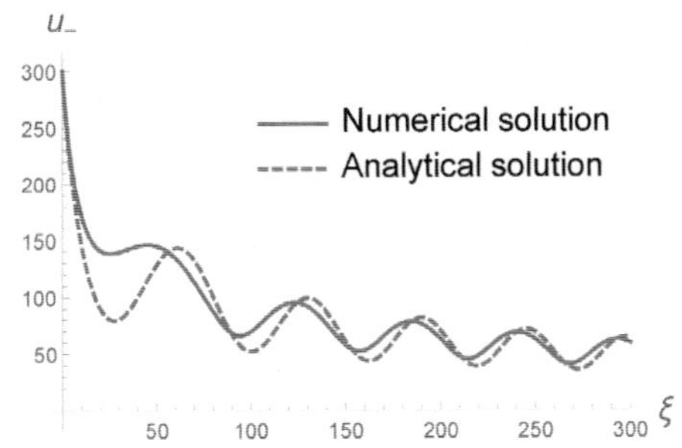

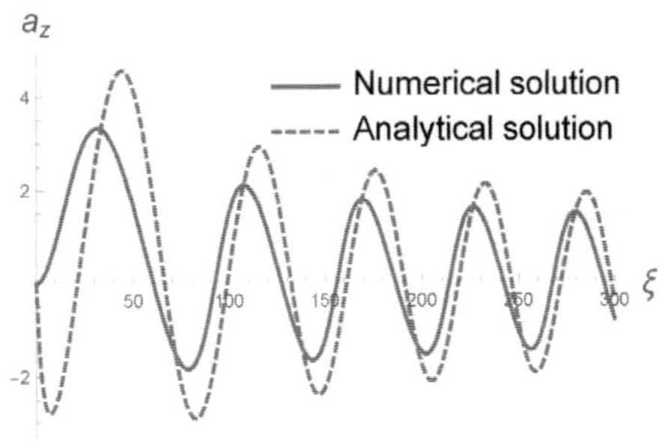

Figure 3. Left panel: numerical solution for $u_-(\xi)$ (solid blue) vs. analytical solution (26) (dashed red). Right panel: numerical solution for $a_z(\xi)$ (solid blue) vs. analytical solution (27) (dashed red). $u_0 = 3 \cdot 10^2$ ($\alpha\sqrt{u_0} \approx 2.1$), $1 - 1/\beta^2 = 10^{-2}$.

Clearly, amplitude of the longitudinal field has very strong scaling with a_0. This, in a view of the novel experimental capabilities [1,2], may be interesting for possible applications to electron acceleration. Besides, the frequency of longitudinal oscillations scales as $\omega_z \sim \omega_p/\sqrt{a_0}$, implying that accelerating particles can spend more time in accelerating phase of the longitudinal field.

3. Conclusion

We examined the impact of radiation reaction on initially transverse circularly polarized plasma wave and showed that radiation reaction could effectively convert transverse waves into longitudinal plasma modes. We obtained analytical approximation for coupled transverse and longitudinal plasma waves. This approximate solution is a self-consistent generalization of the exact solution [8] for a particle moving in a plain wave with account for radiation reaction.

Normally radiation reaction acts like hindrance and inhibits acceleration. However, the effect of conversion of transverse to longitudinal plasma waves induced by radiation reaction could be interesting as a novel alternative mechanism for electron acceleration. Such mechanism may be more effective than conventional LWFA for high laser field amplitudes $a_0 \gg 1$ due to a stronger scaling (a_0^4 vs. a_0^2).

Acknowledgments

The work was supported by the MEPhI Academic Excellence Project (contract No. 02.a03.21.0005, 27.08.2013) and RFBR (Grants No. 16-32-00863mol_a and No. 16-02-00963a).

References

[1] http://www.extreme-light-infrastructure.eu
[2] http://www.stfc.ac.uk/clf/New+Initiatives/ The+Vulcan+10+Petawatt+Project/14684.aspx
[3] Tajima T and Dawson J M 1979 *Phys. Rev. Lett.* **43** 267
[4] Gorbunov L and Kirsanov V 1987 *Sov. Phys. JETP* **66** 40
[5] Bulanov S V, Esirkepov T Zh, Hayashi Y, Kiriyama H, Koga J K, Kotaki H, Mori M and Kando M 2016 *J. Plasma Phys.* **82** 905820308
[6] Akhiezer A and Polovin R 1956 *Sov. Phys. JETP* **3** 696–705
[7] Di Piazza A, Müller C, Hatsagortsyan K Z and Keitel C H 2012 *Rev. Mod. Phys.* **84** 1177
[8] Di Piazza A 2008 *Lett. Math. Phys.* **83** 305-313
[9] Fedotov A M, Elkina N V, Gelfer E G, Narozhny N B and Ruhl H 2014 *Phys. Rev. A.* **90** 053847
[10] Gonoskov A, Bashinov A, Gonoskov I, Harvey C, Ilderton A, Kim A, Marklund M, Mourou G and Sergeev A 2014 *Phys. Rev. Lett.* **113** 014801
[11] Ji L L, Pukhov A, Kostyukov O Yu, Shen B F and Akli K 2014 *Phys. Rev. Lett.* 2014 **112** 145003
[12] Landau L D and Lifshitz E M 1975 *The Classical Theory of Fields* (Oxford: Pergamon Press)